Linear Algebra

Linear Algebra

Vivek Sahai □ Vikas Bist

Alpha Science International Ltd.
Pangbourne

Vivek Sahai
Department of Mathematics and Astronomy
Lucknow University, Lucknow, India

Vikas Bist
Department of Mathematics
Punjab University, Chandigarh, India

Alpha Science International Ltd.
P.O. Box 4067, Pangbourne RG8 8UT, UK

ISBN 1-84265-080-7

Printed in India.

Preface

This book presents a concise and comprehensive introduction to the fundamentals of linear algebra. The subject is developed so that it is accessible to the students of varied backgrounds. A very basic algebra and elementary properties of matrices and determinants are prerequisites. However, almost all the material that is needed for the text is given in the first chapter with reasonable details to make the book as much self contained as possible. A fairly large number of examples are included at the end of each section for conceptual understanding. Examples are also used to illustrate the computational aspect and to explore further related results. Each section ends with a collection of exercises with different levels of difficulty that help the reader to check their understanding and to apply it to concrete problems.

Chapter 2 deals with the elementary concepts like vector space, subspace, basis and dimension. Although the emphasis here is on finite dimensional vector spaces, examples of infinite dimensional vector spaces are also included to highlight the differences between these two classes. Linear transformation, isomorphism theorems, the matrix of a linear transformation, linear forms and dual spaces are also discussed in this chapter.

Chapter 3 is analysis of a single linear operator on a finite dimensional vector space. Characterizations of diagonalizable and triangulable operators appear in this chapter. The concept of generalized eigenvectors is used to obtain an inductive procedure for constructing a Jordan basis for a triangulable linear operator. The last section is about the rational canonical form and again the approach is algorithmic.

Chapter 4 is about finite dimensional inner product spaces. This chapter treats with normal and self adjoint operators and the spectral decomposition. Also it deals with nonnegative operators, isometeries and polar and singular value decomposition.

The last chapter is on bilinear forms and extends the results of inner product spaces to bilinear spaces. The main thrust of this chapter is to classify finite dimensional reflexive bilinear spaces.

Vivek Sahai
Vikas Bist

Contents

CHAPTER 1

Preliminaries

The purpose of this chapter is to give a brief account of a number of useful concepts and facts which are required in the text. The reader may be familiar with the contents of this chapter. Nonetheless, this may serve as a short review, and an introduction to the basic notation.

1. Matrices

We denote by $K^{m\times n}$ the set of all matrices of size $m\times n$, m rows and n columns, whose entries are from the field K. If $A\in K^{m\times n}$, then we denote its (i,j)–th entry by $[A]_{ij}$. A matrix of size $n\times 1$ (respectively, $1\times n$) is called a **column** (respectively, **row**) **vector** of length n. A square matrix is a matrix of size $n\times n$.

The sum of two matrices is allowed whenever the size of the matrices are same; for $A,B\in K^{m\times n}$, $A+B\in K^{m\times n}$ and

$$[A+B]_{ij} = [A]_{ij} + [B]_{ij}\,,\ i=1,\dots,m,\ j=1,\dots,n.$$

The product AB of two matrices is defined only if the number of columns of A is equal to the number of rows of B; for $A\in K^{m\times q}$, $B\in K^{q\times n}$, $AB\in K^{m\times n}$ and

$$[AB]_{ij} = \sum_{k=1}^{q} [A]_{ik}\,[B]_{kj}\,,\ i=1,\dots,m,\ j=1,\dots,n.$$

The associative laws hold for addition and multiplication of matrices:

$$(A+B)+C = A+(B+C)\,,\ \text{for all } A,B,C\in K^{m\times n}$$

and

$$(AB)\,C = A\,(BC)\,,\ \text{for all } A\in K^{m\times p}, B\in K^{p\times q}, C\in K^{q\times n}.$$

But the matrix addition is commutative and the matrix multiplication is not. We also have distributive laws:

$$A\,(B+C) = AB+AC,\ \text{for all } A\in K^{m\times q},\ B,C\in K^{q\times n},$$

and

$$(A+B)\,C = AC+BC,\ \text{for all } A,B\in K^{m\times q}, C\in K^{q\times n}.$$

The elements of the field K are called **scalars.** Multiplication of an $m\times n$ matrix A by a scalar λ is again a matrix λA of size $m\times n$ such that

$$[\lambda A]_{ij} = \lambda\,[A]_{ij}\,,\ i=1,\dots,m,\ j=1,\dots,n.$$

The following are basic properties of scalar multiplication. The reader can easily verify them.

$$\begin{aligned}(\lambda+\mu)A &= \lambda A+\mu A,\\ \lambda(A+B) &= \lambda A+\lambda B,\\ (\lambda\mu)A &= \lambda(\mu A),\\ 1A &= A,\end{aligned}$$

for all $\lambda, \mu \in K$ and $A, B \in K^{m\times n}$.

A sum of the form $\sum_{i=1}^{r} \lambda_i A_i$, where $\lambda_1, \ldots, \lambda_r \in K$ and $A_1, \ldots, A_r \in K^{m\times n}$ is called a **linear combination** of matrices. If $A_1, \ldots, A_r$ are column (respectively, row) vectors, then we call it a linear combination of column (respectively, row) vectors.

Let A be an $m \times n$ matrix. A **submatrix** of A is a matrix obtained from A by deleting its some rows or some columns or both. A **partition** of A is its expression into smaller submatrices:

$$A = \begin{bmatrix} A_{11} & \cdots & A_{1s} \\ \vdots & & \vdots \\ A_{r1} & \cdots & A_{rs} \end{bmatrix},$$

where each A_{ij} is an $m_i \times n_j$ submatrix of A, and $m_1+\cdots+m_r = m$, $n_1+\cdots+n_s = n$. A partitioned matrix is also called a **block matrix**.

For example, let

$$A = \begin{bmatrix} 1 & 3 & 2 & 9 & 2 & 5 & 3 \\ 3 & 9 & 7 & 8 & 3 & 6 & 7 \\ 2 & 0 & 0 & 1 & 3 & 2 & 7 \\ 7 & 6 & 5 & 4 & 3 & 2 & 1 \\ 7 & 8 & 7 & 8 & 9 & 3 & 1 \end{bmatrix}.$$

Then $\begin{bmatrix} 3 & 9 & 3 \\ 0 & 1 & 7 \\ 8 & 8 & 1 \end{bmatrix}$ is a submatrix of A obtained by removing rows 2 and 4, and columns 1, 3, 4 and 5.

$A = \begin{bmatrix} A_{11} & A_{12} & A_{13} \\ A_{21} & A_{22} & A_{23} \end{bmatrix}$ is a partition of A, where

$$A_{11} = \begin{bmatrix} 1 & 3 & 2 \\ 3 & 9 & 7 \\ 2 & 0 & 0 \end{bmatrix}, \quad A_{12} = \begin{bmatrix} 9 & 2 \\ 8 & 3 \\ 1 & 3 \end{bmatrix}, \quad A_{13} = \begin{bmatrix} 5 & 3 \\ 6 & 7 \\ 2 & 7 \end{bmatrix},$$

$$A_{21} = \begin{bmatrix} 7 & 6 & 5 \\ 7 & 8 & 7 \end{bmatrix}, \quad A_{22} = \begin{bmatrix} 4 & 3 \\ 8 & 9 \end{bmatrix}, \quad A_{23} = \begin{bmatrix} 2 & 1 \\ 3 & 1 \end{bmatrix}.$$

Let $A, B \in K^{m\times n}$, and let

$$A = \begin{bmatrix} A_{11} & \cdots & A_{1s} \\ \vdots & & \vdots \\ A_{r1} & \cdots & A_{rs} \end{bmatrix}, \quad B = \begin{bmatrix} B_{11} & \cdots & B_{1s} \\ \vdots & & \vdots \\ B_{r1} & \cdots & B_{rs} \end{bmatrix},$$

$A_{ij}, B_{ij} \in K^{m_i \times n_j}$, $m_1 + \cdots + m_r = m$, $n_1 + \cdots + n_s = n$. In this case we say that the block matrices A and B are of the same type. Clearly the sum $A+B$ is obtainable by adding submatrices blockwise:

$$A+B = \begin{bmatrix} A_{11}+B_{11} & \cdots & A_{1s}+B_{1s} \\ \vdots & & \vdots \\ A_{r1}+B_{r1} & \cdots & A_{rs}+B_{rs} \end{bmatrix}.$$

In the case of multiplication of two block matrices, the partition should be such that the multiplication of submatrices make sense. Such a partition is called a **conformable partition** for the product. If $A \in K^{m \times q}$ is a block matrix:

$$A = \begin{bmatrix} A_{11} & \cdots & A_{1s} \\ \vdots & & \vdots \\ A_{r1} & \cdots & A_{rs} \end{bmatrix},$$

with $A_{ij} \in K^{m_i \times q_j}$, $m_1 + \cdots + m_r = m$, $q_1 + \cdots + q_s = q$. If $B \in K^{q \times n}$ then the conformable partition of B with respect to A for AB is

$$B = \begin{bmatrix} B_{11} & \cdots & B_{1t} \\ \vdots & & \vdots \\ B_{s1} & \cdots & B_{st} \end{bmatrix},$$

where $B_{ij} \in K^{q_i \times n_j}$, $n_1 + \cdots + n_t = n$. In this case we obtain

$$AB = \begin{bmatrix} C_{11} & \cdots & C_{1t} \\ \vdots & & \vdots \\ C_{r1} & \cdots & C_{rt} \end{bmatrix},$$

where $C_{ij} = \sum_{k=1}^{s} A_{ik}B_{kj}$.

A square matrix A is called **upper triangular** (respectively, **lower triangular**) if $[A]_{ij} = 0$ for $i > j$ (respectively, $i < j$). A square matrix A is called a **diagonal matrix** if it is upper triangular as well as lower triangular, that is, $[A]_{ij} = 0$ for $i \neq j$.

If

$$A = \begin{bmatrix} A_{11} & \cdots & A_{1r} \\ \vdots & & \vdots \\ A_{r1} & \cdots & A_{rr} \end{bmatrix}$$

is a block matrix with each A_{ii} a square matrix, then:

(i) A is **block diagonal** if $A_{ij} = 0$ for $i \neq j$;
(ii) A is **block upper triangular** if $A_{ij} = 0$ for $i > j$;
(iii) A is **block lower triangular** if $A_{ij} = 0$ for $i < j$.

If A is an $n \times n$ diagonal matrix with diagonal entries $\lambda_1, \ldots, \lambda_n$, then we write $A = \operatorname{diag}(\lambda_1, \ldots, \lambda_n)$. Similarly, $A = \operatorname{diag}(A_1, \ldots, A_r)$, where each A_i is a $n_i \times n_i$ matrix, denotes a block diagonal matrix with diagonal entries $A_1, \ldots, A_r$.

The $n \times n$ matrix $I_n = \operatorname{diag}(1, \ldots, 1)$ is called the **identity matrix**. Verify that if $A \in K^{m \times n}$, then $I_m A = A = A I_n$.

An $n \times n$ matrix A is said to be **invertible** if there exists an $n \times n$ matrix B such that $AB = I_n = BA$. The matrix B is called an **inverse** of A. If A is invertible then it has a unique inverse. Indeed, if B and B' are inverses of A, then $B = BI_n = B(AB') = (BA)B' = I_nB' = B'$, We denote the inverse of A by A^{-1}. Clearly, $(A^{-1})^{-1} = A$.

Note that every square matrix is not invertible. If A is a square matrix whose r–th row is zero, then for any matrix B, the r–th row of AB is also zero, and so A is not invertible. The sum of two invertible matrices need not be an invertible matrix. The reader should supply an example of such matrices. However, if A and B are invertible matrices, then so is AB, and

$$(AB)^{-1} = B^{-1}A^{-1}.$$

Thus, a finite product of invertible matrices is also an invertible matrix.

The **transpose** of an $m \times n$ matrix is an $n \times m$ matrix A^t such that $[A^t]_{ij} = [A]_{ji}$. If $A \in \mathbb{C}^{n\times n}$, then A^* denotes the **conjugate transpose** of A, that is, $[A^*]_{ij} = \overline{[A]}_{ji}$, where bar denotes the conjugate of a complex number. A square matrix A is **symmetric** (respectively, **Hermitian**) if $A^t = A$ (respectively, $A^* = A$), and **skew–symmetric** (respectively, **skew–Hermitian**) if $A^t = -A$ (respectively, $A^* = -A$).

Clearly, the transpose of an upper triangular matrix is lower triangular, and a diagonal matrix is symmetric.

The following proposition gives elementary properties of transpose.

PROPOSITION 1.1. *Let K be a field. If $A, B \in K^{m\times n}$ and $\lambda \in K$, then*

(i) $(A^t)^t = A$;

(ii) $(A + B)^t = A^t + B^t$;

(iii) $(\lambda A)^t = \lambda A^t$.

If $A \in K^{m\times q}$ and $B \in K^{q\times n}$, then

(iv) $(AB)^t = B^tA^t$.

PROOF. (i) $\left[(A^t)^t\right]_{ij} = [A^t]_{ji} = [A]_{ij}$ for all $i = 1, \ldots, m$, $j = 1, \ldots, n$. Hence $(A^t)^t = A$. (ii) and (iii) are easy.

(iv) $\left[(AB)^t\right]_{ij} = [AB]_{ji} = \sum_{k=1}^{q} [A]_{jk}[B]_{ki} = \sum_{k=1}^{q} [B^t]_{ik}[A^t]_{kj} = [B^tA^t]_{ij}$.

Hence $(AB)^t = B^tA^t$. ∎

Similar results hold for conjugate transpose.

PROPOSITION 1.2. *If $A, B \in \mathbb{C}^{m\times n}$, then for $\lambda \in \mathbb{C}$:*

(i) $(A^*)^* = A$;

(ii) $(A + B)^* = A^* + B^*$;

(iii) $(\lambda A)^* = \overline{\lambda}A^*$.

If $A \in \mathbb{C}^{m\times q}$ and $B \in \mathbb{C}^{q\times n}$, then

(iv) $(AB)^* = B^*A^*$.

Let $E(m,n)_{ij}$ be an $m \times n$ matrix whose (i,j)–th entry is 1 and all other entries zero, and let $e(n)_i$ be a column matrix of length n with 1 at i–th row and all other entries zero.

PROPOSITION 1.3. *Let* $A \in K^{m\times n}$.

(i) $Ae(n)_j$ *is the* j*–th column of* A;

(ii) $e(m)_i^t A$ *is the* i*–th row of* A;

(iii) $e(m)_i^t Ae(n)_j = [A]_{ij}$;

(iv) $e(n)_i^t e(n)_j = \delta_{ij}$ *(Kronecker delta)*;

(v) $e(m)_i e(n)_j^t = E(m,n)_{ij}$.

PROOF. Exercise. ∎

We normally do not write labels m and n for $E(m,n)_{ij}$ and $e(n)_i$ as the size of these matrices is generally clear from the context. In $K^{m\times n}$ the mn matrices E_{ij} $(i = 1,\ldots,m,\ j = 1,\ldots,n)$ are called **matrix units.** A matrix $A \in K^{m\times n}$ can be uniquely expressed as a linear combination of these mn matrix units:

$$A = \sum_{j=1}^{n}\sum_{i=1}^{m} [A]_{ij}\, E_{ij}.$$

In $K^{n\times n}$,

$$E_{ij}E_{rs} = \delta_{jr}E_{is}.$$

This identity is a consequence of Proposition 1.3 and the associativity of matrix multiplication: $E_{ij}E_{rs} = \left(e_ie_j^t\right)\left(e_re_s^t\right) = e_i\left(e_j^te_r\right)e_s^t = \delta_{jr}e_ie_s^t = \delta_{jr}E_{is}$.

We now describe a useful function on matrices called the **trace** of a matrix. If A is a square matrix, then the trace of A, denoted by $\operatorname{tr} A$, is the sum of diagonal entries of A. More precisely, if $A \in K^{n\times n}$, then the trace of A is an element of K, defined by:

$$\operatorname{tr} A = \sum_{i=1}^{n} [A]_{ii}\,.$$

The following proposition summarizes the basic properties of the trace function.

PROPOSITION 1.4. *Let* K *be a field, let* $\lambda \in K$ *and* $A, B, P \in K^{n\times n}$*, where* P *is invertible. Then:*

(i) $\operatorname{tr}(A+B) = \operatorname{tr}(A) + \operatorname{tr}(B)$;

(ii) $\operatorname{tr}(\lambda A) = \lambda \operatorname{tr}(A)$;

(iii) $\operatorname{tr}(AB) = \operatorname{tr}(BA)$;

(iv) $\operatorname{tr}\left(P^{-1}AP\right) = \operatorname{tr}(A)$.

PROOF. (i) and (ii) are easy.

(iii)

$$\begin{aligned}\operatorname{tr}(AB) &= \sum_{i=1}^{n} [AB]_{ii} = \sum_{i=1}^{n}\left(\sum_{j=1}^{n} [A]_{ij}[B]_{ji}\right)\\ &= \sum_{j=1}^{n}\left(\sum_{i=1}^{n} [B]_{ji}[A]_{ij}\right) = \sum_{j=1}^{n} [BA]_{jj} = \operatorname{tr}(BA).\end{aligned}$$

(iv) By (iii), $\operatorname{tr}\left(P^{-1}AP\right) = \operatorname{tr}\left((AP)P^{-1}\right) = \operatorname{tr}(A)$. ∎

2. Elementary operations on matrices

The following operations on rows (respectively, columns) of a matrix are called **elementary row** (respectively, **column**) **operations**:

I interchange of two rows (respectively, columns);
II multiplication of a row (respectively, column) by a nonzero scalar;
III adding to a row (respectively, column) a scalar multiple of another row (respectively, column).

We first show that elementary row (respectively, column) operations on an $m \times n$ matrix A can be performed by premultiplying (respectively, postmultiplying) A by a suitable $m \times m$ (respectively, $n \times n$) matrix.

Observe that if E_{rs} is an $m \times m$ matrix unit, then by Proposition 1.3, $E_{rs} = e_r e_s^t$, and the k–th row of $E_{rs}A$ is $e_k^t (E_{rs}A) = (e_k^t e_r) e_s^t A = \delta_{kr} e_s^t A$. Thus, $E_{rs}A$ is an $m \times n$ matrix whose r–th row is identical to the s–th row of A, and all other entries zero.

$$E_{rs}A = \begin{bmatrix} 0 & \cdots & 0 \\ \vdots & & \vdots \\ 0 & \cdots & 0 \\ [A]_{s1} & \cdots & [A]_{sn} \\ 0 & \cdots & 0 \\ \vdots & & \vdots \\ 0 & \cdots & 0 \end{bmatrix} \begin{matrix} \\ \\ \\ \rightarrow r\text{-th row.} \\ \\ \\ \\ \end{matrix}$$

Similarly, if E_{rs} is an $n \times n$ matrix unit, then the k–th column of AE_{rs} is $(AE_{rs}) e_k = Ae_r (e_s^t e_k) = \delta_{sk} Ae_r$. Thus, AE_{rs} has s–th column identical to the r–th column of A, and all other entries zero.

$$AE_{rs} = \begin{bmatrix} 0 & \cdots & 0 & [A]_{1r} & 0 & \cdots & 0 \\ \vdots & & \vdots & \vdots & \vdots & & \vdots \\ 0 & \cdots & 0 & [A]_{mr} & 0 & \cdots & 0 \end{bmatrix}.$$
$$\downarrow$$
$$s\text{-th column}$$

Define the following $n \times n$ square matrices:

1. $\mathcal{E}(r,s) = I_n + E_{rs} + E_{sr} - E_{rr} - E_{ss}$, called an **elementary permutation matrix**;
2. $\mathcal{E}_r(\lambda) = I_n + (\lambda - 1) E_{rr}$, $\lambda \neq 0$, called an **elementary dilation**;
3. $\mathcal{E}_{rs}(\lambda) = I_n + \lambda E_{rs}$, $r \neq s$, called an **elementary transvection**.

These elementary matrices are invertible; and the inverse of an elementary matrix is an elementary matrix of the same type: $\mathcal{E}(r,s)^{-1} = \mathcal{E}(r,s)$, $\mathcal{E}_r(\lambda)^{-1} = \mathcal{E}_r(\lambda^{-1})$, $\mathcal{E}_{rs}(\lambda)^{-1} = \mathcal{E}_{rs}(-\lambda)$. Thus, using the above observations, we can prove the next result.

PROPOSITION 2.1. *Let A be an $m \times n$ matrix.*

1. *$\mathcal{E}(r,s) A$ (respectively, $A\mathcal{E}(r,s)$) is an $m \times n$ matrix which is same as A except that the r–th and s–th rows (respectively, columns) are interchanged, that is, it is obtained from A by elementary row (respectively, column) operation of type I.*

2. $\mathcal{E}_r(\lambda)A$ *(respectively,* $A\mathcal{E}_r(\lambda)$*) is an* $m\times n$ *matrix which is same as* A *except that the* r*–th row (respectively, column) is multiplied by* λ*, that is, it is obtained from* A *by applying elementary row (respectively, column) operation of type II.*
3. $\mathcal{E}_{rs}(\lambda)A$ *(respectively,* $A\mathcal{E}_{rs}(\lambda)$*) is an* $m\times n$ *matrix which is same as* A *except that the* r*–th row (respectively,* s*–th column) is replaced by* r*–th row (respectively,* s*–th column) of* A *plus* λ *times the* s*–th row (respectively,* r*–th column), that is, it is obtained from* A *by applying elementary row (respectively, column) operation of type III.*

Obviously, we are premultiplying A by elementary matrices of size $m\times m$, and postmultiplying A by elementary matrices of size $n\times n$.

An $m\times n$ matrix is said to be a **row reduced echelon matrix** if:

(i) all zero rows are in the bottom position;
(ii) the first nonzero element (from the left), called the leading entry, of a nonzero row is 1;
(iii) if, from the top, first t rows are nonzero, and the leading entry of each i–th row $i=1,\dots,t$, is at the k_i–th column, then all other entries of k_i–th column are zero, and $k_i>k_{i-1}$, $i=2,\dots,t$.

The zero matrix is deemed to be a row reduced echelon matrix. Examples of row reduced echelon matrices are

$$\begin{bmatrix}1&2&3&4\end{bmatrix},\quad \begin{bmatrix}1&6&0&3\\0&0&1&2\\0&0&0&0\end{bmatrix},\quad \begin{bmatrix}1&5&0&0&3&0\\0&0&0&1&3&0\\0&0&0&0&0&1\\0&0&0&0&0&0\end{bmatrix}.$$

Observe that only row reduced echelon invertible matrix of size $n\times n$ is I_n.

Matrices A and B of the same size are said to be **row equivalent** if one can be obtained from the other by elementary row operations. Thus, A and B are row equivalent if and only if there are elementary matrices $P_1,\dots,P_k$ such that $B=P_k\cdots P_1A$.

THEOREM 2.2. *An* $m\times n$ *matrix* A *is row equivalent to a row reduced echelon matrix of size* $m\times n$.

PROOF. Clearly, we can assume that A is a nonzero matrix. Among all nonzero rows of A select a row whose leading entry is at the left most column j_1. Thus, all the columns of A which are on the left of column j_1 (if any) are zero. Interchange this row with the first row of A (if required), and we have the matrix A_1 of the following form:

$$A_1=\begin{bmatrix}0&\cdots&0&a_1&*&\cdots&*\\ \vdots&&\vdots&\vdots&\vdots&&\vdots\\ 0&\cdots&0&a_m&*&\cdots&*\end{bmatrix},\quad a_1\neq 0.$$

$$\underset{j_1\text{-th column}}{\downarrow}$$

$*$ denotes the entries of A_1 whose precise values are of no concern.

Multiply the first row by a_1^{-1} so that the leading entry of the first row is 1. Then apply elementary row operations of type III to get A_2:

$$A_2 = \begin{bmatrix} 0 & \cdots & 0 & 1 & * & \cdots & * \\ 0 & \cdots & 0 & 0 & * & \cdots & * \\ \vdots & & \vdots & \vdots & \vdots & & \vdots \\ 0 & \cdots & 0 & 0 & * & \cdots & * \end{bmatrix}.$$
$$\downarrow$$
$$j_1\text{-th column}$$

If there are no nonzero rows below the first row of A_2 then it is a row reduced echelon matrix. Otherwise, among all nonzero rows of A_2 which are below the first row select a row whose leading entry is at the left most column j_2. Thus, all columns of A_2 which are on the left of j_2-th column, except the j_1-th column, are zero. Interchange this row with the second row of A_2, and we have:

$$A_3 = \begin{bmatrix} 0 & \cdots & 0 & 1 & * & \cdots & * & b_1 & * & \cdots & * \\ 0 & \cdots & 0 & 0 & 0 & \cdots & 0 & b_2 & * & \cdots & * \\ \vdots & & \vdots & \vdots & \vdots & & \vdots & \vdots & \vdots & & \vdots \\ 0 & \cdots & 0 & 0 & 0 & \cdots & 0 & b_m & * & \cdots & * \end{bmatrix}, \ b_2 \neq 0.$$
$$\downarrow \qquad\qquad \downarrow$$
$$j_1\text{-th column} \qquad j_2\text{-th column}$$

Next, multiply the second row by b_2^{-1} so that the leading entry of the second row is 1. Then apply elementary row operations of type III to get:

$$A_4 = \begin{bmatrix} 0 & \cdots & 0 & 1 & * & \cdots & * & 0 & * & \cdots & * \\ 0 & \cdots & 0 & 0 & 0 & \cdots & 0 & 1 & * & \cdots & * \\ 0 & \cdots & 0 & 0 & 0 & \cdots & 0 & 0 & * & \cdots & * \\ \vdots & & \vdots & \vdots & \vdots & & \vdots & \vdots & \vdots & & \vdots \\ 0 & \cdots & 0 & 0 & 0 & \cdots & 0 & 0 & * & \cdots & * \end{bmatrix}.$$
$$\downarrow \qquad\qquad \downarrow$$
$$j_1\text{-th column} \qquad j_2\text{-th column}$$

If all the rows below the second row of A_4 are zero then A_4 is a row reduced echelon matrix. Otherwise, we continue this procedure to get to reduce A to a row reduced echelon matrix. ∎

It is clear from the proof of Theorem 2.2 that to a matrix A there corresponds a unique row reduced echelon matrix. The number of nonzero rows of a row reduced echelon matrix corresponding to A is called the **rank** of A, denoted by rank A.

EXAMPLE 2.1. *We illustrate the above procedure by reducing*

$$A = \begin{bmatrix} 0 & 2 & 2 & -1 & 3 \\ 1 & 2 & 5 & 0 & 7 \\ 1 & 4 & 7 & 0 & 13 \\ 3 & -2 & 7 & 2 & 3 \end{bmatrix}$$

to row reduced echelon matrix.

Since, there is a nonzero entry at the first column and the second row, we interchange the first and second rows:

$$A_1 = \mathcal{E}(1,2)A = \begin{bmatrix} 1 & 2 & 5 & 0 & 7 \\ 0 & 2 & 2 & -1 & 3 \\ 1 & 4 & 7 & 0 & 13 \\ 3 & -2 & 7 & 2 & 3 \end{bmatrix}.$$

The leading entry of the first row is already 1. Now we convert all other entries of the first column to zero. We multiply the first row by -3 and add to the fourth row, and multiply the first row by -1 and add to the third row. Thus, we have:

$$A_2 = \mathcal{E}_{31}(-1)\mathcal{E}_{41}(-3)A_1 = \begin{bmatrix} 1 & 2 & 5 & 0 & 7 \\ 0 & 2 & 2 & -1 & 3 \\ 0 & 2 & 2 & 0 & 6 \\ 0 & -8 & -8 & 2 & -18 \end{bmatrix}.$$

For the next step, we can select any of the rows below the first row. Since all elements of the third row are multiple of 2, the leading entry, computations are easier with this row. Therefore, we prefer to interchange the second and the third rows:

$$A_3 = \mathcal{E}(2,3)A_2 = \begin{bmatrix} 1 & 2 & 5 & 0 & 7 \\ 0 & 2 & 2 & 0 & 6 \\ 0 & 2 & 2 & -1 & 3 \\ 0 & -8 & -8 & 2 & -18 \end{bmatrix}.$$

Dividing the second row by 2:

$$A_4 = \mathcal{E}_2\left(\frac{1}{2}\right) A_3 = \begin{bmatrix} 1 & 2 & 5 & 0 & 7 \\ 0 & 1 & 1 & 0 & 3 \\ 0 & 2 & 2 & -1 & 3 \\ 0 & -8 & -8 & 2 & -18 \end{bmatrix}.$$

Using elementary row operations of type III, we convert all other entries of the second column to 0.

$$A_5 = \mathcal{E}_{42}(8)\,\mathcal{E}_{32}(-2)\,\mathcal{E}_{12}(-2)\,A_4 = \begin{bmatrix} 1 & 0 & 3 & 0 & 1 \\ 0 & 1 & 1 & 0 & 3 \\ 0 & 0 & 0 & -1 & -3 \\ 0 & 0 & 0 & 2 & 6 \end{bmatrix}.$$

Next, we select the third row and multiply it by -1 to make the leading entry 1.

$$A_6 = \mathcal{E}_3(-1)\,A_5 = \begin{bmatrix} 1 & 0 & 3 & 0 & 1 \\ 0 & 1 & 1 & 0 & 3 \\ 0 & 0 & 0 & 1 & 3 \\ 0 & 0 & 0 & 2 & 6 \end{bmatrix}$$

Finally, we multiply the third row by -2 and add to the fourth row to get the required row reduced echelon matrix.

$$A_7 = \mathcal{E}_{43}(-2)\,A_6 = \begin{bmatrix} 1 & 0 & 3 & 0 & 1 \\ 0 & 1 & 1 & 0 & 3 \\ 0 & 0 & 0 & 1 & 3 \\ 0 & 0 & 0 & 0 & 0 \end{bmatrix}.$$

Collecting all these reductions, we have $A_7 = \mathcal{E}_{43}(-2)\,\mathcal{E}_3(-1)\,\mathcal{E}_{42}(8)\,\mathcal{E}_{32}(-2)$ $\mathcal{E}_{12}(-2)\,\mathcal{E}_2\left(\frac{1}{2}\right)\mathcal{E}(2,3)\,\mathcal{E}_{31}(-1)\,\mathcal{E}_{41}(-3)\,\mathcal{E}(1,2)\,A$, that is,

$$\begin{bmatrix} 1 & 0 & 3 & 0 & 1 \\ 0 & 1 & 1 & 0 & 3 \\ 0 & 0 & 0 & 1 & 3 \\ 0 & 0 & 0 & 0 & 0 \end{bmatrix} = \begin{bmatrix} 0 & 2 & -1 & 0 \\ 0 & -\frac{1}{2} & \frac{1}{2} & 0 \\ -1 & -1 & 1 & 0 \\ 2 & -5 & 2 & 1 \end{bmatrix} \begin{bmatrix} 0 & 2 & 2 & -1 & 3 \\ 1 & 2 & 5 & 0 & 7 \\ 1 & 4 & 7 & 0 & 13 \\ 3 & -2 & 7 & 2 & 3 \end{bmatrix}.$$

PROPOSITION 2.3. *An $n \times n$ matrix A is invertible if and only if A is a finite product of elementary matrices.*

PROOF. If A is a finite product of elementary matrices, then A is clearly invertible. Conversely, suppose that A is invertible. Let $P_1, \dots, P_k$ be elementary matrices such that $P_k \cdots P_1 A = E$, where E is a row reduced echelon matrix (Theorem 2.2). Since $P_k \cdots P_1 A$ is also invertible, $E = I_n$. Therefore, $A = P_1^{-1} \cdots P_k^{-1}$. Since the inverse of an elementary matrix is also elementary, A is a finite product of elementary matrices. ∎

COROLLARY 2.4. *Let A and B be matrices of size $m \times n$. Then A and B are row equivalent if and only if there exists an invertible matrix P such that $B = PA$.*

We have reduced matrices to simpler forms by application of elementary row operations. If elementary column operations are also performed then a further reduction is possible.

THEOREM 2.5. *Let A be a nonzero matrix of size $m \times n$. Then by elementary row and column operations, it can be reduced to an $m \times n$ matrix of one of the following forms:*

$$I_r, \quad \begin{bmatrix} I_r & 0 \\ 0 & 0 \end{bmatrix}, \quad \begin{bmatrix} I_r & 0 \end{bmatrix}, \quad \begin{bmatrix} I_r \\ 0 \end{bmatrix}.$$

PROOF. By row and column interchanges bring a nonzero entry of A to the (1,1)-th position, and then divide the first row (or column) by this entry so that in the transformed matrix (1,1)-th entry is 1. Thus A is transformed to

$$A_1 = \begin{bmatrix} 1 & * & \cdots & * \\ * & & & \\ \vdots & & A_1' & \\ * & & & \end{bmatrix}.$$

Now by elementary row and column operations of type III reduce all other entries of first row and first column to zero. Thus, we have

$$A_2 = \begin{bmatrix} 1 & 0 & \cdots & 0 \\ 0 & & & \\ \vdots & & A_2' & \\ 0 & & & \end{bmatrix}.$$

If $A_2' = 0$, then we have the required form. Otherwise, we repeat the procedure on a smaller matrix A_2'. This proves the theorem. ∎

Matrices A and B of the same size are said to be **equivalent** if either of these matrices can be obtained from the other by elementary row and column operations. It then follows from Proposition 2.1 and 2.3 that A and B are equivalent if and only if there are invertible matrices P and Q, of suitable sizes, such that $B = PAQ$. Thus, Theorem 2.5 can be restated as

THEOREM 2.6. *An $m \times n$ matrix is equivalent to an $m \times n$ matrix of one of the following forms:*

$$I_r, \quad \begin{bmatrix} I_r & 0 \\ 0 & 0 \end{bmatrix}, \quad \begin{bmatrix} I_r & 0 \end{bmatrix}, \quad \begin{bmatrix} I_r \\ 0 \end{bmatrix}.$$

Clearly, the number r appearing in the statements of Theorems 2.5 and 2.6 is the rank of the matrix A. Thus A is an $n \times n$ invertible matrix if and only if A is equivalent to the $n \times n$ identity matrix. Also, two equivalent matrices are of the same rank.

EXAMPLE 2.2. *We again consider the matrix*

$$A = \begin{bmatrix} 0 & 2 & 2 & -1 & 3 \\ 1 & 2 & 5 & 0 & 7 \\ 1 & 4 & 7 & 0 & 13 \\ 3 & -2 & 7 & 2 & 3 \end{bmatrix}$$

to find a matrix in the simplest form which is equivalent to A.

First we interchange the first and second rows to bring a nonzero entry at the (1,1)-th position:

$$A_1 = \mathcal{E}(1,2)A = \begin{bmatrix} 1 & 2 & 5 & 0 & 7 \\ 0 & 2 & 2 & -1 & 3 \\ 1 & 4 & 7 & 0 & 13 \\ 3 & -2 & 7 & 2 & 3 \end{bmatrix}.$$

We perform the elementary row and column operations of type III to reduce all other entries of the first row and the first column zero.

$$A_2 = \mathcal{E}_{31}(-1)\mathcal{E}_{41}(-3)A_1\mathcal{E}_{12}(-2)\mathcal{E}_{13}(-5)\mathcal{E}_{15}(-7) = \begin{bmatrix} 1 & 0 & 0 & 0 & 0 \\ 0 & 2 & 2 & -1 & 3 \\ 0 & 2 & 2 & 0 & 6 \\ 0 & -8 & -8 & 2 & -18 \end{bmatrix}.$$

Now the (2,2)-th entry is already nonzero, we divide the second column by 2:

$$A_3 = A_2\mathcal{E}_2\left(\frac{1}{2}\right) = \begin{bmatrix} 1 & 0 & 0 & 0 & 0 \\ 0 & 1 & 2 & -1 & 3 \\ 0 & 1 & 2 & 0 & 6 \\ 0 & -4 & -8 & 2 & -18 \end{bmatrix},$$

(Why did we prefer dividing the second column by 2, not the row?). Then the elementary row and column operations of type III reduce A_3 to:

$$A_4 = \mathcal{E}_{42}(4)\mathcal{E}_{32}(-1)\,A_3\mathcal{E}_{23}(-2)\mathcal{E}_{24}(1)\mathcal{E}_{25}(-3) = \begin{bmatrix} 1 & 0 & 0 & 0 & 0 \\ 0 & 1 & 0 & 0 & 0 \\ 0 & 0 & 0 & 1 & 3 \\ 0 & 0 & 0 & -2 & -6 \end{bmatrix}.$$

Next, we interchange the third and fourth columns to have a nonzero entry at (3,3)-th position:

$$A_5 = A_4\mathcal{E}(3,4) = \begin{bmatrix} 1 & 0 & 0 & 0 & 0 \\ 0 & 1 & 0 & 0 & 0 \\ 0 & 0 & 1 & 0 & 3 \\ 0 & 0 & -2 & 0 & -6 \end{bmatrix}.$$

Finally,

$$A_6 = \mathcal{E}_{43}(2)\, A_5 \mathcal{E}_{35}(-3) = \begin{bmatrix} 1 & 0 & 0 & 0 & 0 \\ 0 & 1 & 0 & 0 & 0 \\ 0 & 0 & 1 & 0 & 0 \\ 0 & 0 & 0 & 0 & 0 \end{bmatrix}.$$

Hence,

$$\begin{bmatrix} I_3 & 0 \\ 0 & 0 \end{bmatrix} = PAQ,$$

where $P = \mathcal{E}_{43}(2)\,\mathcal{E}_{42}(4)\,\mathcal{E}_{32}(-1)\,\mathcal{E}_{31}(-1)\,\mathcal{E}_{41}(-3)\,\mathcal{E}(1,2)$, a 4×4 matrix, and Q is a 5×5 matrix given by $Q = \mathcal{E}_{12}(-2)\,\mathcal{E}_{13}(-5)\,\mathcal{E}_{15}(-7)\,\mathcal{E}_2\left(\frac{1}{2}\right)\mathcal{E}_{23}(-2)\,\mathcal{E}_{24}(1)\,\mathcal{E}_{25}(-3)\,\mathcal{E}(3,4)\mathcal{E}_{35}(-3)$.

3. Determinants

Let K be a field and let $A \in K^{n \times n}$. The **determinant** of A is an element of K defined by

$$\det A = \sum_{\sigma \in S_n} \operatorname{sgn} \sigma \prod_{i=1}^{n} [A]_{i\sigma(i)}.$$

Here, S_n denotes the set (in fact, a group) of all bijections from the set of integers $\{1, \ldots, n\}$ to itself and $\operatorname{sgn} \sigma$ is the **signature** of σ, defined by

$$\operatorname{sgn} \sigma = \prod_{1 \le i < j \le n} \frac{\sigma(j) - \sigma(i)}{j - i} \quad \text{(which is } +1 \text{ or } -1).$$

Note that S_n has $n!$ elements.

Let $\sigma \in S_n$ and $\sigma(i) = k_i$, $i = 1, \ldots, n$. We will denote σ by an ordered n-tuple $(k_1, \ldots, k_n)$. Since σ is a bijection, $(k_1, \ldots, k_n)$ is just a permutation of integers $1, \ldots, n$. For example, a 5-tuple $(2,4,3,1,5)$ denotes an element of S_5 which maps 1 to 2, 2 to 4, 3 to 3, 4 to 1 and 5 to 5. Also

$$\begin{aligned} \operatorname{sgn}(2,4,3,1,5) &= \frac{4-2}{2-1} \cdot \frac{3-2}{3-1} \cdot \frac{1-2}{4-1} \cdot \frac{5-2}{5-1} \cdot \frac{3-4}{3-2} \cdot \frac{1-4}{4-2} \\ &\quad \cdot \frac{5-4}{5-2} \cdot \frac{1-3}{4-3} \cdot \frac{5-3}{5-3} \cdot \frac{5-1}{5-4} \\ &= 1. \end{aligned}$$

Now the reader can verify that

$$\begin{aligned} \det \begin{bmatrix} a_{11} & a_{12} \\ a_{21} & a_{22} \end{bmatrix} &= \operatorname{sgn}(1,2)\, a_{11}a_{22} + \operatorname{sgn}(2,1)\, a_{12}a_{21} \\ &= a_{11}a_{22} - a_{12}a_{21}; \end{aligned}$$

$$\det\begin{bmatrix} a_{11} & a_{12} & a_{13} \\ a_{21} & a_{22} & a_{23} \\ a_{31} & a_{32} & a_{33} \end{bmatrix} = \operatorname{sgn}(1,2,3)\, a_{11}a_{22}a_{33} + \operatorname{sgn}(2,3,1)\, a_{12}a_{23}a_{31}$$

$$+\operatorname{sgn}(3,1,2)\, a_{13}a_{21}a_{32} + \operatorname{sgn}(3,2,1)\, a_{13}a_{22}a_{31}$$
$$+\operatorname{sgn}(2,1,3)\, a_{12}a_{21}a_{33} + \operatorname{sgn}(1,3,2)\, a_{11}a_{23}a_{32}$$
$$= a_{11}a_{22}a_{33} + a_{12}a_{23}a_{31} + a_{13}a_{21}a_{32}$$
$$-a_{13}a_{22}a_{31} - a_{12}a_{21}a_{33} - a_{11}a_{23}a_{32}.$$

We now establish some basic properties of determinants. First, we collect some facts which are required to prove these properties.

Let U be the set of all ordered pairs defined by $U = \{(i,j) \mid 1 \leq i < j \leq n\}$. For $\tau \in S_n$ and for $(i,j) \in U$, either $\tau(j) > \tau(i)$ or $\tau(j) < \tau(i)$. Let $U_1 = \{(\tau(i), \tau(j)) \mid (i,j) \in U, \tau(i) < \tau(j)\}$, and $U_2 = \{(\tau(j), \tau(i)) \mid (i,j) \in U, \tau(i) > \tau(j)\}$. Clearly $U_1 \cup U_2 \subseteq U$. If $(r,s) \in U$, then $\tau(r') = r$, $\tau(s') = s$ for some $r', s' \in \{1, \ldots, n\}$. If $(r', s') \in U$, then $(r,s) \in U_1$, and if $(s', r') \in U$, then $(r,s) \in U_2$. Thus, $(r,s) \in U_1 \cup U_2$. Hence $U = U_1 \cup U_2$.

Now for any $\sigma \in S_n$,

$$\prod_{(i,j)\in U} \frac{\sigma\tau(j) - \sigma\tau(i)}{\tau(j) - \tau(i)} = \prod_{\tau(j)>\tau(i)} \frac{\sigma\tau(j) - \sigma\tau(i)}{\tau(j) - \tau(i)} \prod_{\tau(j)<\tau(i)} \frac{\sigma\tau(j) - \sigma\tau(i)}{\tau(j) - \tau(i)}$$
$$= \prod_{(\tau(i),\tau(j))\in U_1} \frac{\sigma\tau(j) - \sigma\tau(i)}{\tau(j) - \tau(i)} \prod_{(\tau(j),\tau(i))\in U_2} \frac{\sigma\tau(i) - \sigma\tau(j)}{\tau(i) - \tau(j)}$$
$$= \prod_{(i',j')\in U} \frac{\sigma(j') - \sigma(i')}{j' - i'} = \operatorname{sgn}\sigma.$$

Hence, for a given $\tau \in S_n$,

$$\operatorname{sgn}\sigma = \prod_{(i,j)\in U} \frac{\sigma\tau(j) - \sigma\tau(i)}{\tau(j) - \tau(i)} \text{ for all } \sigma \in S_n.$$

Thus, for $\sigma, \tau \in S_n$,

$$\operatorname{sgn}\sigma\tau = \prod_{(i,j)\in U} \frac{\sigma\tau(j) - \sigma\tau(i)}{j - i}$$
$$= \prod_{(i,j)\in U} \frac{\sigma\tau(j) - \sigma\tau(i)}{\tau(j) - \tau(i)} \prod_{(i,j)\in U} \frac{\tau(j) - \tau(i)}{j - i}$$
$$= \operatorname{sgn}\sigma \operatorname{sgn}\tau.$$

Here $\sigma\tau$ denotes the composition of mappings σ and τ.

If $i \in S_n$ is the identity mapping, then clearly $\operatorname{sgn} i = 1$, and $1 = \operatorname{sgn}(\sigma\sigma^{-1}) = \operatorname{sgn}\sigma \operatorname{sgn}\sigma^{-1}$. Thus

$$\operatorname{sgn}\sigma = \operatorname{sgn}\sigma^{-1}.$$

Another fact which we require is that if $\sigma \in S_n$, then $\sigma S_n = \{\sigma\theta \mid \theta \in S_n\} = S_n$. This is easy, since the composition of two bijections is also a bijection.

PROPOSITION 3.1. *Let K be a field and let $A \in K^{n\times n}$. Then*

$$\det A = \det A^t.$$

PROOF. For $\sigma \in S_n$,

$$\prod_{i=1}^{n}[A]_{i\sigma(i)} = \prod_{i=1}^{n}[A^t]_{\sigma(i)i} = \prod_{i=1}^{n}[A^t]_{\sigma(i)\sigma^{-1}(\sigma(i))} = \prod_{i'=1}^{n}[A^t]_{i'\sigma^{-1}(i')}.$$

Thus,

$$\begin{aligned}\det A &= \sum_{\sigma\in S_n} \operatorname{sgn}\sigma \prod_{i=1}^{n}[A]_{i\sigma(i)} = \sum_{\sigma\in S_n} \operatorname{sgn}\sigma \prod_{i=1}^{n}[A^t]_{i\sigma^{-1}(i)} \\ &= \sum_{\sigma^{-1}\in S_n} \operatorname{sgn}\sigma^{-1} \prod_{i=1}^{n}[A^t]_{i\sigma^{-1}(i)} = \det A^t.\end{aligned}$$

∎

Proposition 3.1 shows an equivalence between the rows and columns concerning the properties of determinants. This allows the derivation of properties of determinants involving columns from the corresponding properties involving rows of the transposed matrix and vice versa. Taking advantage of this, subsequent proofs will be only for properties concerning the rows of a matrix.

Let $A \in K^{n\times n}$ and let $\sigma \in S_n$. The product $\prod_{i=1}^{n}[A]_{i\sigma(i)}$ is such that exactly one element from each row of A and exactly one element from each column of A appears as a factor. This observation leads to the following properties of determinants.

PROPOSITION 3.2. *Let $A \in K^{n\times n}$. If B is the $n\times n$ matrix obtained from A by multiplying its i–th row (respectively, i–th column) by a scalar λ, then* $\det B = \lambda \det A$. *Thus,* $\det(\lambda A) = \lambda^n \det A$.

PROPOSITION 3.3. *Let K be a field and let $A \in K^{n\times n}$. If B is the $n\times n$ matrix obtained from A by interchanging its r–th and s–th rows (respectively, columns), then* $\det B = -\det A$.

PROOF. Without any loss, we can assume that $r < s$. Let τ be a bijection which fixes all elements of $\{1,\dots,n\}$ except r and s, and $\tau(r) = s$, $\tau(s) = r$ (such a bijection is called a transposition).

Since $[B]_{ij} = [A]_{ij}$, $i \neq r, s$, $[B]_{rj} = [A]_{sj}$ and $[B]_{sj} = [A]_{rj}$, for all j, so for any $\sigma \in S_n$:

$$\prod_{i=1}^{n}[B]_{i\sigma(i)} = \prod_{i=1,\, i\neq r,s}^{n}[A]_{i\sigma(i)}\,[A]_{r\sigma(s)}\,[A]_{s\sigma(r)} = \prod_{i=1}^{n}[A]_{i\sigma\tau(i)}.$$

Now,

$$\begin{aligned}\det B &= \sum_{\sigma\in S_n} \operatorname{sgn}\sigma \prod_{i=1}^{n}[B]_{i\sigma(i)} = \sum_{\sigma\in S_n} \operatorname{sgn}\sigma \prod_{i=1}^{n}[A]_{i\sigma\tau(i)} \\ &= \operatorname{sgn}\tau \sum_{\sigma\in S_n} \operatorname{sgn}(\sigma\tau) \prod_{i=1}^{n}[A]_{i\sigma\tau(i)} = \operatorname{sgn}\tau \,\det A.\end{aligned}$$

Thus, to establish the result it is enough to show that $\operatorname{sgn}\tau = -1$. This can be verified:

$$\begin{aligned}\operatorname{sgn}\tau &= \prod_{1\le i<j\le n} \frac{\tau(j)-\tau(i)}{j-i} = \prod_{j=r+1}^{s} \frac{\tau(j)-\tau(r)}{j-r} \prod_{i=r+1}^{s-1} \frac{\tau(s)-\tau(i)}{s-i} \\ &= \prod_{j=r+1}^{s-1} \frac{j-s}{j-r} \prod_{i=r+1}^{s-1} \frac{r-i}{s-i}\cdot\frac{r-s}{s-r} = -1.\end{aligned}$$

∎

PROPOSITION 3.4. *Let K be a field and let $A, B, C \in K^{n\times n}$. If the j-th row (respectively, column) of A is the sum of j–th rows (respectively, columns) of B and C, and all other rows (respectively, columns) of A, B and C are identical, then* $\det A = \det B + \det C$.

PROOF.

$$\begin{aligned}\det A &= \sum_{\sigma\in S_n} \operatorname{sgn}\sigma \prod_{i=1}^{n} [A]_{i\sigma(i)} = \sum_{\sigma\in S_n} \operatorname{sgn}\sigma \prod_{i\ne j} [A]_{i\sigma(i)}\, [A]_{j\sigma(j)} \\ &= \sum_{\sigma\in S_n} \operatorname{sgn}\sigma \prod_{i\ne j} [A]_{i\sigma(i)} \left([B]_{j\sigma(j)} + [C]_{j\sigma(j)}\right) \\ &= \sum_{\sigma\in S_n} \operatorname{sgn}\sigma \prod_{i\ne j} [A]_{i\sigma(i)}\, [B]_{j\sigma(j)} + \sum_{\sigma\in S_n} \operatorname{sgn}\sigma \prod_{i\ne j} [A]_{i\sigma(i)}\, [C]_{j\sigma(j)} \\ &= \det B + \det C.\end{aligned}$$

∎

For $A = [a]$, a matrix of size 1×1, clearly, $\det A = a$. The following result is an inductive procedure of computing the determinant of a matrix.

PROPOSITION 3.5. ***Laplace expansion.*** *Let K be a field, $A \in K^{n\times n}$ and let $A(i,j)$ be a submatrix of A obtained by deleting the i–th row and the j–th column of A. Then*

$$\det A = \sum_{j=1}^{n} (-1)^{i+j}\, [A]_{ij}\, \det A(i,j), \ \textit{for any } i,$$

(expansion by minors with respect to the i-th row)

and

$$\det A = \sum_{i=1}^{n} (-1)^{i+j}\, [A]_{ij}\, \det A(i,j), \ \textit{for any } j.$$

(expansion by minors with respect to the j-th column)

PROOF. Since every summand of $\det A = \sum_{\sigma\in S_n} \operatorname{sgn}\sigma \prod_{r=1}^{n} [A]_{r\sigma(r)}$ contains an element of the r–th row. Therefore collecting together all the terms containing $[A]_{ij}$ $(j = 1, \dots, n)$, we obtain

$$(*) \qquad \det A = [A]_{i1}A_{i1} + \cdots + [A]_{in}A_{in}.$$

Thus, to establish the result, it is sufficient to show that $A_{ij} = (-1)^{i+j}\det A(i,j)$.

Consider, first, the case $i = 1 = j$. In this case, we show that $A_{11} = \det A(1,1)$. Indeed,

$$A_{11} = \sum_{\sigma \in S_n} \operatorname{sgn} \sigma \prod_{k=2}^{n} [A]_{k\sigma(k)},$$

where the summation is over all $\sigma \in S_n$ such that $\sigma(1) = 1$. Hence, by definition, it follows that $A_{11} = \det A(1,1)$.

To prove (*) for general case, we shift $[A]_{ij}$ to (1,1)–th position by means of $i-1$ successive interchange of adjacent rows and $j-1$ successive interchange of adjacent columns. Let the resulting matrix be B. Since the relative positions of rows and columns is the same in A and B, $A(i,j) = B(1,1)$. Thus,

$$\begin{aligned} \det B &= [B]_{11} \det B(1,1) + \text{terms not involving } [B]_{11} \\ &= [A]_{ij} \det A(i,j) + \text{terms not involving } [A]_{ij}. \end{aligned}$$

By Proposition 3.3, it follows that $\det B = (-1)^{i-1}(-1)^{j-1} \det A$. Therefore,

$$\det A = (-1)^{i+j}[A]_{ij} \det A(i,j) + \text{terms not involving } [A]_{ij},$$

and the result now follows on comparison with (*). Equation can now be obtained using equation and the fact that $\det A = \det A^t$. ∎

An easy consequence of this expansion formula is the following result.

PROPOSITION 3.6. *If A and B are square matrices, then*

$$\det \begin{bmatrix} A & C \\ 0 & B \end{bmatrix} = \det A \det B.$$

Thus, if the square matrix A is block triangular with blocks A_i, $i = 1, \dots, k$ on the main diagonal, then $\det A = \det A_1 \cdots \det A_k$. In particular, if A is an upper triangular matrix then $\det A$ is the product of diagonal entries.

PROPOSITION 3.7. *If an $n \times n$ matrix A has two identical rows (respectively, columns), then* $\det A = 0$.

PROOF. By induction on n. If $n = 2$ and A has identical rows, then obviously $\det A = 0$. Assume that for any matrix of size $(n-1) \times (n-1)$ with two identical rows the determinant is zero. Let $A \in K^{n \times n}$ with identical r–th and s–th rows. Choose an i–th row of A so that $i \neq r, s$ and write the Laplace expansion of $\det A$ with respect to the i–th row:

$$\det A = \sum_{j=1}^{n} (-1)^{i+j} [A]_{ij} \det A(i,j).$$

Clearly, each $A(i,j)$ has identical rows. By induction hypothesis, $\det A(i,j) = 0$ for all j. Hence, $\det A = 0$. ∎

If $A \in K^{n \times n}$ and has identical r–th and s–th rows, then the interchange of identical rows (respectively, columns) does not change the matrix. On the other hand, by Proposition 3.3, the determinant must change sign. Thus, $\det A = -\det A$, hence $\det A = 0$. This is another proof of Proposition 3.7, but this is valid provided $\operatorname{char} K \neq 2$.

PROPOSITION 3.8. *Let $A \in K^{n\times n}$ and let B be a matrix obtained by A by adding the elements of i–th row (respectively, column) to the corresponding elements of j–th row (respectively, column) multiplied by a scalar λ. Then* $\det B = \det A$.

PROOF. Follows from Propositions 3.4 and 3.7. ∎

Let K be a field, and let $A \in K^{n\times n}$. The (classical) **adjoint** of A, written as $\operatorname{adj} A$, is an $n \times n$ matrix whose (i, j)–th entry is $(-1)^{i+j} \det A(j, i)$.

Now consider the (r, s)–th entry of $A(\operatorname{adj} A)$.

$$[A(\operatorname{adj} A)]_{rs} = \sum_{k=1}^{n} (-1)^{k+s} [A]_{rk} \det A(s, k).$$

If $r = s$, then by Proposition 3.5, $[A(\operatorname{adj} A)]_{rr} = \det A$. If $r \neq s$, then again by Proposition 3.5, this is the determinant of an $n \times n$ matrix B, whose r–th row is identical to the r–th row of A, and $B(r, k) = A(s, k)$ for $k = 1, \ldots, n$. But then B has identical r–th and s–th rows, and so $\det B = 0$ (Proposition 3.7). Thus, $A(\operatorname{adj} A) = \det A\, I_n$. Similar arguments show that $(\operatorname{adj} A)A = \det A\, I_n$. Hence,

$$A(\operatorname{adj} A) = (\operatorname{adj} A)A = \det A\, I_n.$$

In particular, if A is not invertible, then $\det A = 0$.

If we make use of elementary matrices, we have proved that for $A \in K^{n\times n}$ and $\lambda \in K$:

$\det(\mathcal{E}_r(\lambda) A) = \lambda \det A = \det(A\mathcal{E}_r(\lambda))$ (Proposition 3.2);
$\det(\mathcal{E}(r, s) A) = -\det A = \det(A\mathcal{E}(r, s))$ (Proposition 3.3); and
$\det(\mathcal{E}_{rs}(\lambda) A) = \det A = \det(A\mathcal{E}_{rs}(\lambda))$ (Proposition 3.8);

where $r, s = 1, \ldots, n$ and $r \neq s$.

Using the above results and the fact that an invertible matrix is a product of elementary matrices, we now prove that the determinant function is multiplicative.

PROPOSITION 3.9. *For $A, B \in K^{n\times n}$, K a field,* $\det AB = \det A \det B$.

PROOF. If A is an elementary matrix, then using Proposition 3.4 and 3.8, the reader can easily verify that $\det AB = \det A \det B$.

In particular, it follows that if $P_1, \ldots, P_k$ are elementary matrices, then

$$\det(P_1, \ldots, P_k) = \det P_1 \cdots \det P_k.$$

Since the determinant of elementary matrices of all the three types is nonzero, by Proposition 2.3, it follows that the determinant of an invertible matrix is nonzero.

Now if A is invertible, then by Proposition 2.3, $A = Q_1 \cdots Q_l$, where each Q_i is an elementary matrix. Therefore,

$$\begin{aligned} \det AB &= \det Q_1 (Q_2 \cdots Q_l B) = \det Q_1 \det(Q_2 \cdots Q_l B) = \ldots \\ &= \det Q_1 \cdots \det Q_l \det B = \det A \det B. \end{aligned}$$

Finally, if A is not invertible, then $\det A = 0$. Also there exists an invertible matrix P such that $PA = E$, where E is a row reduced echelon matrix (Theorem 2.2). The last row of E must be zero. Otherwise, by definition of row reduced echelon matrix, $E = I_n$, and so $A = P^{-1}$, an invertible matrix. Thus, $PAB = EB$, and the last row of EB is zero. But then $\det(PAB) = \det(EB) = 0$. Now $\det(PAB) = \det P \det(AB)$ and $\det P \neq 0$. Hence, $\det(AB) = 0 = \det A \det B$. ∎

An immediate consequence of the above equation is the following result.

COROLLARY 3.10. *A square matrix is invertible if and only if its determinant is a nonzero scalar.*

4. Systems of linear equations

Consider the following system of m linear equations in n variables (unknowns) over a field K:

$$\begin{aligned} a_{11}x_1 + \cdots + a_{1n}x_n &= b_1 \\ &\vdots \\ a_{m1}x_1 + \cdots + a_{mn}x_n &= b_m \end{aligned} \tag{1}$$

$a_{ij}, b_i \in K$, $i = 1, \ldots, m$, $j = 1, \ldots, n$. A column vector $\begin{bmatrix} \alpha_1 & \cdots & \alpha_n \end{bmatrix}^t \in K^n$ is a solution of (1) if, on substituting $x_j = \alpha_j$, $j = 1, \ldots, n$, the above equations hold. A system of linear equations is said to be **consistent**, if it has a solution; otherwise, it is **inconsistent**. Let

$$A = \begin{bmatrix} a_{11} & \cdots & a_{1n} \\ \vdots & & \vdots \\ a_{m1} & \cdots & a_{mn} \end{bmatrix}, \; x = \begin{bmatrix} x_1 \\ \vdots \\ x_n \end{bmatrix} \text{ and } b = \begin{bmatrix} b_1 \\ \vdots \\ b_m \end{bmatrix}.$$

Then using matrix multiplication, the above system (1) can be represented by

$$Ax = b. \tag{2}$$

The $m \times n$ matrix A is called the coefficient matrix and the $m \times (n+1)$ matrix $[A \mid b]$ is called the augmented matrix of the system (2). Two systems $Ax = b$ and $A'x = b'$, $A, A' \in K^{m \times n}$, $b, b' \in K^m$, are said to be **equivalent** if they have the same set of solutions.

We now discuss a general procedure of finding solution(s) of the system (2).

By elementary row operations on A, we can reduce A to a row reduced echelon matrix. Equivalently, there exists an invertible $m \times m$ matrix P such that PA is a row reduced echelon matrix. Clearly, $PAx = Pb$ and $Ax = b$ are equivalent systems. Note that the system $PAx = Pb$ is consistent if and only if any row of PA, which is zero, the same row of Pb must be zero. In other words, the rank of PA must be equal to the rank of augmented $m \times (n+1)$ matrix $[PA \mid Pb]$ $(= P[A \mid b])$. Since A and PA, and $[A \mid b]$ and $P[A \mid b]$ are equivalent matrices, we have the following criterion to decide whether a given system is consistent.

PROPOSITION 4.1. *A system of m linear equations in n variables (unknowns) over a field K: $Ax = b$ is consistent if and only if* $\operatorname{rank} A = \operatorname{rank} [A \mid b]$.

Now we assume that the system (2) is consistent and A is in the row reduced echelon form with $\operatorname{rank} A = r$. Thus, the first r rows of A are nonzero. By repeated interchanges of columns of A, if required, we convert A to the following form

$$A' = \begin{bmatrix} 1 & & 0 & b_{1r+1} & \cdots & b_{1n} \\ & \ddots & & \vdots & & \vdots \\ 0 & & 1 & b_{rr+1} & \cdots & b_{rn} \\ 0 & \cdots & 0 & 0 & \cdots & 0 \\ \vdots & & \vdots & \vdots & & \vdots \\ 0 & \cdots & 0 & 0 & \cdots & 0 \end{bmatrix}.$$

Thus, there exists an $n \times n$ matrix E which is product of elementary permutation matrices such that $A' = AE$. Further, if $\alpha \in K^n$ is a solution of $Ax = b$, then $E^{-1}\alpha$ is a solution of $A'y = b$. Since $A'(E^{-1}\alpha) = (AE)(E^{-1}\alpha) = A\alpha = b$. Thus, $A'y = b$ is consistent if and only if the system $Ax = b$ is consistent. Therefore, we consider the following consistent system:

$$\begin{bmatrix} 1 & & 0 & b_{1r+1} & \cdots & b_{1n} \\ & \ddots & & \vdots & & \vdots \\ 0 & & 1 & b_{rr+1} & \cdots & b_{rn} \\ 0 & \cdots & 0 & 0 & \cdots & 0 \\ \vdots & & \vdots & \vdots & & \vdots \\ 0 & \cdots & 0 & 0 & \cdots & 0 \end{bmatrix} \begin{bmatrix} y_1 \\ \vdots \\ \vdots \\ \vdots \\ y_n \end{bmatrix} = \begin{bmatrix} b_1 \\ \vdots \\ \vdots \\ \vdots \\ b_m \end{bmatrix}. \tag{3}$$

Since (3) is a consistent system, $b_{r+1} = \cdots = b_m = 0$. Also, we have

$$y_i + \sum_{k=r+1}^{n} b_{ik}y_k = b_i, \ i = 1, \ldots, r.$$

Thus we have determined first r variables in terms of remaining $n - r$ variables; and we have solutions of the form

$$p = \begin{bmatrix} b_1 - \sum_{k=r+1}^{n} b_{1k}y_k \\ \vdots \\ b_r - \sum_{k=r+1}^{n} b_{rk}y_k \\ y_{r+1} \\ \vdots \\ y_n \end{bmatrix},$$

where $y_{r+1}, \ldots, y_n$ are arbitrary elements of K. Finally, $E^{-1}p$ is a solution of (2).

EXAMPLE 4.1. *Solve the following system of equations:*

$$\begin{aligned} 2x_2 + 2x_3 - x_4 + 3x_5 &= 6 \\ x_1 + 2x_2 + 5x_3 + 7x_5 &= 12 \\ x_1 + 4x_2 + 7x_3 + 13x_5 &= 22 \\ 3x_1 - 2x_2 + 7x_3 + 2x_4 + 3x_5 &= 4. \end{aligned} \tag{4}$$

In the matrix form, we can write the above system as

$$\begin{bmatrix} 0 & 2 & 2 & -1 & 3 \\ 1 & 2 & 5 & 0 & 7 \\ 1 & 4 & 7 & 0 & 13 \\ 3 & -2 & 7 & 2 & 3 \end{bmatrix} \begin{bmatrix} x_1 \\ x_2 \\ x_3 \\ x_4 \\ x_5 \end{bmatrix} = \begin{bmatrix} 6 \\ 12 \\ 22 \\ 4 \end{bmatrix}.$$

We have seen that premultiplying the coefficient matrix by

$$P = \begin{bmatrix} 0 & 2 & -1 & 0 \\ 0 & -1/2 & 1/2 & 0 \\ -1 & -1 & 1 & 0 \\ 2 & -5 & 2 & 1 \end{bmatrix}$$

will reduce the coefficient matrix to a row reduced echelon form (Example 2.1). Therefore, premultiplying the system by P, we have an equivalent system of linear equations:

$$\begin{bmatrix} 1 & 0 & 3 & 0 & 1 \\ 0 & 1 & 1 & 0 & 3 \\ 0 & 0 & 0 & 1 & 3 \\ 0 & 0 & 0 & 0 & 0 \end{bmatrix} \begin{bmatrix} x_1 \\ x_2 \\ x_3 \\ x_4 \\ x_5 \end{bmatrix} = \begin{bmatrix} 2 \\ 5 \\ 4 \\ 0 \end{bmatrix}.$$

By Proposition 4.1, the above system and hence (4) is a consistent system of linear equations. Next, we interchange 3rd and 4th columns to write this in the form as in (3). Thus, we have $E = \mathcal{E}(3,4)$, and we now have

$$\begin{bmatrix} 1 & 0 & 0 & 3 & 1 \\ 0 & 1 & 0 & 1 & 3 \\ 0 & 0 & 1 & 0 & 3 \\ 0 & 0 & 0 & 0 & 0 \end{bmatrix} \begin{bmatrix} y_1 \\ y_2 \\ y_3 \\ y_4 \\ y_5 \end{bmatrix} = \begin{bmatrix} 2 \\ 5 \\ 4 \\ 0 \end{bmatrix},$$

that is, $y_1 = 2 - 3y_4 - y_5$, $y_2 = 5 - y_4 - 3y_5$, $y_3 = 4 - 3y_5$, and y_4, y_5 are arbitrary elements of K. Thus, a typical solution vector is

$$p = \begin{bmatrix} 2 - 3s - t \\ 5 - s - 3t \\ 4 - 3t \\ s \\ t \end{bmatrix}, \quad s, t \in K.$$

Finally, solution of (4) is Ep, that is,

$$\begin{bmatrix} 2 - 3s - t \\ 5 - s - 3t \\ s \\ 4 - 3t \\ t \end{bmatrix}, \quad s, t \in K.$$

5. Polynomials

A **polynomial** of degree n over a field K with an indeterminate x is an expression of the form:

$$a_0 + a_1 x + \cdots + a_n x^n,$$

$a_0, \ldots, a_n$ are in K, $a_n \neq 0$. The addition and multiplication of polynomials over K are defined in the usual manner:

$$\left(a_0 + a_1 x + a_2 x^2 + \cdots\right) + \left(b_0 + b_1 x + b_2 x^2 + \cdots\right)$$
$$= (a_0 + b_0) + (a_1 + b_1)\,x + (a_2 + b_2)\,x^2 + \cdots,$$

and

$$\left(a_0 + a_1 x + a_2 x^2 + \cdots\right)\left(b_0 + b_1 x + b_2 x^2 + \cdots\right)$$
$$= (a_0 b_0) + (a_1 b_0 + a_0 b_1)\,x + (a_2 b_0 + a_1 b_1 + a_0 b_2)\,x^2 + \cdots.$$

We will denote by $K[x]$, the set of all polynomials over K with an indeterminate x. The degree of a polynomial $p(x)$ will be denoted by $\deg p(x)$. Elements of the field K are called the **constant polynomials**. Non zero elements of K are the polynomials of degree zero; the degree of $0 \in K$ is defined to be $-\infty$. Polynomials of degree 1 are called the **linear polynomials**. For $f(x), g(x) \in K[x]$:

$$\deg(f(x) + g(x)) \leq \max\{\deg f(x), \deg g(x)\}$$

and

$$\deg(f(x)g(x)) = \deg f(x) + \deg g(x).$$

If $f(x) = a_0 + a_1x + \cdots + a_nx^n$, a polynomial of degree n over K, the coefficient a_n is called the **leading coefficient** of $f(x)$. A polynomial with leading coefficient 1 is called a **monic polynomial**. Clearly, 1 is the monic polynomial of degree zero.

PROPOSITION 5.1. ***Division Algorithm.*** *For $f(x)$, $g(x) \in K[x]$, $g(x) \neq 0$, there are polynomials $q(x)$, $r(x)$ in $K[x]$ with*

$$f(x) = q(x)g(x) + r(x)$$

*such that either $r(x) = 0$ or $\deg r(x) < \deg g(x)$. The polynomials $q(x)$ and $r(x)$ are uniquely determined by these conditions, called the **quotient** and the **remainder** when $g(x)$ is divided by $f(x)$.*

PROOF. If $\deg f(x) < \deg g(x)$, then let $q(x) = 0$ and $r(x) = g(x)$. Now assume that $\deg f(x) \geq \deg g(x)$. Let $f(x) = a_0 + a_1x + \cdots + a_nx^n$, $a_n \neq 0$ and $g(x) = b_0 + b_1x + \cdots + b_mx^m$, $b_m \neq 0$. If $f_1(x) = f(x) - a_nb_m^{-1}x^{n-m}g(x)$, then $f_1(x) \in K[x]$ and $\deg f_1(x) < \deg f(x)$. Therefore, by induction on $\deg f(x)$, we conclude that there are polynomials $q_1(x)$ and $r(x)$ in $K[x]$ with $f_1(x) = q_1(x)g(x) + r(x)$ such that either $r(x) = 0$ or $\deg r(x) < \deg g(x)$. Hence, we have $f(x) = q(x)g(x) + r(x)$ with $q(x) = a_nb_m^{-1}x^{n-m} + q_1(x)$.

Let there be $q_1(x)$, $q_2(x)$, $r_1(x)$, $r_2(x)$ in $K[x]$ with

$$f(x) = q_i(x)g(x) + r_i(x)$$

such that either $r_i(x) = 0$ or $\deg r_i(x) < \deg g(x)$, $i = 1, 2$. Then

$$(q_1(x) - q_2(x))g(x) = r_2(x) - r_1(x).$$

The above equation clearly gives a contradiction if $r_1(x) \neq r_2(x)$ as then the degrees of the polynomials appearing on the left and right sides are different. Hence $r_1(x) = r_2(x)$ and $q_1(x) = q_2(x)$ as $g(x) \neq 0$. ∎

The above proof is essentially the long division process which we learn in high school algebra.

Let $f(x)$ and $g(x)$ be two nonzero polynomials in $K[x]$. Then $g(x)$ is said to **divide** $f(x)$ if there exists a polynomial $h(x)$ in $K[x]$ such that $f(x) = g(x)h(x)$. In this case we also say that $g(x)$ is a **divisor** or a **factor** of $f(x)$. If $g(x)$ divides $f(x)$, we write $g(x) | f(x)$. Clearly, for any $\alpha \in K \setminus \{0\}$, $\alpha | f(x)$ and $f(x) | \alpha f(x)$. The reader can also verify that $f(x) | g(x)$ and $g(x) | f(x)$ if and only if $g(x) = \alpha f(x)$ for some $\alpha \in K \setminus \{0\}$.

A monic polynomial $d(x) \in K[x]$ is called the **greatest common divisor** (gcd) of $f(x)$ and $g(x)$, written as $d(x) = \gcd(f(x), g(x))$ if $d(x) | f(x)$ and $d(x) | g(x)$ and further if $h(x) \in K[x]$ such that $h(x) | f(x)$ and $h(x) | g(x)$, then $h(x) | d(x)$. If $\gcd(f(x), g(x)) = 1$, the polynomials $f(x)$ and $g(x)$ are called **relatively prime.** Note that since gcd of two polynomials is monic, gcd will always be unique, whenever it exists. In fact, we now show that gcd of two polynomials always exists.

The division algorithm works in $K[x]$ in the same manner as in the set of integers $\mathbb{Z}$. Thus, we can have a similar method for finding gcd of two polynomials. Let $f(x)$ and $g(x)$ be two nonconstant polynomials in $K[x]$ and let $\deg g(x) \leq \deg f(x)$. We divide $f(x)$ by $g(x)$. If $g(x)$ is a factor of $f(x)$, then $\gcd(f(x), g(x)) = \alpha^{-1} g(x)$, where α is the leading coefficient of $g(x)$. Otherwise, let $f_1(x) = g(x)$ and $g_1(x)$ be the remainder obtained on dividing $f(x)$ by $g(x)$. Then $\deg g_1(x) \leq \deg f_1(x)$. At subsequent k–th step: $\deg g_k(x) \leq \deg f_k(x)$. If $g_k(x)$ is a factor of $f_k(x)$, then $\gcd(f(x), g(x)) = \alpha_k^{-1} g_k(x)$, where α_k is the leading coefficient of $g_k(x)$. Otherwise, let $f_{k+1}(x) = g_k(x)$ and $g_{k+1}(x)$ be the remainder when $f_k(x)$ is divided by $g_k(x)$. Continuing similarly, we have the gcd of $f(x)$ and $g(x)$. Since $f_{k+1}(x) = g_k(x)$, $g_{k+1}(x) = f_k(x) - q_k(x) g_k(x)$, $q_k(x) \in K[x]$, and the process stops if $f_p(x) = 0$ for some p. Substituting backwards we can now write:

$$\gcd(f(x), g(x)) = a(x) f(x) + b(x) g(x),$$

for some $a(x), b(x) \in K[x]$. This gives the following result.

PROPOSITION 5.2. *Let $f(x)$ and $g(x)$ be two nonconstant polynomials in $K[x]$. Then* $\gcd(f(x), g(x))$ *exists and is unique. Further,*

$$\gcd(f(x), g(x)) = a(x) f(x) + b(x) g(x),$$

for some $a(x), b(x) \in K[x]$. In particular, $f(x)$ and $g(x)$ are relatively prime if and only if $a(x) f(x) + b(x) g(x) = 1$ for some $a(x), b(x) \in K[x]$.

Let $f(x)$ be a nonconstant polynomial in $K[x]$. An element $\alpha \in K$ is called a **root** of $f(x)$ if $f(\alpha) = 0$. The division algorithm gives the following criterion for the root of a polynomial.

PROPOSITION 5.3. *An element α in K is a root of a nonconstant polynomial $f(x)$ in $K[x]$ if and only if $x - \alpha$ is a factor of $f(x)$.*

PROOF. If $x - \alpha$ is a factor of $f(x)$, then $f(x) = (x - \alpha) q(x)$, for some $q(x) \in K[x]$, and so $f(\alpha) = (\alpha - \alpha) q(\alpha) = 0$.

Conversely, by division algorithm $f(x) = (x - \alpha) q(x) + r(x)$, such that $r(x) = 0$ or $\deg r(x) < 1$. If $r(x) \neq 0$, then $\deg r(x) < 1$ implies that $r(x) = r \in K$. Since α is a root of $f(x)$, $f(\alpha) = 0$, and so $r = (\alpha - \alpha) q(\alpha) + r = f(\alpha) = 0$, a contradiction. Hence, $x - \alpha$ is a factor of $f(x)$. ∎

COROLLARY 5.4. *A polynomial of degree n (≥ 1) over K can not have more than n roots.*

Note that a polynomial of degree n over K may not have any root in K. For example, $x^4 + 1 \in \mathbb{Q}[x]$ has no roots in $\mathbb{Q}$.

A nonconstant polynomial $p(x)$ in $K[x]$ is called **reducible** in $K[x]$ if there are polynomials $q_1(x)$, $q_2(x)$ in $K[x]$ such that $p(x) = q_1(x) q_2(x)$, $\deg q_i(x) \geq 1$, $i = 1, 2$. Otherwise $p(x)$ is called **irreducible**. For example, $x^2 - 2$ is irreducible in $\mathbb{Q}[x]$, but is reducible in $\mathbb{R}[x]$. Clearly, any polynomial of degree 1 is irreducible. If $f(x)$ is irreducible, then so is $\alpha f(x)$, for any $\alpha \in K \setminus \{0\}$. Thus, if $p(x)$ is irreducible and $q(x)$ $(\neq 0)$ divides $p(x)$, then $q(x) = \alpha p(x)$, for some $\alpha \in K$. Irreducible polynomials actually behave like prime numbers in $\mathbb{Z}$. Note that if $p(x)$ is irreducible, and $p(x)$ divides $f(x) g(x)$ in $K[x]$, then either $p(x) \mid f(x)$ or $p(x) \mid g(x)$. Since, if $p(x) \nmid f(x)$, then as $p(x)$ is irreducible $\gcd(p(x), f(x)) = 1$.

Thus, there are $a(x), b(x) \in K[x]$ such that $a(x)p(x) + b(x)f(x) = 1$, and so $g(x) = a(x)p(x)g(x) + b(x)f(x)g(x)$. Since the right side has a factor $p(x)$, we conclude that $p(x) | g(x)$.

Another property is the decomposition of every polynomials as a product of irreducible polynomials.

PROPOSITION 5.5. ***Unique factorization of polynomials.*** *Let $p(x)$ be a nonconstant polynomial in $K[x]$. Then $p(x)$ can be written as*

$$p(x) = \alpha q_1(x) \cdots q_k(x),$$

where α is the leading coefficient of $p(x)$ and $q_1(x), \ldots, q_k(x)$ are monic irreducible polynomials in $K[x]$. Moreover, this factorization is unique up to the order in which these irreducible factors appear.

PROOF. By induction on $\deg p(x)$. If $p(x)$ is irreducible, in particular if $p(x)$ is of degree 1, then $p(x) = \alpha q(x)$, α is the leading coefficient of $p(x)$, and it is obvious that $q(x)$ is monic and irreducible. If $p(x)$ is reducible then $p(x) = p_1(x)p_2(x)$, for $p_1(x), p_2(x) \in K[x]$ such that $\deg p_i(x) \geq 1$, $i = 1, 2$, by induction hypothesis, $p_1(x) = \alpha_1 q_1(x) \cdots q_l(x)$ and $p_2(x) = \alpha_2 q_{l+1}(x) \cdots q_k(x)$, α_i is the leading coefficient of $p_i(x)$, $i = 1, 2$, and each $q_i(x)$ is monic irreducible polynomial in $K[x]$. Therefore, $p(x) = \alpha q_1(x) \cdots q_k(x)$, and $\alpha = \alpha_1\alpha_2$ is the leading coefficient of $p(x)$.

For uniqueness, let $p(x) = \alpha q_1(x) \cdots q_k(x)$ and $p(x) = \alpha q_1'(x) \cdots q_r'(x)$ be two factorizations of $p(x)$ into irreducibles in $K[x]$. Then, $q_1(x) \cdots q_k(x) = q_1'(x) \cdots q_r'(x)$. Since $q_1(x)$ divides $q_1'(x) \cdots q_r'(x)$, $q_1(x)$ divides $q_i'(x)$ for some i, and as both $q_1(x)$ and $q_i'(x)$ are both monic, $q_1(x) = q_i'(x)$. Cancelling these two factors and repeating this process until all $q_1(x), \ldots, q_k(x)$ are cancelled, leads us to conclude that all factors are the same up to order. ∎

A nonconstant polynomial $p(x)$ in $K[x]$ is said to **split** in $K[x]$ if all its irreducible factors in $K[x]$ are linear polynomials, that is, if

$$p(x) = \alpha(x - \alpha_1) \cdots (x - \alpha_n),$$

where $\alpha, \alpha_1, \ldots, \alpha_n \in K$, α is the leading coefficient of $p(x)$ and $n = \deg p(x)$.

A field K is called **algebraically closed** if every polynomial over K splits in $K[x]$. For example, $\mathbb{C}$ is an algebraically closed field while $\mathbb{R}$ is not as $x^2 + 1$ does not split in $\mathbb{R}[x]$.

For more details, see [**3**].

CHAPTER 2

Vector Spaces

1. Definition and examples

A **vector space** over a field K is a nonempty set V together with two operations: $V \times V \to V$, usually denoted as addition, $(u, v) \mapsto u + v$; and $K \times V \to V$, usually denoted as multiplication, $(\alpha, u) \mapsto \alpha u$, which satisfy the following conditions:

1. $u + (v + w) = (u + v) + w$, for all $u, v, w \in V$ (associativity);
2. $u + v = v + u$, for all $u, v \in V$ (commutativity);
3. there exists an element 0 in V such that $u + 0 = u$, for all $u \in V$ (existence of identity element);
4. for each $u \in V$, there exists $-u \in V$ such that $u + (-u) = 0$ (existence of additive inverse);
5. $1u = u$, for all $u \in V$, where 1 is the multiplicative identity of K;
6. $\alpha(\beta u) = (\alpha\beta)u$, for all $\alpha, \beta \in K$ and $u \in V$;
7. $\alpha(u + v) = \alpha u + \alpha v$, for all $\alpha \in K$ and $u, v \in V$;
8. $(\alpha + \beta)u = \alpha u + \beta u$, for all $\alpha, \beta \in K$ and $u \in V$.

The elements of V are called **vectors**, the elements of K are called **scalars**, and the multiplication of a vector by a scalar is called **scalar multiplication**. The conditions 1 – 4 state that V is an abelian group under addition. It then follows from the definition of groups that identity element 0 of V is unique and each element of V has a unique additive inverse. Also conditions 7 and 8 signify that scalar multiplication is distributive over addition of vectors as well as over addition of scalars.

Note that the + sign appearing on the left side of the equality in 8 is for the addition of the field elements of K, while on the right side, + is for the addition of vectors. In spite of this usage, there is no chance of confusion as it is clear from the context what we are actually adding. Similar remarks hold for multiplication; in the parentheses of the left side of the equality in 6 is scalar multiplication, and in the parentheses of the right side, it is multiplication of field elements.

Throughout this book, the letter K denotes a field.

We list some straightforward consequences of the above definition.

PROPOSITION 1.1. *Let V be a vector space over K. Then:*

(i) $0v = 0$, *for all* $v \in V$;
(ii) $\alpha 0 = 0$, *for all* $\alpha \in K$;
(iii) $(-1)v = -v$, *for all* $v \in V$;
(iv) *for* $\alpha \in K$, $v \in V$, $\alpha v = 0$ *implies that either* $\alpha = 0$ *or* $v = 0$.

PROOF. (i) $0v = (0 + 0)v = 0v + 0v$. Therefore, $0 = 0v + (-0v) = 0v + 0v + (-0v) = 0v$.

(ii) Similarly, $\alpha 0 = \alpha(0+0) = \alpha 0 + \alpha 0$, that is, $\alpha 0 = 0$.
(iii) $v + (-1)v = 1v + (-1)v = (1 + (-1))v = 0v = 0$. Thus, $(-1)v = -v$.
(iv) If $\alpha \neq 0$ and $\alpha v = 0$, then $v = 1v = (\alpha^{-1}\alpha)\, v = \alpha^{-1}(\alpha v) = \alpha^{-1}0 = 0$. ∎

1.1. Examples

1. In high school geometry, a vector in a space is represented by a directed line segment, that is, a line segment on which a direction is specified by distinguishing its end points; its length is the magnitude of the vector. Two directed line segments of equal length and in the same direction represent the same vector. The sum of two vectors is obtained by parallelogram rule. The elements of $\mathbb{R}$ are called scalars. If $\lambda \in \mathbb{R}$ and v is a vector, then λv is a vector parallel to v whose magnitude is changed by the factor $|\lambda|$; negative λ means the reversal of direction. The reader can verify that the set of all such vectors form a vector space over $\mathbb{R}$.

2. $K^{m\times n}$, the set of all matrices of size $m \times n$ with entries from K is a vector space over K under the usual addition and the scalar multiplication of matrices, that is, for $A, B \in K^{m\times n}$ and $\lambda \in K$:
$$[A+B]_{ij} = [A]_{ij} + [B]_{ij},$$
and
$$[\lambda A]_{ij} = \lambda [A]_{ij}, \text{ for } i = 1, \ldots, m;\ j = 1, \ldots, n.$$
In particular, $K^{1\times n}$, the set of column vectors of length n is a vector space over K: for $x = \begin{bmatrix} x_1 \\ \vdots \\ x_n \end{bmatrix}$, $y = \begin{bmatrix} y_1 \\ \vdots \\ y_n \end{bmatrix} \in K^{1\times n}$ and $\lambda \in K$, the addition and the scalar multiplication are given by:
$$x + y = \begin{bmatrix} x_1 + y_1 \\ \vdots \\ x_n + y_n \end{bmatrix} \text{ and } \lambda x = \begin{bmatrix} \lambda x_1 \\ \vdots \\ \lambda x_n \end{bmatrix}.$$
We will denote the vector space $K^{1\times n}$ by K^n.

3. $K[x]$, the set of all polynomials in an indeterminate x with coefficients from K is a vector space over K under the usual addition of polynomials, and for $p(x) = a_0 + a_1 x + \cdots + a_n x^n \in K[x]$, $\lambda \in K$, the scalar multiplication is defined by $\lambda p(x) = \lambda a_0 + \lambda a_1 x + \cdots + \lambda a_n x^n$.

4. K^∞, the set of all sequences with entries from K is a vector space over K with componentwise addition and scalar multiplication, that is, for $a = \{a_n\}_{n=1}^\infty$, $b = \{b_n\}_{n=1}^\infty$ in K^∞ and for $\lambda \in K$:
$$a + b = \{a_n + b_n\}_{n=1}^\infty \text{ and } \lambda a = \{\lambda a_n\}_{n=1}^\infty.$$

5. Let S be an arbitrary nonempty set and let K^S be the set of all mappings from S to K. Then K^S is a vector space over K with pointwise addition and scalar multiplication, that is, for $f, g \in K^S$ and $\lambda \in K$:
$$(f+g)(s) = f(s) + g(s),$$
$$(\lambda f)(s) = \lambda f(s) \text{ for all } s \in S.$$

Note that, if $S = \mathbb{Z}^+$, then $K^\infty = K^{\mathbb{Z}^+}$.

6. Let I be an open interval of $\mathbb{R}$ and let

$$C(I) = \{f : I \to \mathbb{R} \mid f \text{ is continuous on } I\}.$$

Then $C(I)$ is a vector space over $\mathbb{R}$ with pointwise addition and scalar multiplication.

7. Let F be a field and let K be its subfield. Then F is a vector space over K under the usual addition and multiplication of F, that is, for $\alpha \in K$ and $x \in F$, the scalar multiplication αx is the multiplication of α and x in F.

   Thus, $\mathbb{C}$ can be considered as a vector space over $\mathbb{R}$ (or over $\mathbb{Q}$) and $\mathbb{R}$ can be considered as a vector space over $\mathbb{Q}$.

8. It is customary to denote the operations of a vector space by addition $(+)$ and by juxtaposition. The convenience of this notation is evident from the examples discussed above. This example shows that 'addition' and 'scalar multiplication' in a vector space can be entirely different.

   Let $V = \mathbb{R}^+$, the set of all positive real numbers. Then V can be regarded as an abelian group with respect to the multiplication of real numbers. Also for $x \in V$ and $\alpha \in \mathbb{R}$, $x^\alpha \in V$. It is easy to verify that V is a vector space over $\mathbb{R}$ with addition of two vectors in V defined to be the usual multiplication of real numbers; and the scalar multiplication is the exponentiation, that is, $\alpha \in \mathbb{R}$, and $x \in V$, αx is replaced by x^α. Now the zero vector in V is actually 1, and the condition pertaining to scalar multiplication $1x = x$ is replaced by $x^1 = x$; $\alpha(\beta x) = (\alpha\beta)x$ is replaced by $(x^\beta)^\alpha = x^{\alpha\beta}$; $(\alpha + \beta)x = \alpha x + \beta x$ is replaced by $x^{\alpha+\beta} = x^\alpha x^\beta$; $\alpha(x + y) = \alpha x + \alpha y$ is replaced by $(xy)^\alpha = x^\alpha y^\alpha$, where $x, y \in V$ and $\alpha, \beta \in \mathbb{R}$.

1.2. Exercises

1. Let V be a vector space over K. Show that, if for $\alpha \in K$ and $x \in V$, $\alpha x = x$ then $\alpha = 1$ or $x = 0$.

2. Let $\mathbb{Q}\left[\sqrt{3}\right] = \left\{a + b\sqrt{3} \mid a, b \in \mathbb{Q}\right\}$. Prove that $\mathbb{Q}\left[\sqrt{3}\right]$ is a vector space over $\mathbb{Q}$ with respect to the usual addition of real numbers, and for $r \in \mathbb{Q}$, $a+b\sqrt{3} \in \mathbb{Q}\left[\sqrt{3}\right]$, scalar multiplication is given by $r\left(a + b\sqrt{3}\right) = ra + rb\sqrt{3}$.

3. Let I be an open interval in $\mathbb{R}$. Let V be the set of all real valued functions defined on I which have at most finite number of points of discontinuity. Prove that V is a vector space over $\mathbb{R}$ under pointwise addition and scalar multiplication of functions.

4. Let V be a vector space over a field F and let K be a subfield of F. Prove that V is a vector space over K.

5. Let $V = K^2$. Define coordinatewise addition on V, and scalar multiplication:

$$\alpha \begin{bmatrix} x \\ y \end{bmatrix} = \begin{bmatrix} \alpha x \\ 0 \end{bmatrix}; \alpha \in K, \begin{bmatrix} x \\ y \end{bmatrix} \in K^2.$$

   Is V a vector space with respect to these operations?

6. Let $V = \{t_a : \mathbb{R} \to \mathbb{R} \mid t_a(x) = x + a,\ a \in \mathbb{R}\}$. Is V a vector space over $\mathbb{R}$ if the addition on V is the composition of mappings and scalar multiplication on V is: $\alpha t_a = t_{\alpha a}$, $\alpha \in \mathbb{R}$?

7. Let V be a vector space over $\mathbb{C}$. Define another scalar multiplication $*$ on V: $\alpha * x = \text{Re}(\alpha)x$, $\alpha \in \mathbb{C}$, $x \in V$, where $\text{Re}(\alpha)$ is the real part of α. Is V a vector space with respect to original addition and scalar multiplication $*$?

8. Same question as in Problem 7, except that $*$ is now defined by: $\alpha * x = \overline{\alpha}x$, $\alpha \in \mathbb{C}$, $x \in V$, where $\overline{\alpha}$ is the complex conjugate of α.

9. Let V be a vector space over a field K. Define a scalar multiplication $*$ on V: $\alpha * x = \alpha^p x$, $\alpha \in K$, $x \in V$, where p is a prime. Discuss when is V a vector space over K with respect to the original addition and new scalar multiplication $*$?

2. Subspaces

Let V be a vector space over K. A subset W of V is a **subspace** of V if W itself is a vector space with respect to the addition and the scalar multiplication of V. It is obvious that a subspace must contain the zero vector. The following is an easy criterion for a subset of a vector space to be a subspace.

PROPOSITION 2.1. *Let V be a vector space over K. A nonempty subset W of V is a subspace of V if and only if $\alpha u + \beta v \in W$ for all $\alpha, \beta \in K$, whenever $u, v \in W$.*

PROOF. If W is a subspace of V and $u, v \in W$, then clearly $\alpha u + \beta v \in W$ for all $\alpha, \beta \in K$.

Conversely, assume that W is a nonempty subset of V, and $\alpha u + \beta v \in W$, for all $\alpha, \beta \in K$, whenever $u, v \in W$. Since W is nonempty, choose $w \in W$. Then $0 = 1w + (-1)w \, (\alpha = 1, \beta = -1)$, and so $0 \in W$. Also, for any $u \in W$, $-u = 1 \cdot 0 + (-1)u \in W$. Thus, W has additive identity and every element of W has additive inverse. It is easy to verify that all other properties of addition and scalar multiplication hold in W. Hence, W is a vector space over K. ∎

For a vector space V, clearly, the whole space V and $\{0\}$ are subspaces of V. We call $\{0\}$ as the zero subspace of V. A subspace W of V is said to be proper if $W \neq V$; and nonzero if $W \neq \{0\}$.

PROPOSITION 2.2. *Let V be a vector space and let $\{W_i \mid i \in I\}$ be a collection of subspaces of V. Then $\bigcap_{i \in I} W_i$ is a subspace of V.*

PROOF. Exercise. ∎

If W_1 and W_2 are subspaces of a vector space V, then $W_1 \cup W_2$ need not be a subspace of V (Exercise 7). In fact, we have the following result.

PROPOSITION 2.3. *Let W_1 and W_2 be subspaces of a vector space V. Then $W_1 \cup W_2$ is a subspace of V if and only if $W_1 \subseteq W_2$ or $W_2 \subseteq W_1$.*

PROOF. Assume that $W_1 \cup W_2$ is a subspace of V and $W_1 \not\subseteq W_2$. Let $w_2 \in W_2$ and choose $w_1 \in W_1 \setminus W_2$. Since $W_1 \cup W_2$ is a subspace of V, $w_1 + w_2 \in W_1 \cup W_2$. If $w_1 + w_2 \in W_2$, then $w_1 = (w_1 + w_2) - w_2 \in W_2$, a contradiction. Therefore, $w_1 + w_2 \in W_1$, and $w_2 \in W_1$. Hence $W_2 \subseteq W_1$.

The converse is obvious. ∎

Let V be a vector space over K, and let $S = \{x_i \mid i \in I\}$ be a nonempty subset of V. By a **linear combination** (or K-linear combination) of elements of S, we mean an expression $\sum_{i \in I} \alpha_i x_i$, where each $\alpha_i \in K$ and $\alpha_i = 0$ except for finitely many indices $i \in I$. Thus, a linear combination is essentially the sum of a finite number of nonzero vectors. Denote by $\langle S \rangle$, the set of all (K-)linear combination of elements of S. Then $\langle S \rangle$ is a subspace of V, called the **subspace generated by** S. We let $\langle \emptyset \rangle = \{0\}$. Note that $\langle S \rangle = \cap W$, where the set intersection is over all those subspaces of V which contain S. If $S = \{x_1, \ldots, x_n\}$, a finite set, then we write $\langle x_1, \ldots, x_n \rangle$ for $\langle \{x_1, \ldots, x_n\} \rangle$.

If $W_1, \ldots, W_n$ are finitely many subspaces of V, then we write $\sum_{i=1}^{n} W_i$ or $W_1 + \cdots + W_n$ for $\langle W_1 \cup \cdots \cup W_n \rangle$.

2.1. Examples

1. Consider the vector space $K^{m \times n}$ over K (Example 2, p.26). Let $I = \{i_1, \ldots, i_k\}$, and $J = \{j_1, \ldots, j_l\}$ be subsets of ordered sets $\{1, \ldots, m\}$ and $\{1, \ldots, n\}$, and let
$$K^{m \times n}(I \times J) = \{A \in K^{m \times n} \mid [A]_{ij} = 0 \text{ for all } (i,j) \in I \times J\}.$$
Then $K^{m \times n}(I \times J)$ is a subspace of $K^{m \times n}$.

2. Let $A \in K^{n \times n}$ and let $N = \{x \in K^n \mid Ax = 0\}$. Then N is a subspace of K^n.

3. Let $n \in \mathbb{N}$, and let $K_n[x]$ be the set of all those polynomials in $K[x]$ whose degree is at most n. Then $K_n[x]$ is a subspace of $K[x]$ (Example 3, p.26). Verify that $K_n[x] = \langle 1, x, \ldots, x^n \rangle$.

4. Let $V = \mathbb{R}^\infty$, vector space of all sequences of real numbers. Then W_1, the set of all convergent sequences of real numbers, is a subspace of V. Also, $W_2 = \{\{a_n\}_{n=1}^{\infty} \mid \sum a_i^2 < \infty\}$, the set of all square summable sequences, is a subspace of W_1, and hence also of V.

5. Let I be an open interval in $\mathbb{R}$. Then the following are subspaces of vector space $C(I)$ over $\mathbb{R}$ (Example 6, p.27).
$$C^n(I) = \{f \in C(I) \mid f \text{ is } n \text{ times continuosly differentiable, } n \in \mathbb{N}\};$$
$$C^\infty(I) = \{f \in C(I) \mid f \text{ is infinitely continuosly differentiable, } n \in \mathbb{N}\};$$
$$C^w(I) = \{f \in C(I) \mid f \text{ is analytic}\}.$$
In fact,
$$C^w(I) \subseteq C^\infty(I) \subseteq C^{n+1}(I) \subseteq C^n(I) \subseteq C(I),$$
and each is a subspace of its superset.

6. Let I be an open interval in $\mathbb{R}$. The set of all solutions of n–th order homogeneous linear differential equation:
$$y^{(n)}(x) + a_{n-1}(x)y^{(n-1)}(x) + \cdots + a_1(x)y^{(1)}(x) + a_0(x)y(x) = 0,$$
where $a_0(x), \ldots, a_n(x) \in C(I)$, is a subspace of $C(I)$.

   For example, the equation $y^{(3)}(x) - 2y^{(2)}(x) - y^{(1)}(x) + 2y(x) = 0$ has general solution $\alpha e^{-x} + \beta e^{x} + \gamma e^{2x}$, $\alpha, \beta, \gamma \in \mathbb{R}$. Thus, the solution space W of this equation is a subspace of $C^3(\mathbb{R})$ (here $I = \mathbb{R}$), and $W = \langle e^{-x}, e^{x}, e^{2x} \rangle$.

7. Let S be an arbitrary set; then K^S is a vector space over K (Example 5, p.26). Let T be a subset of S. Define

$$K_T^S = \{f \in K^S \mid f(t) = 0 \text{ for all } t \in T\}.$$

Verify that K_T^S is a subspace of K^S.

Also, $K^{[S]} = \{f \in K^S \mid f(s) = 0 \text{ except for finitely many } s \in S\}$, is a subspace of K^S.

For each $s \in S$, define $\delta_s : S \to K$ given by

$$\delta_s(t) = \begin{cases} 0, & t \neq s \\ 1, & t = s \end{cases}.$$

Then $K^{[S]} = \langle\{\delta_s \mid s \in S\}\rangle$. Indeed, if $f \in K^{[S]}$, and $s_1, \ldots, s_n \in S$ such that $f(s_i) \neq 0$ and $f(s) = 0$ for all $s \in S \setminus \{s_1, \ldots, s_n\}$, then $f = \sum_{i=1}^{n} f(s_i)\delta_{s_i}$.

Clearly, if S is a finite set, then $K^{[S]} = K^S$.

2.2. Exercises

1. Let $V = \mathbb{R}^n$, and let W be a subset of V consisting of vectors $\begin{bmatrix} x_1 & \cdots & x_n \end{bmatrix}^t$ such that
   (a) $x_n = 0$;
   (b) $x_1 + x_n = x_{[(n+1)/2]}$, $[\,]$ denotes the greatest integer function;
   (c) $x_1 + \cdots + x_n = 1$;
   (d) $x_1 + \cdots + x_n = 0$;
   (e) $x_1 + \cdots + x_n \geq 0$;
   (f) $a_1x_1 + \cdots + a_nx_n \geq 0$, $a_1, \ldots, a_n \in \mathbb{R}$;
   (g) $x_1^2 + \cdots + x_n^2 \geq 1$;
   (h) $x_{2k+1} = x_1$, for all $k = 1, \ldots, [(n-1)/2]$;
   (i) $x_i = x_{n+1-i}$, for all $i = 1, \ldots, n$.

   In which of these conditions, is W a subspace of V?

2. Let $V = \mathbb{C}^2$, and let $W = \left\{\begin{bmatrix} x_1 \\ x_2 \end{bmatrix} \mid x_1 \in \mathbb{R}, x_2 \in \mathbb{C}\right\}$. Is W a subspace of the complex vector space V? Is it a subspace of V, when V is considered as a vector space over $\mathbb{R}$?

3. Consider the vector space $\mathbb{R}[x]$ over $\mathbb{R}$. Which of the following subsets of $\mathbb{R}[x]$ are subspaces:
   (a) $\{p(x) \mid p(x) = p(1-x) \text{ for all } x\}$;
   (b) $\{p(x) \mid p(1) \geq 0\}$;
   (c) $\{p(x) \mid p(x) = p(-x)\}$;
   (d) $\{p(x) \mid p^{(1)}(0) + p(0) = 0\}$, $p^{(1)}(x)$ is the derivative of $p(x)$;
   (e) $\{p(x) \mid p(x) \text{ has a root in the interval } [-1,1]\}$?

4. Consider the vector space $K^{n\times n}$ over K. Prove that the following subsets of $K^{n\times n}$ are subspaces:
   (a) the set of all symmetric matrices;
   (b) the set of all skew-symmetric matrices;
   (c) the set of all diagonal matrices;
   (d) the set of all upper triangular matrices;
   (e) the set of all matrices with trace zero.

5. Express the vector $\begin{bmatrix} 3 \\ 5 \\ 2 \\ 5 \end{bmatrix}$ as a $\mathbb{Q}$-linear combination of elements $\begin{bmatrix} 1 \\ 1 \\ 1 \\ 1 \end{bmatrix}$, $\begin{bmatrix} 1 \\ 1 \\ 0 \\ 1 \end{bmatrix}$, $\begin{bmatrix} 0 \\ 1 \\ 0 \\ 1 \end{bmatrix}$. Is $\mathbb{Q}^4 = \left\langle \begin{bmatrix} 1 \\ 1 \\ 1 \\ 1 \end{bmatrix}, \begin{bmatrix} 1 \\ 1 \\ 0 \\ 1 \end{bmatrix}, \begin{bmatrix} 0 \\ 1 \\ 0 \\ 1 \end{bmatrix} \right\rangle$?

6. Can a vector $\begin{bmatrix} \sqrt{3} \\ 2 \\ 3/5 \end{bmatrix}$ be expressed as a $\mathbb{Q}$-linear combination of vectors $a = \begin{bmatrix} 1 \\ 1 \\ 1 \end{bmatrix}$, $b = \begin{bmatrix} 0 \\ 1 \\ 1 \end{bmatrix}$ and $c = \begin{bmatrix} 0 \\ 0 \\ 1 \end{bmatrix}$? Can it be expressed as a $\mathbb{R}$-linear combination of vectors a, b and c?

7. Consider the vector space $K_n[x]$ over K. Let W_1 and W_2 be subspaces of $K_n[x]$ defined by:
$$W_1 = \{p(x) \in K_n[x] \mid p(0) = 0\}, \text{ and}$$
$$W_2 = \{p(x) \in K_n[x] \mid p(1) = 0\}.$$
Find subsets S_1 and S_2 such that $W_1 \cap W_2 = \langle S_1 \rangle$, and $W_1 + W_2 = \langle S_2 \rangle$. Verify that the set $W_1 \cup W_2$ is not a subspace.

8. Let $a = \begin{bmatrix} 0 \\ 1 \\ -1 \end{bmatrix}$, $b = \begin{bmatrix} 0 \\ 1 \\ 1 \end{bmatrix}$, $c = \begin{bmatrix} -1 \\ 1 \\ 0 \end{bmatrix}$ and $d = \begin{bmatrix} 1 \\ 0 \\ 1 \end{bmatrix}$ be vectors in $\mathbb{R}^3$. Let $W_1 = \langle a, b \rangle$, $W_2 = \langle b, c \rangle$ and $W_3 = \langle c, d \rangle$. Show that $W_2 = W_3$, $W_1 \cap W_2 = \langle b \rangle$. Also identify the subspace $W_1 + W_2$. Is $W_1 \cup W_2$ a subspace of $\mathbb{R}^3$?

9. Let X be a subset of a vector space V. Show that $\langle X \rangle = \langle \langle X \rangle \rangle$.

10. Let X and Y be subsets of a vector space V. Show that $\langle X \cup Y \rangle = \langle X \rangle + \langle Y \rangle$.

11. Let $W_1, \ldots, W_n$ are subspaces of a vector space V and $V = \cup_{i=1}^n W_i$. Show that $V = W_r$ for some r.

12. Let V be a vector space over K and let $x_1, \ldots, x_n$ be nonzero vectors such that $V = \langle x_1, \ldots, x_n \rangle$. Prove that if for some $i = 1, \ldots, n$, $x_i = \sum_{j=1, j \neq i}^n \alpha_j x_j$, $\alpha_j \in K$, then $V = \langle x_1, \ldots, x_{i-1}, x_{i+1}, \ldots x_n \rangle$.

13. Let V be a vector space over a finite field K. Prove that if S is a finite subset of V, then $|\langle S \rangle|$ is finite.

14. Let V be a vector space over an infinite field K. Show that if S is a countable subset of V, then $|\langle S \rangle| = |K|$.

3. Basis and dimension

A vector space V over a field K is said to be **finite dimensional** if there are finite number of elements $x_1, \ldots, x_n$ in V such that $V = \langle x_1, \ldots, x_n \rangle$. If no such finite number of elements exist in V, V is called **infinite dimensional**.

The vector space $K^{m\times n}$ of $m \times n$ matrices with entries from K is finite dimensional. Since, for $A \in K^{m\times n}$, $A = \sum_{i=1}^{m} \sum_{j=1}^{n} [A]_{ij} E_{ij}$, and so

$$K^{m\times n} = \langle E_{ij},\ i = 1, \dots, m,\ j = 1, \dots, n\rangle .$$

In particular, K^n is a finite dimensional vector space over K. On the other hand, the vector space $K[x]$ of polynomials over K in an indeterminate x is infinite dimensional. Since, if there are polynomials $p_1(x), \dots, p_n(x)$ in $K[x]$ such that $K[x] = \langle p_1(x), \dots, p_n(x)\rangle$, then any polynomial which is a K-linear combination of these polynomials can not be of degree more than $\max\{\deg p_1(x), \dots, \deg p_n(x)\}$.

Let V be a vector space over K. A finite subset $\{x_1, \dots, x_n\}$ of V is a **linearly dependent** set (or a K–linearly dependent set, whenever we want to specify the underlying field K) if there are scalars $\alpha_1, \dots, \alpha_n$ in K, not all of which are zero, such that

$$\alpha_1 x_1 + \cdots + \alpha_n x_n = 0.$$

Otherwise, $\{x_1, \dots, x_n\}$ is called a **linearly independent** set.

Thus, $\{x_1, \dots, x_n\}$ is a linearly independent set if and only if for $\alpha_1, \dots, \alpha_n$ in K, $\alpha_1 x_1 + \cdots + \alpha_n x_n = 0$ implies that $\alpha_1 = \cdots = \alpha_n = 0$.

If $\{x_1, \dots, x_n\}$ is a linearly independent (respectively, linearly dependent) subset of a vector space V, then we also write this as: the vectors $x_1, \dots, x_n$ are linearly independent (respectively, linearly dependent) in V.

An arbitrary subset X (not necessarily finite) of a vector space V is said to be linearly independent if every finite subset of X is linearly independent; otherwise X is said to be linearly dependent. Clearly, if X is a linearly independent subset of V, then every subset of X is also linearly independent; and if X is linearly dependent, then every subset of V which contains X is also linearly dependent. Since for the zero vector 0, $1 \cdot 0 = 0$; 0 is linearly dependent. Proposition 1.1 states that every nonzero vector is linearly independent. It is vacuously true that the empty set is linearly independent.

Let V be a finite dimensional vector space over K. Define $\mu(V)$, the least number of elements of V which are required to generate V. Clearly, $\mu(V) = 0$ if and only if $V = \{0\}$; and $\mu(V) \geq 1$ if $V \neq \{0\}$.

PROPOSITION 3.1. *Let V be a finite dimensional vector space over K, and let $\mu(V) = n$. If $V = \langle x_1, \dots, x_n\rangle$, then $x_1, \dots, x_n$ are linearly independent vectors.*

PROOF. Clearly, $x_1, \dots, x_n$ are nonzero distinct elements of V. If $x_1, \dots, x_n$ are linearly dependent, then there are scalars $\alpha_1, \dots, \alpha_n$ in K, not all of which are zero, such that $\alpha_1 x_1 + \cdots + \alpha_n x_n = 0$. Let $\alpha_k \neq 0$; then

$$x_k = \sum_{i=1,\, i\neq k}^{n} \left(-\alpha_k^{-1}\alpha_i\right) x_i,$$

and so $V = \langle x_1, \dots, x_{k-1}, x_{k+1}, \dots, x_n\rangle$. This contradicts that $\mu(V) = n$. Hence, $x_1, \dots, x_n$ are linearly independent. ∎

Let V be a vector space over a field K. A subset $\mathcal{B}$ of V is a **basis** of V if:

1. $V = \langle \mathcal{B}\rangle$, and
2. $\mathcal{B}$ is a linearly independent subset of V.

Proposition 3.1 shows that a finite dimensional vector space has a basis. In fact, an infinite dimensional vector space also has a basis. In this case, Zorn's Lemma is required to show the existence of a basis. Further, any two bases of a vector space have the same cardinality. Again we will prove this result for finite dimensional vector spaces. For arbitrary vector spaces, the reader can see [2].

PROPOSITION 3.2. *Let V be a vector space over K. A finite subset $\mathcal{B}$ of V is a basis of V if and only if every element of V is a unique linear combination of elements of $\mathcal{B}$.*

PROOF. Let $\mathcal{B} = \{x_1, \dots, x_n\}$ be a basis of V. Suppose that $v = \sum_{i=1}^{n} \alpha_i x_i$, $\alpha_i \in K$, and also that $v = \sum_{i=1}^{n} \beta_i x_i$, $\beta_i \in K$. Then $\sum_{i=1}^{n} (\alpha_i - \beta_i) x_i = 0$. By linear independence of $\mathcal{B}$, it follows that $\alpha_i = \beta_i$ for all i.

Conversely, it is clear that $V = \langle \mathcal{B} \rangle$. Let $\alpha_1, \dots, \alpha_n$ be in K such that $\alpha_1 x_1 + \cdots + \alpha_n x_n = 0$. Since $0 = 0x_1 + \cdots + 0x_n$; by uniqueness of representation of elements of V as linear combination of $\mathcal{B}$, $\alpha_1 = \cdots = \alpha_n = 0$. Hence, $\mathcal{B}$ is a basis of V. ∎

THEOREM 3.3. *Let V be a finite dimensional vector space over K, and let X and Y be finite subsets of V. If Y is linearly independent and $V = \langle X \rangle$, then $|Y| \leq |X|$.*

PROOF. Let $X = \{x_1, \dots, x_m\}$ and $Y = \{y_1, \dots, y_n\}$. Since $V = \langle X \rangle$, $y_n = \sum_{i=1}^{m} \alpha_i x_i$, $\alpha_i \in K$. Linear independence of Y implies that $y_n \neq 0$. Let k be the first index such that $\alpha_k \neq 0$. Then

$$x_k = \alpha_k^{-1} y_n + \sum_{i=1,\, i \neq k}^{m} \left(-\alpha_k^{-1} \alpha_i\right) x_i .$$

Therefore, $V = \langle \{x_i, i = 1, \dots, m, i \neq k\} \cup \{y_n\} \rangle$. In particular, $y_{n-1} = \beta y_n + \sum_{i=1,\, i \neq k}^{m} \beta_i x_i$, $\beta, \beta_i \in K$. Since y_n and y_{n-1} are linearly independent, $y_{n-1} - \beta y_n \neq 0$. Let j be the first index such that $\beta_j \neq 0$. Then

$$x_j = \beta_j^{-1} y_{n-1} + \left(-\beta_j^{-1} \beta\right) y_n + \sum_{i=1,\, i \neq j,k}^{m} \left(-\beta_j^{-1} \beta_i\right) x_i,$$

and so $V = \langle \{x_i, i = 1, \dots, m, i \neq j, k\} \cup \{y_{n-1}, y_n\} \rangle$.

This process of adding an element of Y to X and removing an element of X can be repeated. If $n > m$, after m steps, we conclude that $V = \langle y_n, y_{n-1}, \dots, y_{n-m+1} \rangle$. But then $y_n, y_{n-1}, \dots, y_{n-m+1}, y_{n-m}$ are linearly independent, a contradiction. Hence, $m \leq n$. ∎

COROLLARY 3.4. *Let V be a finite dimensional vector space over K. If X is a linearly independent subset of V, then $|X| \leq \mu(V)$. In particular, every linearly independent subset of V is finite.*

COROLLARY 3.5. *If X and Y are bases of a finite dimensional vector space V, then $|X| = |Y| = \mu(V)$.*

Let V be a vector space over a field K. The **dimension** of V, denoted by $\dim V$, is the cardinality of a basis of V. We also denote it by $\dim_K V$, whenever we want to specify the field K.

It is immediate from Corollary 3.4, that a vector space V is of infinite dimension if and only if V has a linearly independent subset of infinite cardinality.

PROPOSITION 3.6. *Let V be a finite dimensional vector space over K. Then the following statements are equivalent for a subset $\mathcal{B}$ of V:*

1. *$\mathcal{B}$ is a basis;*
2. *$\mathcal{B}$ is a minimal generating set, that is no subset of $\mathcal{B}$ can generate V.*
3. *$\mathcal{B}$ is a maximal linearly independent set,*

PROOF. ($1 \Rightarrow 2$) If $X \subseteq \mathcal{B}$ such that $V = \langle X \rangle$, then by Theorem 3.3, $|\mathcal{B}| \leq |X|$. Hence, $\mathcal{B} = X$.

($2 \Rightarrow 3$) Let Y be a linearly independent subset of V and let $\mathcal{B} \subseteq Y$. Since $V = \langle \mathcal{B} \rangle$, by Theorem 3.3, $|Y| \leq |\mathcal{B}|$. Since $\mathcal{B}$ is a minimal generating set, $\mathcal{B} = Y$.

($3 \Rightarrow 1$) Let $v \in V$, and let $\mathcal{B} = \{x_1, \ldots, x_n\}$, a maximal linearly independent set. If $v \notin \mathcal{B}$, then $\mathcal{B} \cup \{v\}$ is a linearly dependent set. Thus there are $\alpha, \alpha_1, \ldots, \alpha_n$ in K, not all of which are zero, such that $\alpha v + \alpha_1 x_1 + \cdots + \alpha_n x_n = 0$. If $\alpha = 0$, the linear independence of $\mathcal{B}$ implies that $\alpha_1 = \cdots = \alpha_n = 0$. Thus, $\alpha \neq 0$ and $v = \sum_{i=1}^{n} \left(-\alpha^{-1}\alpha_i\right) x_i$. Hence, $v \in \langle \mathcal{B} \rangle$, and $\mathcal{B}$ is a basis of V. ∎

THEOREM 3.7. *A linearly independent subset of a finite dimensional vector space can be extended to form a basis of the vector space.*

PROOF. Let V be a finite dimensional vector space over a field K, and let X be a linearly independent subset of V. By Corollary 3.4, $|X|$ is finite. Write $X = \{x_1, \ldots, x_m\}$. If $V = \langle X \rangle$, then by definition, X is a basis of V. If $V \neq \langle X \rangle$, then choose $x_{m+1} \in V \setminus \langle X \rangle$, and let $X_1 = X \cup \{x_{m+1}\}$. Then X_1 is a linearly independent set. Indeed, if $\sum_{i=1}^{m+1} \alpha_i x_i = 0$, $\alpha_i \in K$, and if $\alpha_{m+1} \neq 0$, then $x_{m+1} = \sum_{i=1}^{m} \left(-\alpha_{m+1}^{-1}\alpha_i\right) x_i \in \langle X \rangle$, a contradiction, and so $\alpha_{m+1} = 0$, which further implies that $\alpha_1 = \cdots = \alpha_m = 0$. Next, if $V \neq \langle X_1 \rangle$, then choose $x_{m+2} \in V \setminus \langle X_1 \rangle$, and $X_2 = X_1 \cup \{x_{m+2}\}$. Then X_2 is a linearly independent set. Inductively, if $V \neq \langle X_i \rangle$, then choose $x_{m+i+1} \in V \setminus \langle X_i \rangle$, and then let $X_{i+1} = X_i \cup \{x_{m+i+1}\}$, a linearly independent set. Since V is finite dimensional, $V = \langle X_k \rangle$, for some k. Hence, X_k is a basis of V containing X. ∎

COROLLARY 3.8. *If V is a vector space of dimension n and if X is a linearly independent set of V with n elements, then X is a basis of V.*

PROPOSITION 3.9. *Let V be a finite dimensional vector space and let W be a subspace of V. Then $\dim W \leq \dim V$; and $V = W$ if and only if $\dim V = \dim W$.*

PROOF. Let $\mathcal{B}_0$ be a basis of W. Then it can be extended to form a basis $\mathcal{B}$ of V (Theorem 3.7). Therefore, $|\mathcal{B}_0| \leq |\mathcal{B}|$, and $\dim W \leq \dim V$. The second part is a direct consequence of Corollary 3.8. ∎

Thus, every subspace of a finite dimensional vector space is also finite dimensional. If a vector space has a subspace of infinite dimension, then the vector space itself is of infinite dimension.

3.1. Examples

1. The vector space $K^{m\times n}$ has a basis $\{E_{ij} \mid i = 1, \ldots, m;\ j = 1, \ldots, n\}$. Thus, $\dim K^{m\times n} = mn$. In particular, K^n is an n-dimensional vector space over K (Recall that $K^n = K^{n\times 1}$). The set

$$\left\{ e_1 = \begin{bmatrix} 1 \\ 0 \\ \vdots \\ 0 \end{bmatrix}, e_2 = \begin{bmatrix} 0 \\ 1 \\ \vdots \\ 0 \end{bmatrix}, \ldots, e_n = \begin{bmatrix} 0 \\ 0 \\ \vdots \\ 1 \end{bmatrix} \right\}$$

is a basis of K^n, called the **standard basis** of K^n.

2. The vector space $K[x]$ is an infinite dimensional vector space over K, and $\mathcal{B} = \{1, x, \dots, x^n, \dots\}$ is a basis of $K[x]$. The subspace $K_n[x]$ of $K[x]$ is an $(n+1)$-dimensional vector space over K, and $\{1, x, \dots, x^n\}$ is its basis. Further,
$$K\left[x^2\right] = \left\{a_0 + a_1x^2 + \cdots + a_kx^{2k} \mid k \in \mathbb{N}, a_0, \dots, a_k \in K\right\}$$
is a proper subspace of $K[x]$ and its basis is $\mathcal{B}' = \left\{1, x^2, \dots, x^{2n}, \dots\right\}$. Since the mapping $x^r \mapsto x^{2r}$, $r \in \mathbb{N}$, is a bijection from $\mathcal{B}$ to $\mathcal{B}'$, $\dim K[x] = \dim K\left[x^2\right]$.

   Thus, in an infinite dimensional vector space V, a proper subspace can have dimension equal to $\dim V$. Proposition 3.9 shows that this can not happen in a finite dimensional vector space.

3. The vector space $K^{[S]}$ over K has a basis $\mathcal{B} = \{\delta_s \mid s \in S\}$ (see Example 7, p.30). Thus, $\dim K^{[S]} = |S|$. In particular, K^S is finite dimensional if and only if S is a finite set.

4. Let K be a finite field with q elements, and let V be a vector space over K of dimension n. If $\mathcal{B} = \{x_1, \dots, x_n\}$ is a basis of V, then each element of V is of the form $\alpha_1x_1 + \cdots + \alpha_nx_n$, where $\alpha_1, \dots, \alpha_n \in K$. Therefore, $|V| = q^n$.

   Now we find the number of distinct ordered bases of V. If $x_1 \in V \setminus \{0\}$, then x_1 itself is linearly independent. Now $x_2 \in V$ such that x_1, x_2 are linearly independent if and only if $x_2 \notin \langle x_1 \rangle$, that is, $x_2 \neq \alpha x_1$, for any $\alpha \in K$. Thus, there are $q^n - q$ choices for x_2 so that x_1, x_2 are linearly independent. Therefore, we have $(q^n - 1)(q^n - q)$ distinct ordered linearly independent subsets of V with two elements. If $n > 2$, then $\{x_1, x_2, x_3\}$ is linearly independent if and only if $x_3 \notin \langle x_1, x_2 \rangle$, where x_1, x_2 are already linearly independent. Since there are q^2 elements of the form $\alpha_1x_1 + \alpha_2x_2$, $\alpha_1, \alpha_2 \in K$, there are $q^n - q^2$ choices for x_3. Thus, there are $(q^n - 1)(q^n - q)\left(q^n - q^2\right)$ distinct ordered linearly independent subsets of V with three elements. Continuing similarly, we conclude that there are $(q^n - 1) \cdots \left(q^n - q^{n-1}\right)$ distinct ordered bases of V.

   Note that we have found the number of ordered bases of V, not the number of distinct bases of V. If $\{x_1, \dots, x_n\}$ is a basis of V, then a permutation of these elements is the same basis but a different ordered basis. The number of distinct (unordered) basis of V is $\frac{(q^n-1)\cdots\left(q^n-q^{n-1}\right)}{n!}$.

5. The rank of a matrix $A \in K^{m \times n}$ is the number of linearly independent rows of A. For $A \in K^{m \times n}$, let $\langle \text{rows}\, A \rangle$ be the space generated by the rows of A, called the **row space** of A. This is clearly a subspace of $K^{1 \times n}$. Similarly, let $\langle \text{cols}\, A \rangle$ be the space generated by the columns of A, called the **column space** of A; a subspace of K^m. Observe that $\dim \langle \text{rows}\, A^t \rangle = \dim \langle \text{cols}\, A \rangle$ and $\dim \langle \text{cols}\, A^t \rangle = \dim \langle \text{rows}\, A \rangle$.

   We now prove that for any matrix A,
$$\text{rank}\, A = \dim \langle \text{rows}\, A \rangle = \dim \langle \text{cols}\, A \rangle\,.$$
It is enough to verify that $\dim \langle \text{rows}\, A \rangle = \dim \langle \text{cols}\, A \rangle$. Let $\dim \langle \text{cols}\, A \rangle = k$ and let $c_1, \dots, c_k$ be linearly independent columns of A. Then the j–th

column of A is

$$Ae_j = b_{1j}c_1 + \cdots + b_{kj}c_k, \text{ for some } b_{1j}, \ldots, b_{kj} \in K \text{ and } j = 1, \ldots, n.$$

Let $C = (c_{ij})$, a $m \times k$ matrix whose k columns are $c_1, \ldots, c_k$ respectively, and let $B = (b_{ij})$, a $k \times n$ matrix. Then it is easy to verify that $A = CB$. Now the i–th row of A is

$$\begin{aligned} e_i^t A &= \left(e_i^t C\right) B \\ &= \left(c_{i1} e_1^t + \cdots + c_{ik} e_k^t\right) B \\ &= c_{i1} \left(e_1^t B\right) + \cdots + c_{ik} \left(e_k^t B\right), \text{ for } i = 1, \ldots, m, \text{ and } e_i \in K^m. \end{aligned}$$

Thus each row of A is a linear combination of k rows of B, and so

$$(*) \qquad \dim \langle \text{rows } A \rangle \le k = \dim \langle \text{cols } A \rangle$$

Next, using $(*)$ for A^t, we have

$$\dim \langle \text{cols } A \rangle = \dim \left\langle \text{rows } A^t \right\rangle \le \dim \left\langle \text{cols } A^t \right\rangle = \dim \langle \text{rows } A \rangle .$$

Hence, $\dim \langle \text{rows } A \rangle = \dim \langle \text{cols } A \rangle$.

Immediate consequences of the above result are: $\operatorname{rank} A^t = \operatorname{rank} A$ and $\operatorname{rank} A \le \min(m, n)$, $A \in K^{m \times n}$.

We prove another observation: if $A \in K^{m \times n}$ and P is an invertible $m \times m$ matrix, then $\operatorname{rank} A = \operatorname{rank} PA$.

The i–th row of PA is:

$$e_i^t (PA) = \left(e_i^t P\right) A = [P]_{i1} \left(e_1^t A\right) + \cdots + [P]_{im} \left(e_m^t A\right).$$

Thus, $\dim \langle \text{rows } PA \rangle \le \dim \langle \text{rows } A \rangle$.

Now using this inequality for P^{-1}, we have

$$\dim \langle \text{rows } A \rangle = \dim \left\langle \text{rows } P^{-1} (PA) \right\rangle \le \dim \langle \text{rows } PA \rangle .$$

Hence, $\dim \langle \text{rows } PA \rangle = \dim \langle \text{rows } A \rangle$, that is, $\operatorname{rank} PA = \operatorname{rank} A$.

According to Proposition 1.2.2, every matrix is row equivalent to an echelon matrix. The nonzero rows of an echelon matrix are precisely its linearly independent rows. Therefore, the rank of a matrix is also equal to the number of nonzero rows in the row reduced echelon form of A.

6. Let V be a subspace of $\mathbb{R}^5$ generated by

$$v_1 = \begin{bmatrix} 1 \\ 3 \\ -2 \\ 2 \\ 3 \end{bmatrix}, \quad v_2 = \begin{bmatrix} 2 \\ 7 \\ -5 \\ 6 \\ 5 \end{bmatrix}, \quad v_3 = \begin{bmatrix} 3 \\ 6 \\ -3 \\ 0 \\ 13 \end{bmatrix},$$

and let W be a subspace generated by

$$w_1 = \begin{bmatrix} 1 \\ 3 \\ 0 \\ 2 \\ 1 \end{bmatrix}, \quad w_2 = \begin{bmatrix} 5 \\ 16 \\ -3 \\ 12 \\ 6 \end{bmatrix}, \quad w_3 = \begin{bmatrix} 3 \\ 8 \\ 3 \\ 4 \\ 2 \end{bmatrix}.$$

We find a basis for $V + W$ and for $V \cap W$.

The following observations will be used, the reader is advised to verify these statements.

Let $u_1, \dots, u_m \in K^n$ and let A be an $n \times m$ matrix whose i–th column is u_i. If A' is a matrix obtained by applying elementary column operations on A, then the subspace generated by columns of A' is the same as the subspace generated by columns of A. Also the rank of A is equal to the dimension of subspace $\langle u_1, \dots, u_m \rangle$, and the linearly independent columns of A form a basis of $\langle u_1, \dots, u_m \rangle$.

Recall that if $V = \langle X \rangle$ and $W = \langle Y \rangle$, then $V+W = \langle X \cup Y \rangle$. Therefore, $V + W = \langle v_1, v_2, v_3, w_1, w_2, w_3 \rangle$. Let A be a matrix whose columns are generators of $V + W$:

$$A = \begin{bmatrix} 1 & 2 & 3 & 1 & 5 & 3 \\ 3 & 7 & 6 & 3 & 16 & 8 \\ -2 & -5 & -3 & 0 & -3 & 3 \\ 2 & 6 & 0 & 2 & 12 & 4 \\ 3 & 5 & 13 & 1 & 6 & 2 \end{bmatrix}.$$

Applying elementary column transformations, we can reduce A to

$$A' = \begin{bmatrix} 1 & 0 & 0 & 0 & 0 & 0 \\ 2 & 1 & 0 & 0 & 0 & 0 \\ 0 & 0 & 1 & 0 & 0 & 0 \\ 0 & 2 & 0 & 0 & 0 & 0 \\ 0 & 0 & 0 & 1 & 0 & 0 \end{bmatrix}.$$

Here our strategy is to reduce A to a matrix which is like a column reduced echelon form, but we will avoid division by scalars if it changes integer entries to fractions.

Clearly, rank $A' = 4$, so $\dim(V + W) = 4$. The nonzero columns of A' give a basis of $V + W$, that is,

$$\mathcal{B} = \left\{ u_1 = \begin{bmatrix} 1 \\ 2 \\ 0 \\ 0 \\ 0 \end{bmatrix}, u_2 = \begin{bmatrix} 0 \\ 1 \\ 0 \\ 2 \\ 0 \end{bmatrix}, u_3 = \begin{bmatrix} 0 \\ 0 \\ 1 \\ 0 \\ 0 \end{bmatrix}, u_4 = \begin{bmatrix} 0 \\ 0 \\ 0 \\ 0 \\ 1 \end{bmatrix} \right\}$$

is a basis of $V + W$.

Next, to find a basis of $V \cap W$, we express generators of V and W as a linear combination of elements of $\mathcal{B}$:

$$\begin{aligned} v_1 &= u_1 + u_2 - 2u_3 + 3u_4; \\ v_2 &= 2u_1 + 3u_2 - 5u_3 + 5u_4; \\ v_3 &= 3u_1 - 3u_3 + 13u_4; \\ w_1 &= u_1 + u_2 + u_4; \\ w_2 &= 5u_1 + 6u_2 - 3u_3 + 6u_4; \\ w_3 &= 3u_1 + 2u_2 + 3u_3 + 2u_4. \end{aligned}$$

Note that $V \cap W$ is a subspace of V and W, so $\dim V \cap W$ can not exceed $\min(\dim V, \dim W)$. Since the rank of matrix with columns w_1, w_2, w_3 is 2, $\dim V \cap W \le 2$. Further, if $\dim V \cap W = 2$, then $W = V \cap W \subseteq V$. This means that every element of W is a linear combination of v_1, v_2, v_3. But this is not the case as a matrix with columns v_1, v_2, v_3 and w_1 is of rank 4.

Thus, $\dim V \cap W = 1$. Since we are required to find a basis of $V \cap W$, we use a direct argument.

If $v \in V \cap W$, then $v = \alpha_1 v_1 + \alpha_2 v_2 + \alpha_3 v_3$, $\alpha_i \in \mathbb{R}$ and also $v = \beta_1 w_1 + \beta_2 w_2 + \beta_3 w_3$, $\beta_i \in \mathbb{R}$. Expressing v_i's and w_i's in terms of u_i's, we get

$$\begin{aligned} 0 &= (\alpha_1 + 2\alpha_2 + 3\alpha_3 - \beta_1 - 5\beta_2 - 3\beta_3)\, u_1 \\ &\quad + (\alpha_1 + 3\alpha_2 - \beta_1 - 6\beta_2 - 2\beta_3)\, u_2 \\ &\quad + (-2\alpha_1 - 5\alpha_2 - 3\alpha_3 + 3\beta_2 - 3\beta_3)\, u_3 \\ &\quad + (3\alpha_1 + 5\alpha_2 + 13\alpha_3 - \beta_1 - 6\beta_2 - 2\beta_3)\, u_4. \end{aligned}$$

This gives the following system of linear equations:

$$\begin{aligned} \alpha_1 + 2\alpha_2 + 3\alpha_3 - \beta_1 - 5\beta_2 - 3\beta_3 &= 0, \\ \alpha_1 + 3\alpha_2 - \beta_1 - 6\beta_2 - 2\beta_3 &= 0, \\ -2\alpha_1 - 5\alpha_2 - 3\alpha_3 + 3\beta_2 - 3\beta_3 &= 0, \\ 3\alpha_1 + 5\alpha_2 + 13\alpha_3 - \beta_1 - 6\beta_2 - 2\beta_3 &= 0. \end{aligned}$$

Reducing the coefficient matrix of the above equation to row echelon form , we have

$$\begin{bmatrix} 1 & 0 & 0 & 0 & 1 & -1 \\ 0 & 1 & 0 & 0 & -1 & 1 \\ 0 & 0 & 1 & 0 & 0 & 0 \\ 0 & 0 & 0 & 1 & 4 & 4 \end{bmatrix}.$$

This means that $\alpha_1 = \beta_3 - \beta_2$, $\alpha_2 = -\alpha_1$, $\alpha_3 = 0$ and $\beta_1 = -4(\beta_2 + \beta_3)$. Thus, $\alpha_1 + \alpha_2 = 0$, $\alpha_3 = 0$, so that v is of the form $\alpha(v_2 - v_1)$, $\alpha \in \mathbb{R}$. Hence, $V \cap W = \left\langle \begin{bmatrix} 1 & 4 & -3 & 4 & 2 \end{bmatrix}^t \right\rangle$. Verify that using the relations $\beta_1 = -4(\beta_2 + \beta_3)$ for w_1, w_2, w_3, we get the same basis vector for $V \cap W$.

3.2. Exercises

1. Let $W = \left\{ \begin{bmatrix} x_1 & x_2 & x_3 & x_4 \end{bmatrix}^t \in \mathbb{R}^4 \mid 2x_1 + 3x_2 = 4x_3 + x_4 \right\}$. Show that W is a subspace of $\mathbb{R}^4$. Find a basis of W and extend it to form a basis of $\mathbb{R}^4$. Do the same if $W = \left\{ \begin{bmatrix} x_1 & x_2 & x_3 & x_4 \end{bmatrix}^t \in \mathbb{R}^4 \mid x_1 + x_2 = 0, x_3 - x_4 = 0 \right\}$.

2. A matrix $A \in K^{m \times n}$ is invertible if and only if the columns of A form a basis of K^n.

3. Using the above exercise, find a basis of $\mathbb{R}^4$ containing vector $\begin{bmatrix} 1 & 0 & 1 & 0 \end{bmatrix}^t$ and $\begin{bmatrix} 0 & 1 & 0 & 1 \end{bmatrix}^t$.

4. Find a two dimensional subspace of $\mathbb{R}^4$ which does not contain vectors $\begin{bmatrix} 1 & 3 & 2 & 5 \end{bmatrix}^t$ and $\begin{bmatrix} 2 & 4 & 3 & 1 \end{bmatrix}^t$.

5. Let V be finite dimensional vector space over K and let W be a subspace of V. Show that if $W \neq V$, then there exists a subspace W' such that $W \cap W' = \{0\}$ and $V = W + W'$.

6. Let $W = \{p(x) \in K[x] \mid x^2 p^{(2)}(x) - 4xp^{(1)}(x) + 6p(x) = 0\}$. Show that W is a finite dimensional subspace of $K[x]$. Verify that $2x^2 + 3x^3 \in W$, and find a basis which contains $2x^2 + 3x^3$.

7. Show that $W = \{p(x) \in K_n[x] \mid p(x) = p(-x)\}$, $n \geq 2$, is a subspace of $K_n[x]$. What is the dimension of W? For $n = 7$, find a basis containing $1 + x^2 + x^4$, $1 + x^4 + x^6$.

8. Let $W_1 = \left\langle \begin{bmatrix} 2 & 0 & 3 & 1 & 1 \end{bmatrix}^t, \begin{bmatrix} 1 & 0 & 2 & 1 & 1 \end{bmatrix}^t, \begin{bmatrix} 2 & 0 & 3 & 1 & 3 \end{bmatrix}^t \right\rangle$ and $W_2 = \left\langle \begin{bmatrix} 2 & 1 & 1 & 0 & 1 \end{bmatrix}^t, \begin{bmatrix} 3 & 2 & 3 & 2 & 3 \end{bmatrix}^t, \begin{bmatrix} 1 & 1 & 1 & 1 & 1 \end{bmatrix}^t \right\rangle$ be subspaces of $\mathbb{R}^5$. Find a basis for $W_1 + W_2$ and a basis for $W_1 \cap W_2$.

9. Let $\{v_1, \ldots, v_n\}$ be a basis of a vector space over K and let for $j = 1, \ldots, n$, $w_j = \sum_{i=1}^{n} \alpha_{ij} v_i$, $\alpha_{ij} \in K$. Show that $\{w_1, \ldots, w_n\}$ is a basis of V if and only if $\det(\alpha_{ij}) \neq 0$.

10. Let $A \in K^{n \times n}$, $n > 1$ and let $C(A) = \{X \in K^{n \times n} \mid AX = XA\}$. Prove that $\dim C(A) \geq 2$.

11. Verify that the following sets are bases corresponding to given subspaces of $K^{n \times n}$:
    (a) $\{E_{ij} + E_{ji}, i \neq j, i, j = 1, \ldots, n\} \cup \{E_{11}, \ldots, E_{nn}\}$ for the space of symmetric matrices $\{A \in K^{n \times n} \mid A^t = A\}$.
    (b) $\{E_{ij} - E_{ji}, i < j, i, j = 1, \ldots, n\}$ for the space of skew symmetric matrices $\{A \in K^{n \times n} \mid A^t = -A\}$.
    (c) $\{E_{ij}, i \neq j, i, j = 1, \ldots, n\} \cup \{E_{ii} - E_{11}, i = 2, \ldots, n\}$ for the space of traceless matrices $\{A \in K^{n \times n} \mid \operatorname{tr} A = 0\}$.
    (d) $\{E_{ij}, i \leq j, i, j = 1, \ldots, n\}$ for the space of upper triangular matrices $\left\{A \in K^{n \times n} \mid [A]_{ij} = 0 \text{ for } i > j\right\}$.
    (e) $\{E_{ii}, i = 1, \ldots, n\}$ for the space of diagonal matrices
    $$\left\{A \in K^{n \times n} \mid [A]_{ij} = 0 \text{ for } i \neq j\right\}.$$

12. Verify that $\mathbb{C}$ is a 2–dimensional vector space over $\mathbb{R}$, and is an infinite dimensional vector space over $\mathbb{Q}$.

4. Linear transformations

Let V and V' be vector spaces over K. A mapping $T : V \to V'$ is a **linear transformation** if for all $u, v \in V$ and $\alpha, \beta \in K$:

$$T(\alpha u + \beta v) = \alpha T(u) + \beta T(v).$$

Further, if T is a bijective mapping, then T is called an **isomorphism**. We say that V is **isomorphic** to V', written as $V \simeq V'$ if there is an isomorphism from V to V'.

If $T : V \to V'$ is an isomorphism, then its inverse mapping $T^{-1} : V' \to V$ is also an isomorphism. Since, for $\alpha, \beta \in K$, and $x, y \in V'$,

$$\begin{aligned} T^{-1}(\alpha x + \beta y) &= T^{-1}\left(\alpha T\left(T^{-1}(x)\right) + \beta T\left(T^{-1}(y)\right)\right) \\ &= T^{-1}\left(T\left(\alpha T^{-1}(x) + \beta T^{-1}(y)\right)\right) \\ &= \alpha T^{-1}(x) + \beta T^{-1}(y). \end{aligned}$$

Thus, $V \simeq V'$ if and only if $V' \simeq V$, and so $V \simeq V'$ states that V and V' are isomorphic. In fact, an easy computation shows that the relation 'is isomorphic to' is an equivalence relation.

The following statements are first consequences of the above definition.

PROPOSITION 4.1. *Let V and V' be vector spaces over K, and let $T : V \to V'$ be a linear transformation. Then:*

(i) $T(0) = 0$;
(ii) $T(-v) = -T(v)$, $v \in V$;
(iii) *$T(U)$ is a subspace of V', whenever U is a subspace of V;*
(iv) *$T^{-1}(U')$ is a subspace of V, whenever U' is a subspace of V'.*

PROOF. (i) We remark that this statement means that a linear transformation maps the zero of the vector space V (on the left of the expression) to the zero of the vector space V' (on the right of the expression). Since $T(0) = T(0+0) = T(0) + T(0)$, so $T(0) = 0$.

(ii) $T(-v) = T(-v) + 0 = T(-v) + (T(v) + (-T(v))) = (T(-v) + T(v)) + (-T(v)) = T((-v) + v) + (-T(v)) = T(0) + (-T(v)) = -T(v)$.

(iii) If $x, y \in T(U)$, then for $u, v \in U$, $x = T(u)$ and $y = T(v)$. If U is a subspace of V, then $\alpha u + \beta v \in U$ for all $\alpha, \beta \in K$, and so $\alpha x + \beta y = \alpha T(u) + \beta T(v) = T(\alpha u + \beta v) \in T(U)$. Hence, $T(U)$ is a subspace of V'.

(iv) If $u, v \in T^{-1}(U')$, then $T(u), T(v) \in U'$. If U' is a subspace of V', then for all $\alpha, \beta \in K$, $\alpha T(u) + \beta T(v) = T(\alpha u + \beta v) \in U'$, that is, $\alpha u + \beta v \in T^{-1}(U')$. Hence, $T^{-1}(U')$ is a subspace of V. ∎

Let $T : V \to V'$ be a linear transformation. Since $\{0\}$ is a subspace of V', by the above Proposition, $T^{-1}(\{0\})$ is a subspace of V. We denote this subspace by $\ker T$, called the **kernel** of T. Therefore,

$$\ker T = \{v \in V \mid T(v) = 0\}.$$

PROPOSITION 4.2. *Let $T : V \to V'$ be a linear transformation. Then T is injective if and only if $\ker T = \{0\}$.*

PROOF. If T is injective, and $v \in \ker T$, then $T(v) = 0 = T(0)$ implies that $v = 0$. Hence, $\ker T = \{0\}$.

Conversely, let $\ker T = \{0\}$. For $u, v \in V$ if $T(u) = T(v)$, then $T(u - v) = 0$, that is, $u - v \in \ker T$. Hence, $u = v$, and T is injective. ∎

It follows from the above proposition that vector spaces V and V' are isomorphic if and only if there is a linear transformation $T : V \to V'$ with $\ker T = \{0\}$ and $\operatorname{im} T = V'$. In particular, if $T : V \to V'$ is an injective linear transformation, then $V \simeq \operatorname{im} T$.

Let V and V' be vector spaces over a field K. Denote by $L(V, V')$ the set of all linear transformations from V to V'. Our next result states that $L(V, V')$ is also a vector space over K. We write $L(V)$ for $L(V, V)$ and call its elements **linear operators** on V.

PROPOSITION 4.3. *Let V, V' and V'' be vector spaces over K.*

(i) *If $S, T \in L(V, V')$, then $S + T \in L(V, V')$, where $S + T : V \to V'$ is defined by $(S+T)(v) = S(v) + T(v)$.*
(ii) *If $T \in L(V, V')$, then for $\alpha \in K$, $\alpha T \in L(V, V')$, where $\alpha T : V \to V'$ is defined by $(\alpha T)(v) = \alpha T(v)$.*
(iii) *$L(V, V')$ is a vector space over K.*
(iv) *If $T \in L(V, V'')$ and $S \in L(V'', V')$, then the composite of S and T, $S \circ T \in L(V, V')$.*

PROOF. (i) and (ii) are easy to verify. (iii) $L(V,V')$ is a vector space over K with addition and scalar multiplication as defined in (i) and (ii) respectively. For (iv), let $\alpha,\beta \in K$ and $u,v \in V$. Then $S \circ T(\alpha u + \beta v) = S(T(\alpha u + \beta v)) = S(\alpha T(u) + \beta T(v)) = \alpha S(T(u)) + \beta S(T(v)) = \alpha S \circ T(u) + \beta S \circ T(v)$. Hence, $S \circ T \in L(V,V')$. ∎

If V is a vector space over a field K, then $L(V)$ is a vector space over K together with a binary operation, the composition of mappings, which is associative. Also, the identity mapping $I : V \to V$ is linear, and for all $T \in L(V)$, $T \circ I = T = I \circ T$. A vector space over a field K together with an associative binary operation is called a K–**algebra**. Thus, $L(V)$ is a K–algebra with identity. We denote by $GL(V)$ (or $GL_K(V)$, whenever we want to specify the field K), the set of all isomorphisms from V to itself. Clearly, $GL(V)$ is a group with the composition of mappings as binary operation.

PROPOSITION 4.4. *Let V and V' be vector spaces over K, and let $\mathcal{B}$ be a finite basis of V. If $f : \mathcal{B} \to V'$ is an arbitrary mapping, then there exists a unique $T \in L(V,V')$ such that $T|_{\mathcal{B}} = f$.*

PROOF. Let $\mathcal{B} = \{x_1, \ldots, x_n\}$. Then $v \in V$ can be written uniquely as $v = \sum_{i=1}^n \alpha_i x_i$, $\alpha_i \in K$ (Proposition 3.2). Define $T(v) = \sum_{i=1}^n \alpha_i f(x_i)$. It is easy to verify that $T \in L(V,V')$.

Next, for uniqueness, let $S \in L(V,V')$ such that $S|_{\mathcal{B}} = f$. Then for any $v = \sum_{i=1}^n \alpha_i x_i$ in V,

$$S(v) = \sum_{i=1}^{n} \alpha_i S(x_i) = \sum_{i=1}^{n} \alpha_i f(x_i) = T(v).$$

Hence, $S \equiv T$. ∎

This proposition implies that a linear transformation is completely determined by its action on a basis. Thus, two linear transformations are identical if and only if they agree on a basis of the domain space. An important consequence of the above result is the following proposition.

PROPOSITION 4.5. *Let V and V' be finite dimensional vector spaces over K of dimensions n and m respectively. Then* $\dim L(V,V') = nm$.

PROOF. Let $\mathcal{B} = \{x_1, \ldots, x_n\}$ and $\mathcal{B}' = \{y_1, \ldots, y_m\}$ be bases of V and V' respectively. Then for each $i = 1, \ldots, m$ and $j = 1, \ldots, n$, let $T_{ij} \in L(V,V')$ whose action on $\mathcal{B}$ is given by

$$T_{ij}(x_k) = \begin{cases} y_i, \text{ for } k = j \\ 0, \text{ for } k \neq j \end{cases}, \text{ that is, } T_{ij}(x_k) = \delta_{jk} y_i.$$

We now show that $\{T_{ij} \mid i = 1, \ldots, m,\ j = 1, \ldots, n\}$ is a basis of $L(V,V')$. Hence, $\dim L(V,V') = nm$.

Let $T \in L(V,V')$. Then as $T(x_l) \in W$, so $T(x_l) = \sum_{i=1}^m \alpha_{il} y_i$, $\alpha_{il} \in K$. It then follows that $T = \sum_{i=1}^m \sum_{j=1}^n \alpha_{ij} T_{ij}$. Since for each $x_k \in \mathcal{B}$,

$$\left(\sum_{i=1}^{m} \sum_{j=1}^{n} \alpha_{ij} T_{ij} \right)(x_k) = \sum_{i=1}^{m} \sum_{j=1}^{n} \alpha_{ij} T_{ij}(x_k) = \sum_{i=1}^{m} \alpha_{ik} y_i = T(x_k).$$

Finally, for linear independence, let $\sum_{i=1}^{m}\sum_{j=1}^{n}\beta_{ij}T_{ij}=0$, $\beta_{ij}\in K$. (Note that 0 on the right side is the zero linear transformation). Then for each $k=1,\ldots,n$:

$$0=0(x_k)=\sum_{i=1}^{m}\sum_{j=1}^{n}\beta_{ij}T_{ij}(x_k)=\sum_{i=1}^{m}\beta_{ik}y_i,$$

and so, $\beta_{ik}=0$ for all $i=1,\ldots,m$. ∎

We now prove that two finite dimensional vector spaces are isomorphic if and only if they have the same dimensions. We start with the following lemma.

LEMMA 4.6. *Let V and V' be vector spaces over K, and let $T:V\to V'$ be an injective linear transformation. If $x_1,\ldots,x_n$ are linearly independent elements of V, then $T(x_1),\ldots,T(x_n)$ are linearly independent elements of V'.*

PROOF. Since T is injective, $T(x_1),\ldots,T(x_n)$ are distinct elements of V'. Suppose that $\alpha_1T(x_1)+\cdots+\alpha_nT(x_n)=0$, $\alpha_1,\ldots,\alpha_n\in K$. Then

$$T(\alpha_1x_1+\cdots+\alpha_nx_n)=0,$$

and so, $\alpha_1x_1+\cdots+\alpha_nx_n=0$. Hence, $\alpha_1=\cdots=\alpha_n=0$, and $T(x_1),\ldots,T(x_n)$ are linearly independent elements of V'. ∎

THEOREM 4.7. *Let V and V' be finite dimensional vector spaces over K. Then $V\simeq V'$ if and only if $\dim V=\dim V'$.*

PROOF. If $V\simeq V'$, then there is a bijective linear transformation $T:V\to V'$. If $\mathcal{B}$ is a basis of V, then $T(\mathcal{B})$ is linearly independent subset of V' (Lemma 4.6). Thus, $|T(\mathcal{B})|\le\dim V'$ (Theorem 3.3). Since T is bijective, $|\mathcal{B}|=|T(\mathcal{B})|$. Therefore, $\dim V\le\dim V'$. Similar arguments on the isomorphism $T^{-1}:V'\to V$ show that $\dim V'\le\dim V$. Hence, $\dim V=\dim V'$.

Conversely, if $\dim V=\dim V'$, then for any bases $\mathcal{B}$ and $\mathcal{B}'$ of V and V', $|\mathcal{B}|=|\mathcal{B}'|$, that is, there is a bijection $f:\mathcal{B}\to\mathcal{B}'$. Let $T\in L(V,V')$ such that $T|_{\mathcal{B}}=f$, and let $S\in L(V',V)$ such that $S|_{\mathcal{B}'}=f^{-1}$. Then it is easy to verify that $T\circ S:V'\to V'$ and $S\circ T:V\to V$ are the identity linear operators. Hence, T is an isomorphism and $V\simeq V'$. ∎

COROLLARY 4.8. *If V is a vector space over K of finite dimension n, then $V\simeq K^n$.*

4.1. Examples

1. Let $A\in K^{m\times n}$ and let $\tau_A:K^n\to K^m$ defined by $\tau_A(u)=Au$. Then τ_A is linear transformation. A vector $u=[\alpha_1\cdots\alpha_n]^t\in K^n$ is in $\ker\tau_A$ if and only if $Au=0$, that is, $\alpha_1Ae_1+\cdots+\alpha_nAe_n=0$, where $\{e_1,\ldots,e_n\}$ is the standard basis of K^n. Therefore, $\ker\tau_A=\{0\}$ if and only if $\{Ae_1,\ldots,Ae_n\}$ is a linearly independent subset of K^m. Since Ae_i is the i–th column of A, we have

   $$\tau_A \text{ is injective if and only if } m\ge n \text{ and } \operatorname{rank}A=n.$$

   On the other hand, τ_A is surjective if and only if for each $v\in K^m$, there is $u=[\alpha_1\cdots\alpha_n]^t\in K^n$ such that $v=Au=\alpha_1Ae_1+\cdots+\alpha_nAe_n$. Equivalently, $K^m=\langle Ae_1,\ldots,Ae_n\rangle$. Since $\dim K^m=m$, we have

   $$\tau_A \text{ is surjective if and only if } m\le n \text{ and } \operatorname{rank}A=m.$$

It then follows that τ_A is an isomorphism if and only if $m = n$ and A is invertible.

2. The transpose mapping $t : K^{m\times n} \to K^{n\times m}$ defined by $t(A) = A^t$ is clearly a bijective linear transformation. Thus, $K^{m\times n} \simeq K^{n\times m}$. In particular, $K^{1\times n} \simeq K^n$.

   Also, the mapping $\phi : K^{m\times n} \to K^{mn}$ defined by
$$\phi(A) = ([A]_{11}, \ldots, [A]_{1n}, [A]_{21}, \ldots, [A]_{m1}, \ldots, [A]_{mn})$$
   is an isomorphism.

3. The trace mapping $\mathrm{tr} : K^{n\times n} \to K$ defined by $\mathrm{tr}(A) = \sum_{i=1}^n [A]_{ii}$ is a linear transformation (Proposition 1.1.4). Since, for $\alpha \in K$, $\mathrm{tr}(\alpha E_{11}) = \alpha$, the mapping is surjective. It can be easily verified that the trace map is not injective.

   Let $\phi : K^{n\times n} \to K$ be a linear transformation. Let X be an $n \times n$ matrix defined by $[X]_{ij} = \phi(E_{ji})$, $(i, j = 1, \ldots, n)$. Then for $A \in K^{n\times n}$
$$\begin{aligned}\phi(A) &= \phi\left(\sum_{j=1}^n\sum_{i=1}^n [A]_{ij} E_{ij}\right) = \sum_{j=1}^n\sum_{i=1}^n [A]_{ij}\,\phi(E_{ij})\\ &= \sum_{j=1}^n\sum_{i=1}^n [A]_{ij}[X]_{ji} = \sum_{j=1}^n [XA]_{jj} = \mathrm{tr}(XA).\end{aligned}$$
   It follows from Exercise 18, that X is uniquely determined. Hence, if $\phi \in L(K^{n\times n}, K)$, then there exists unique $X \in K^{n\times n}$ such that $\phi(A) = \mathrm{tr}(XA)$, for all $A \in K^{n\times n}$.

4. For $A \in K^{m\times n}$, the mapping $\lambda_A : K^{n\times p} \to K^{m\times p}$ defined by $\lambda_A(X) = AX$ is a linear transformation. Similarly, for $B \in K^{p\times q}$, the mapping $\rho_B : K^{n\times p} \to K^{n\times q}$ defined by $\rho_B(X) = XB$ is also a linear transformation.

   Now let $A \in K^{n\times n}$. Consider the linear operators λ_A and ρ_A on $K^{n\times n}$. By Proposition 4.3, $\phi_A = \lambda_A - \rho_A$ is also a linear operator on $K^{n\times n}$. Since $\phi_A(A) = 0$, ϕ_A is not injective. Further, if $\mathrm{Char}\, K$ does not divide n, then ϕ_A is also not surjective. Indeed, there is no matrix B such that $\phi_A(B) = I_n$. Otherwise, in K, $n = \mathrm{tr}(I_n) = \mathrm{tr}(\phi_A(B)) = \mathrm{tr}(AB - BA) = 0$. In particular, if $K = \mathbb{Q}$ or $\mathbb{R}$ or $\mathbb{C}$, then ϕ_A is not surjective.

5. The evaluation mapping $\varepsilon_u : K[x] \to K$, $u \in K$, defined by
$$\varepsilon_u(a_0 + a_1x + \cdots + a_nx^n) = a_0 + a_1u + \cdots + a_nu^n$$
   is a surjective linear transformation because $\varepsilon_u(x - u + a) = a$ for any $a \in K$; and $\ker \varepsilon_u = \{f(x) \in K[x] \mid f(u) = 0\}$.

6. The mapping $T : K[x] \to K^\infty$, defined by $T(a_0 + a_1x + \cdots + a_nx^n) = (a_0, a_1, \ldots, a_n, 0, \ldots)$ is a linear transformation which is injective but not surjective because there is no polynomial $p(x) \in K[x]$ such that $T(p(x)) = (1, 1, \ldots, 1, \ldots)$.

7. Let I be an open interval of $\mathbb{R}$ containing 0. Let $\mathcal{D} : C^1(I) \to C^0(I)$ be the differential mapping, $\mathcal{D}(f) = f^{(1)}(x)$. Then $\mathcal{D}$ is a linear transformation, and its kernel is the set of all constant functions defined on I.

The integral mapping $\mathcal{I} : C^0(I) \to C^1(I)$ defined by

$$\mathcal{I}(f)(x) = \int_0^x f(t)\, dt,$$

for all $x \in I$ is also linear. Verify that $\mathcal{D}\mathcal{I}(f) = f$, for all $f \in C^0(I)$. In particular, $\mathcal{D}$ is surjective, and $\mathcal{I}$ is injective. Also, note that for $f(x) = ax + b$, $a, b \in \mathbb{R}$, $\mathcal{I}\mathcal{D}(f)(x) = ax \neq f(x)$.

4.2. Exercises

1. Determine which of the following mappings are linear.

(a) $T : \mathbb{R}^2 \to \mathbb{R}^3$, $T\begin{bmatrix} x \\ y \end{bmatrix} = \begin{bmatrix} x \\ y \\ 0 \end{bmatrix}$.

(b) $T : \mathbb{R}^2 \to \mathbb{R}^3$, $T\begin{bmatrix} x \\ y \end{bmatrix} = \begin{bmatrix} x \\ y \\ 1 \end{bmatrix}$.

(c) $T : \mathbb{R}^2 \to \mathbb{R}^3$, $T\begin{bmatrix} x \\ y \end{bmatrix} = \begin{bmatrix} x \\ y \\ y \end{bmatrix}$.

(d) $T : \mathbb{R}^3 \to \mathbb{R}^2$, $T\begin{bmatrix} x \\ y \\ z \end{bmatrix} = \begin{bmatrix} x \\ y \end{bmatrix}$.

(e) $T : \mathbb{R}^3 \to \mathbb{R}^2$, $T\begin{bmatrix} x \\ y \\ z \end{bmatrix} = \begin{bmatrix} x + y \\ z \end{bmatrix}$.

(f) $T : \mathbb{R}^3 \to \mathbb{R}^2$, $T\begin{bmatrix} x \\ y \\ z \end{bmatrix} = \begin{bmatrix} xy \\ yz \end{bmatrix}$.

(g) $T : \mathbb{R}^3 \to \mathbb{R}^3$, $T\begin{bmatrix} x \\ y \\ z \end{bmatrix} = \begin{bmatrix} z \\ x \\ y \end{bmatrix}$.

(h) $T : \mathbb{R}^3 \to \mathbb{R}^3$, $T\begin{bmatrix} x \\ y \\ z \end{bmatrix} = \begin{bmatrix} x + 1 \\ y \\ z \end{bmatrix}$.

2. Let $T : \mathbb{R}^3 \to \mathbb{R}^3$, $T\begin{bmatrix} x \\ y \\ z \end{bmatrix} = \begin{bmatrix} 2x + 2y - z \\ 2x + 5y - 2z \\ 4x + 4y - 2z \end{bmatrix}$. Verify that T is a linear operator on $\mathbb{R}^3$. Find the kernel and the image of T.

3. Verify that $T : \mathbb{R}_n[x] \to \mathbb{R}^{(n+1)\times(n+1)}$, $n \geq 1$, given by

$$T\left(a_0 + a_1 x + \cdots + a_n x^n\right) = \begin{bmatrix} a_0 & a_1 & \cdots & a_n \\ & \ddots & \ddots & \vdots \\ & & \ddots & a_1 \\ 0 & & & a_0 \end{bmatrix}$$

is an injective linear transformation.

4. Find the image and the kernel of the linear transformation $S : \mathbb{C}^{n\times n} \to \mathbb{C}^{n\times n}$ given by $S(A) = A - A^*$.

5. Let $\mathcal{B}$ and $\mathcal{B}'$ be bases of vector spaces V and V' respectively, and let $T \in L(V, V')$. Prove that $T : V \to V'$ is an isomorphism if and only if $T|_{\mathcal{B}}$ is a bijection from $\mathcal{B}$ to $\mathcal{B}'$.

6. Let W be a subspace of a vector space $V$and let $T \in L(W, V')$. Show that T can be extended to a linear transformation from V to V'.

7. Let V and V' be vector spaces over K and let $\{v_1, \ldots, v_n\}$ be a basis of V. Let $T \in L(V, V')$. Prove the following statements.
   (a) T is injective if and only if $Tv_1, \ldots, Tv_n$ are linearly independent. Thus, if T is injective then $\dim V \leq \dim V'$.
   (b) T is surjective if and only if $V' = \langle Tv_1, \ldots, Tv_n\rangle$. Thus, if T is surjective then $\dim V' \leq \dim V$.
   (c) T is bijective if and only if $\{Tv_1, \ldots, Tv_n\}$ is a basis of V'. Thus, if T is bijective then $\dim V = \dim V'$.

8. Let V_1, V_2, V_1' and V_2' be vector spaces over K. Let $\theta : V_1 \to V_2$ and $\psi : V_1' \to V_2'$ be isomorphisms. Show that $\Gamma : L(V_1, V_1') \to L(V_2, V_2')$ defined by $\Gamma(\phi) = \psi \circ \phi \circ \theta^{-1}$ is an isomorphism.

9. Let U, V and W be vector spaces over K, let $T \in L(V, W)$ and let $S \in L(U, V)$. Show that $\ker S \subseteq \ker TS$ and $\operatorname{im} TS \subseteq \operatorname{im} T$.

10. Let T be a linear operator on a vector space V, let $v \in V$ and let m be a positive integer such that $T^m v = 0$ and $T^{m-1}v \neq 0$. Then $v, Tv, \ldots, T^{m-1}v$ are linearly independent.

11. A linear operator T on a vector space V is called **nilpotent** if $T^k = 0$ for some positive integer k. The positive integer m such that $T^m = 0$ and $T^{m-1} \neq 0$ is called the **nilpotency index** of T. Prove that if m is the nilpotency index of T, then m can not exceed the dimension of V.

12. If T is a nilpotent linear operator on a vector space V over K, then for $\alpha \in K$, $I - \alpha T$ is an invertible linear operator on V.

13. Define $S : K^n \to K^n$ by $S\begin{bmatrix} x_1 \\ \vdots \\ x_{n-1} \\ x_n \end{bmatrix} = \begin{bmatrix} 0 \\ x_1 \\ \vdots \\ x_{n-1} \end{bmatrix}$. Show that S is a nilpotent linear operator. Using this, conclude that $T : K^n \to K^n$ defined by $T\begin{bmatrix} x_1 \\ x_2 \\ \vdots \\ x_n \end{bmatrix} = \begin{bmatrix} x_1 \\ x_2 - x_1 \\ \vdots \\ x_n - x_{n-1} \end{bmatrix}$ is an invertible linear operator.

14. Let V be a vector space over K, $\operatorname{Char} K = 0$. Let $S, T \in L(V)$ such that $ST - TS = I$, I is the identity operator on V. Prove that

$$S^{n+1}T - TS^{n+1} = (n+1)\, S^n,$$

for $n = 0, 1, \ldots$. Use this identity to establish that in this case $L(V)$ is an infinite dimensional vector space. Hence, if V is a finite dimensional vector space then there do not exist linear operators S, T on V such that $ST - TS = I$.

15. Let $T : \mathbb{R}[x] \to \mathbb{R}[x]$ defined by $T(p(x)) = xp(x)$. Show that T is a linear operator on $\mathbb{R}[x]$. Also if D is the differential operator on $\mathbb{R}[x]$, then $DT - TD = I$.

16. A linear operator T on a vector space V is called an **idempotent** if $T^2 = T$. Let S and T be idempotents on V. Prove the following statements.
    (a) $I - T$ is an idempotent.
    (b) $S + T$ is an idempotent if and only if $ST = 0 = TS$.
    (c) $S - T$ is an idempotent if and only if $ST = T = TS$.

17. The set $\mathbb{R}$ is a vector space over $\mathbb{R}$ of dimension 1, and we have seen in Example 8, p.27, that $\mathbb{R}^+$ is also a vector space over $\mathbb{R}$. Let $f : \mathbb{R} \to \mathbb{R}^+$ defined by $f(x) = e^x$. Prove that f is an isomorphism.

18. Prove that if $X \in K^{n \times n}$ and $X \neq 0$, then there exists $A \in K^{n \times n}$ such that $\operatorname{tr}(XA) \neq 0$.
    Hint: if $X = (x_{rs})$ and $x_{pq} \neq 0$, $\operatorname{tr}(XE_{qp}) \neq 0$.

5. Quotient spaces

Let W be a subspace of a vector space V defined over a field K, and let $x \in V$. Then the set $x + W$ defined by

$$x + W = \{x + w \mid w \in W\}$$

is called a **coset** of W in V.

Note that if $x \in W$, then $x + W = W$, and for $x \notin W$, $x + W$ is not a subspace of V.

PROPOSITION 5.1. *Let W be a subspace of a vector space V. Then:*
(i) *for $x, y \in V$, $x + W = y + W$ if and only if $x - y \in W$;*
(ii) *any two cosets of W in V are either identical or disjoint.*

PROOF. (i) If $x + W = y + W$, then $x = y + w$ for some $w \in W$, and so $x - y = w \in W$. Conversely, if $x - y \in W$, then for $w \in W$, $x + w = y + ((x - y) + w) \in y + W$. Thus, $x + W \subseteq y + W$. Similarly, $y + W \subseteq x + W$.

(ii) Let $x, y \in V$ and $(x + W) \cap (y + W) \neq \emptyset$. Then for $z \in (x + W) \cap (y + W)$, $z - x \in W$ and $z - y \in W$. By (i), $x + W = z + W = y + W$. ∎

THEOREM 5.2. *Let V be a vector space over K and let W be a subspace of V. Let $V/W = \{x + W \mid x \in V\}$, the set of all cosets of W in V. Define the following addition $\boxplus$ and scalar multiplication $\boxdot$ on V/W:*
(i) $(x + W) \boxplus (y + W) = x + y + W$, *and*
(ii) $\alpha \boxdot (x + W) = \alpha x + W$,

for all $x, y \in V$ and $\alpha \in K$. Then V/W, together with these operations, is a vector space over K.

PROOF. First, we need to verify that these operations are well defined. Let $x+W = x'+W$ and $y+W = y'+W$. Then by Proposition 5.1, $x-x'$ and $y-y'$ are in W. Since W is subspace, $(x+x')-(y+y') \in W$, and so $x+y+W = x'+y'+W$. Hence, $(x+W) \boxplus (y+W) = (x'+W) \boxplus (y'+W)$ and the addition $\boxplus$ is well defined. Also, for $\alpha \in K$, $\alpha(x-x') \in W$, and so $\alpha x + W = \alpha x' + W$. Hence, $\alpha \boxdot (x+W) = \alpha \boxdot (x'+W)$ and the scalar multiplication $\boxdot$ is well defined.

The reader should verify that these operations actually satisfy all conditions of a vector space. The identity element of V/W is $0+W \equiv W$, and the additive inverse of $v+W$ is $-v+W$. ∎

The vector space defined in Theorem 5.2 is called the **quotient space** of V by W. From now onwards, we will normally denote $\boxplus$ and $\boxdot$ by ordinary addition $+$ and juxtaposition, whenever there is no chance of confusion.

The quotient space $V/\{0\}$ can clearly be identified with V under the isomorphism $v \mapsto v+\{0\}$. Thus, we write V for $V/\{0\}$. Also, the quotient space V/V is the zero space as $v+V = V$ for all $v \in V$.

Let W be a subspace of a vector space V. The mapping $\pi : V \to V/W$ defined by $\pi(v) = v+W$ is clearly a surjection, and by definition of addition and scalar multiplication on V/W, it is immediate that π is linear. The linear transformation $\pi : V \to V/W$ is called the **canonical epimorphism** of V onto V/W.

THEOREM 5.3. *Let V and V' be vector spaces over K, and let $T \in L(V, V')$.*

(i) *If W and W' are subspaces of V and V' respectively, and $T(W) \subseteq W'$, then T induces a linear transformation $\overline{T} : V/W \to V'/W'$ defined by $\overline{T}(v+W) = T(v) + W'$.*

(ii) ***First Isomorphism Theorem.*** *If T is surjective, then $V/\ker T \simeq V'$.*

PROOF. (i) First we show that $\overline{T}$ is well defined. If $x+W = y+W$ for $x, y \in V$, then $x-y \in W$, and so $T(x) - T(y) \in T(W) \subseteq W'$. Hence, by Proposition 5.1, $T(x) + W' = T(y) + W'$, and $\overline{T}(x+W) = \overline{T}(y+W)$. By linearity of T it follows that $\overline{T}$ is also linear.

(ii) Let $W = \ker T$ and $W' = \{0\}$. Then T induces a linear transformation $\overline{T} : V/\ker T \to V'$ defined by $\overline{T}(v + \ker T) = T(v)$. Since T is surjective, to $y \in V'$, there exists $x \in V$ such that $T(x) = y$, and so $\overline{T}(x + \ker T) = y$. Therefore $\overline{T}$ is surjective. Thus, to prove that $\overline{T}$ is an isomorphism, we require to check that $\overline{T}$ is injective. For $x \in V$, $x + \ker T \in \ker \overline{T}$ if and only if $T(x) = 0$, that is, $x \in \ker T$. Hence, $\overline{T}$ is injective. ∎

COROLLARY 5.4. *Let V and V' be vector spaces over K and let $T \in L(V, V')$. Then $V/\ker T \simeq \operatorname{im} T$.*

COROLLARY 5.5. ***Second Isomorphism Theorem.*** *Let V be a vector space over K and let W_1 and W_2 be subspaces of V. Then*

$$(W_1 + W_2)/W_2 \simeq W_1/W_1 \cap W_2.$$

PROOF. Define $T : W_1 \to (W_1+W_2)/W_2$ by $T(w_1) = w_1 + W_2$. Clearly, T is a well defined surjective linear transformation. Also,

$$\begin{aligned} \ker T &= \{w_1 \in W_1 \mid w_1 + W_2 = W_2\} \\ &= \{w_1 \in W_1 \mid w_1 \in W_2\} = W_1 \cap W_2. \end{aligned}$$

Therefore, by Theorem 5.3(ii), $(W_1+W_2)/W_2 \simeq W_1/W_1 \cap W_2$. ∎

COROLLARY 5.6. ***Third Isomorphism Theorem.*** *Let V be a vector space over K and let U and W be subspaces of V such that $U \subseteq W$. Then*

$$(V/U)\,/\,(W/U) \simeq V/W.$$

PROOF. The mapping $\phi : V/U \to V/W$ given by $\phi(x+U) = x+W$ is a well defined linear transformation which is clearly surjective. Also

$$\begin{aligned} \ker\phi &= \{x+U \in V/U \mid x+W = 0\ (\in V/W)\} \\ &= \{x+U \in V/U \mid x \in W\} = W/U. \end{aligned}$$

Therefore, by Theorem 5.3(ii), $(V/U)\,/\,(W/U) \simeq V/W$. ∎

Our next result is a dimension formula for quotient spaces.

THEOREM 5.7. *Let V be a finite dimensional vector space over K and let W be a subspace of V. Then*

$$\dim V = \dim W + \dim V/W.$$

PROOF. Let $X = \{x_1, \ldots, x_r\}$ be a basis of W and let $Y = \{y_1, \ldots, y_s\}$ be a finite subset of V such that $X \cup Y$ is a basis of V and $X \cap Y = \varnothing$ (Theorem 3.7). We show that the set $\overline{Y} = \{y_1 + W, \ldots, y_s + W\}$ is a basis of V/W. It then follows that $\dim V/W = s$ and $\dim V = r + s = \dim W + \dim V/W$.

Suppose that $\sum_{i=1}^{s} \alpha_i (y_i + W) = 0\ (\in V/W)$, for some $\alpha_1, \ldots, \alpha_s \in K$. Then $\sum_{i=1}^{s} \alpha_i y_i \in W$, and so $\sum_{i=1}^{s} \alpha_i y_i = 0$. Otherwise $\sum_{i=1}^{s} \alpha_i y_i$ is a nontrivial K-linear combination of elements of X which implies that $X \cup Y$ is a linearly dependent set , a contradiction. But $\sum_{i=1}^{s} \alpha_i y_i = 0$ implies that $\alpha_1 = \cdots = \alpha_s = 0$ as Y is a linearly independent set in V. Hence, $\overline{Y}$ is a linearly independent set in V/W.

Finally, $\overline{Y}$ generates V/W. For any $v \in V$, there are scalars $\alpha_1, \ldots, \alpha_r, \beta_1, \ldots, \beta_s$ such that $v = \sum_{i=1}^{r} \alpha_i x_i + \sum_{j=1}^{s} \beta_j y_j$. Thus, $v + W = \sum_{j=1}^{s} \beta_j (y_j + W)$. Hence, $\overline{Y}$ generates V/W. ∎

COROLLARY 5.8. ***Dimension formula for linear transformations.*** *Let V be a finite dimensional vector space over K and let $T \in L(V, V')$ for any vector space V' over K. Then* $\dim V = \dim\ker T + \dim \operatorname{im} T$.

PROOF. By Theorem 5.7, $\dim V = \dim\ker T + \dim(V/\ker T)$. Since $V/\ker T \simeq \operatorname{im} T$ (Corollary 5.4), $\dim(V/\ker T) = \dim\operatorname{im} T$. This completes the proof. ∎

COROLLARY 5.9. ***Dimension formula for subspaces.*** *Let W_1 and W_2 be finite dimensional subspaces of a vector space V. Then*

$$\dim W_1 + \dim W_2 = \dim(W_1 + W_2) + \dim(W_1 \cap W_2).$$

PROOF. Clearly, $W_1 + W_2$ is also a finite dimensional subspace of V. By Corollary 5.5, $\dim(W_1 + W_2)/W_2 = \dim(W_1/W_1 \cap W_2)$. Also, by Theorem 5.7, $\dim W_1 = \dim(W_1 \cap W_2) + \dim(W_1/W_1 \cap W_2)$. Therefore,

$$\begin{aligned} \dim W_1 + \dim W_2 &= (\dim(W_1 \cap W_2) + \dim(W_1/W_1 \cap W_2)) + \dim W_2 \\ &= \dim(W_1 \cap W_2) + (\dim(W_1/W_1 \cap W_2) + \dim W_2) \\ &= \dim(W_1 \cap W_2) + (\dim(W_1 + W_2)/W_2 + \dim W_2) \\ &= \dim(W_1 \cap W_2) + \dim(W_1 + W_2). \end{aligned}$$

∎

Let V be a finite dimensional vector space over a field K, and let $T \in L(V, V')$, where V' is any vector space over K. The dimension of the subspace $\ker T$ is called the **nullity** of T, denoted by $\operatorname{nullity} T$, and the dimension of the subspace $\operatorname{im} T$ is called the **rank** of T, denoted by $\operatorname{rank} T$. Thus, Corollary 5.8 states that

$$\dim V = \operatorname{nullity} T + \operatorname{rank} T.$$

Another important consequence of Theorem 5.7 is the following proposition.

PROPOSITION 5.10. *Let V and V' be finite dimensional vector spaces over K of the same dimensions, and let $T \in L(V, V')$. Then, the following statements are equivalent:*

(i) *T is surjective;*
(ii) *T is injective;*
(iii) *T is bijective.*

PROOF. Obviously, we need to verify equivalence of (ii) and (iii) only.

If T is surjective, then $\operatorname{im} T = V'$, and so $\dim \ker T = 0$. Hence T is injective. Conversely, if T is injective, then $\ker T = \{0\}$, and so $\dim V = \dim \operatorname{im} T$. Since $\dim V = \dim V'$, T is surjective. ∎

An isomorphism $T : V \to V'$ is also called an **invertible** linear transformation, since the inverse mapping $T^{-1} : V' \to V$ is also a linear transformation and $T^{-1}T$ and $TT^{-1}$are identity linear operators on V and V' respectively. By Proposition 5.10, it follows that $T \in L(V, V')$ is invertible if and only if $\operatorname{rank} T = \dim V = \dim V'$.

5.1. Examples

1. If V is a finite dimensional vector space and if W is a nonzero proper subspace of V, then $\dim W < \dim V$ and $\dim V/W < \dim V$ (Theorem 5.7). However, an infinite dimensional vector space V can have a proper subspace W such that $\dim W = \dim V = \dim V/W$.

   Let $V = \mathbb{R}[x]$, the vector space of polynomials with real coefficients. Then V is an infinite dimensional vector space over $\mathbb{R}$, and has a basis $\{1, x, \dots, x^n, \dots\}$. Let

$$W = \{p(x) \in \mathbb{R}[x] \mid p(t) = p(-t) \text{ for all } t \in \mathbb{R}\}.$$

   Then W is a proper nonzero subspace of V which consists of polynomials of the form $a_0 + a_1x^2 + \cdots + a_nx^{2n}$, $n = 0, 1, \dots$. Clearly, W has a basis $\{1, x^2, \dots, x^{2n}, \dots\}$. The quotient space V/W is also infinite dimensional, and it can be easily verified that $\{x + W, x^3 + W, \dots, x^{2n+1} + W, \dots\}$ is a basis of V/W. Now it is clear that $\dim W = \dim V = \dim V/W$.

   Note that if W is a finite dimensional subspace of a vector space V over K such that V/W is finite dimensional, then V is also finite dimensional. Since, if $W = \langle x_1, \dots, x_m \rangle$ and $V/W = \langle y_1 + W, \dots, y_n + W \rangle$, then $V = \langle x_1, \dots, x_m, y_1, \dots, y_n \rangle$. Indeed, for any $v \in V$, $v + W = \sum_{j=1}^{n} \beta_j (y_j + W) = \left(\sum_{j=1}^{n} \beta_j y_j\right) + W$, $\beta_1, \dots, \beta_n \in K$, that is, $v - \sum_{j=1}^{n} \beta_j y_j \in W$, and so $v - \sum_{j=1}^{n} \beta_j y_j = \sum_{i=1}^{m} \alpha_i x_i$ or $v = \alpha_1 x_1 + \cdots + \alpha_m x_m + \beta_1 y_1 + \cdots + \beta_n y_n$.

2. *If X and Y are subspaces of a vector space V such that V/X and V/Y are finite dimensional, then the quotient space $V/(X\cap Y)$ is also finite dimensional.*

   By Corollary 5.6, $(V/(X\cap Y))/(X/(X\cap Y)) \simeq V/X$, and by Corollary 5.5, $X/(X\cap Y) \simeq (X+Y)/Y$. Since $(X+Y)/Y$ is a subspace of finite dimensional vector space V/Y, $(X+Y)/Y$ is finite dimensional, and so $\dim X/(X\cap Y)$ is finite. Now it follows from the discussion of Example 1 above that $\dim V/(X\cap Y)$ is finite.

3. Let T be a linear operator on a vector space V and let W be a subspace of V. If T_1 is a restriction of T to W, then $T_1 : W \to V$ is a linear transformation. Therefore, $W/\ker T_1 \simeq \operatorname{im} T_1$. Since $\ker T_1 = W\cap \ker T$ and $\operatorname{im} T_1 = T(W)$, we have:

$$W/(W\cap \ker T) \simeq T(W).$$

In particular, if W is a finite dimensional subspace of V, then we have the following dimension formula:

$$\dim W = \dim (W\cap \ker T) + \dim T(W).$$

5.2. Exercises

1. Let $W = \left\langle \begin{bmatrix} 1 & 2 & 1 & 0 & 1\end{bmatrix}^t, \begin{bmatrix} 1 & 0 & 1 & 1 & 1\end{bmatrix}^t, \begin{bmatrix} 1 & 2 & 1 & 3 & 1\end{bmatrix}^t \right\rangle$, a subspace of $\mathbb{R}^5$. Find a basis of $\mathbb{R}^5/W$.

2. Let $V = K^n$ and let $W = \left\{ \begin{bmatrix} x_1 & \cdots & x_n \end{bmatrix}^t \in K^n \mid x_1 + \cdots + x_n = 0 \right\}$. Compute the dimension of V/W. Deduce that there is no subspace U of V such that $W \subseteq U \subseteq V$ and $U \neq V, W$.

3. For a positive integer m, let

$$W_m = \left\{ p(x) \in \mathbb{R}[x] \mid p(0) = p^{(1)}(0) = \cdots = p^{(m-1)}(0) = 0 \right\},$$

   Verify that W_m is a subspace of $\mathbb{R}[x]$. Prove that $\dim W_m = \dim \mathbb{R}[x]$ and $\dim(\mathbb{R}[x]/W_m) = m+1$.

4. Consider a linear operator T on $\mathbb{R}^3$ defined by $T\begin{bmatrix} x \\ y \\ z\end{bmatrix} = \begin{bmatrix} x+y \\ 0 \\ z\end{bmatrix}$. Determine an isomorphism $S : \mathbb{R}^3/\ker T \to \operatorname{Im} T$.

5. Let $T : \mathbb{R}[x] \to \mathbb{R}^2$ defined by $Tp(x) = \begin{bmatrix} p(1) \\ p(-1)\end{bmatrix}$. Verify that T is a linear transformation. Find its kernel, and also determine an isomorphism $S : \mathbb{R}[x]/\ker T \to \mathbb{R}^2$.

6. Let V be the space of all $n \times n$ upper triangular matrices with entries from K and let W be the subspace of V consisting of all $n\times n$ upper triangular matrices with zeros on the diagonal. Prove that $V/W \simeq K^n$.

7. Let V and V' be vector spaces over K and let W be a subspace of V. Show that $\Gamma = \{T \in L(V, V') \mid T(w) = 0 \text{ for all } w \in W\}$ is a subspace of $L(V, V')$, and this subspace is isomorphic to $L(V/W, V')$. Deduce that if V and V' are finite dimensional, then $\dim \Gamma = (\dim V - \dim W) \dim V'$.

8. Let V be a finite dimensional vector space and $S, T \in L(V)$. If ST is the identity operator on V, then so is TS.

9. The above statement is not true for infinite dimensional vector spaces. Let $V = K^\infty$. Define $S(c_1, c_2, \dots) = (c_2, c_3, \dots)$ and $T(c_1, c_2, \dots) = (0, c_1, \dots)$. Then S and T are linear operators on V such that ST is the identity operator and TS is not.

10. Let T be a linear operator on a finite dimensional vector space V. Prove that there is a basis $\{v_1, \dots, v_n\}$ of V such that $\{Tv_1, \dots, Tv_k\}$ is a basis of $\operatorname{im} T$ and $\{v_{k+1}, \dots, v_n\}$ is a basis of $\ker T$.

11. Let V_1 and V_2 be finite dimensional vector spaces over K and let $S \in L(V_1, V_2)$. Prove that
    (a) $\ker S \subseteq \ker T$ if and only if $T = PS$ for some $P \in L(V_2)$.
    (b) $\operatorname{im} S \subseteq \operatorname{im} T$ if and only if $S = TQ$ for some $Q \in L(V_1)$.

12. Let V_1 and V_2 be finite dimensional vector spaces over K and let $S \in L(V_1, V_2)$. Prove that there exists $T \in L(V_2, V_1)$ such that $S = STS$ and $T = TST$.

13. Let V be a $2n$–dimensional vector space and let W_1 and W_2 be $(n+1)$–dimensional subspaces of V. Prove that $\dim(W_1 \cap W_2) \geq 2$.

14. Let W_1 and W_2 be finite dimensional subspaces of a vector space such that $\dim(W_1 + W_2) = 1 + \dim(W_1 \cap W_2)$. Show that $W_1 + W_2$ is equal to one of the subspaces and $W_1 \cap W_2$ is equal to the other.

15. Let W_1 and W_2 be subspaces of finite dimensional vector space V. Show that if $\dim W_1 + \dim W_2 > \dim V$, then $W_1 \cap W_2 \neq \{0\}$.

16. Let T be a linear operator on a finite dimensional vector space. Show that if $\operatorname{rank} T = \operatorname{rank} T^2$, then $\operatorname{im} T \cap \ker T = \{0\}$.

17. Let V_1 and V_2 be finite dimensional vector spaces, $T \in L(V_1, V_2)$ and P and Q be bijective linear operators on V_1 and V_2 respectively. Prove that $\operatorname{rank}(PTQ) = \operatorname{rank} T$.

18. Let $S, T \in L(V_1, V_2)$, V_1 and V_2 finite dimensional vector spaces. Show that $\operatorname{im}(S+T) \subseteq \operatorname{im} S + \operatorname{im} T$. Deduce that $\operatorname{rank}(S+T) \leq \operatorname{rank} S + \operatorname{rank} T$.

19. Let V_1, V_2 and V_3 be finite dimensional vector spaces over K and let $T \in L(V_1, V_2)$, $S \in L(V_2, V_3)$. Prove that
    (a) $\operatorname{rank} ST = \operatorname{rank} T - \dim(\operatorname{im} T \cap \ker S)$.
    (b) $\operatorname{rank} ST \leq \min\{\operatorname{rank} S, \operatorname{rank} T\}$.

    Deduce that
$$\operatorname{rank} S + \operatorname{rank} T - \dim V_2 \leq \operatorname{rank} ST \leq \min\{\operatorname{rank} S, \operatorname{rank} T\}.$$

6. Direct sum

Let $V_1, \dots, V_m$ be vector spaces over K. Consider the cartesian product of sets $V_1, \dots, V_m$:

$$V = V_1 \times \cdots \times V_m = \{(v_1, \dots, v_m) \mid v_i \in V_i\}.$$

Then V is a vector space over K under the componentwise addition and scalar multiplication:

$$\begin{aligned}(x_1, \dots, x_m) + (y_1, \dots, y_m) &= (x_1 + y_1, \dots, x_m + y_m),\\ \alpha\,(x_1, \dots, x_m) &= (\alpha x_1, \dots, \alpha x_m),\end{aligned}$$

where $x_i, y_i \in V_i$, $i = 1, \dots, m$ and $\alpha \in K$. This vector space is called the **external direct sum** of vector spaces $V_1, \dots, V_m$, and is denoted by $\oplus_{i=1}^m V_i$ or $V_1 \oplus \cdots \oplus V_m$.

PROPOSITION 6.1. *Let $V_1, \dots, V_m$ be vector spaces over a field K. Then $V = V_1 \oplus \cdots \oplus V_m$ is finite dimensional if and only if each V_i is finite dimensional. In this case*

$$\dim V_1 \oplus \cdots \oplus V_m = \dim V_1 + \cdots + \dim V_m.$$

PROOF. For each i, define $\widehat{V}_i = \{(v_1, \dots, v_m) \mid v_j = 0 \text{ for } j \neq i\}$. Then each $\widehat{V}_i$ is a subspace of V. Also, the reader can verify that the mapping $V_i \to \widehat{V}_i$ defined by $v_i \mapsto (0, \dots, 0, v_i, 0, \dots, 0)$ is an isomorphism, that is, $V_i \simeq \widehat{V}_i$. Since, $(v_1, \dots, v_m) = (v_1, 0, \dots, 0) + \cdots + (0, \dots, 0, v_m)$, $V = \left\langle \widehat{V}_1 \cup \cdots \cup \widehat{V}_m \right\rangle$. Now if each V_i is finite dimensional, then each $\widehat{V}_i$ is finitely generated vector space, and so V is generated by finitely many elements, that is, $\dim V$ is finite.

Conversely, if $\dim V$ is finite then every subspace of V is finite dimensional. Since $\widehat{V}_i$ is a subspace of V and $\widehat{V}_i \simeq V_i$, so $\dim V_i$ is finite.

For the second part, we apply induction on m. If $m = 1$, then the statement is clearly true. Assume that the result is true for $m - 1$. Now, define the mapping $V_1 \oplus \cdots \oplus V_m \to V_1 \oplus \cdots \oplus V_{m-1}$ by $(v_1, \dots, v_m) \mapsto (v_1, \dots, v_{m-1})$. Then the reader can easily verify that this mapping is surjective linear transformation with kernel $\widehat{V}_m$. Therefore, $V_1 \oplus \cdots \oplus V_m / \widehat{V}_m \simeq V_1 \oplus \cdots \oplus V_{m-1}$. Hence, by Proposition 5.7 and by the induction hypothesis:

$$\begin{aligned}\dim V_1 \oplus \cdots \oplus V_m &= \dim V_1 \oplus \cdots \oplus V_{m-1} + \dim \widehat{V}_m\\ &= \dim V_1 + \cdots + \dim V_{m-1} + \dim V_m.\end{aligned}$$

∎

THEOREM 6.2. *Let V be a vector space over K, and let $W_1, \dots, W_m$ be subspaces of V. If*

1. $V = W_1 + \cdots + W_m$, *and*
2. *for each k, $W_k \cap \sum_{i \neq k} W_i = \{0\}$,*

then $V \simeq W_1 \oplus \cdots \oplus W_m$.

PROOF. Define $T : \oplus_{i=1}^m W_i \to V$ by $T\,(w_1, \dots, w_m) = w_1 + \cdots + w_m$. By the definition of T, it is obvious that T is well defined, and by the very definition of addition and scalar multiplication on $\oplus_{i=1}^m W_i$, T is linear. Since $V = W_1 + \cdots + W_m$, T is surjective. Now to establish the result, we need to check the injectivity of T.

Let $v \in \ker T$; then for $v = (v_1, \dots, v_m)$, $v_1 + \cdots + v_m = 0$. But then $v_k = -\sum_{i \neq k} v_i \in W_k \cap \sum_{i \neq k} W_i = \{0\}$. Hence, T is injective and $V \simeq W_1 \oplus \cdots \oplus W_m$. ∎

A vector space V is said to be an **internal direct sum** of $W_1, \ldots, W_m$ if $W_1, \ldots, W_m$ are subspaces of V, which satisfy the conditions of above theorem.

Note that if V is an internal direct sum of $W_1, \ldots, W_m$ then each W_i is a subspace of V and if V is an external direct sum of $W_1, \ldots, W_m$ then W_i is not a subspace of V but $\widehat{W_i}$ is a subspace of V which is isomorphic to W_i.

External direct sum constructs a new vector space from a given family of vector spaces, and internal direct sum decomposes a given vector space into sum of smaller subspaces. We shall see that internal direct sum plays an important role in studying vector spaces and linear transformations. From now onwards, unless otherwise is stated, by direct sum we mean an internal direct sum; and whenever we write $V = \oplus_{i=1}^{m} W_i$, we mean that V is the (internal) direct sum of subspaces $W_1, \ldots, W_m$.

The following criterion for (internal) direct sum is useful.

PROPOSITION 6.3. *Let $W_1, \ldots, W_m$ be subspaces of a vector space V. Then $V = \oplus_{i=1}^{m} W_i$ if and only if each element v of V admits a unique representation as a sum $v = v_1 + \cdots + v_m$, where $v_i \in W_i$, $i = 1, \ldots, m$.*

PROOF. If $V = \oplus_{i=1}^{m} W_i$; and $v = v_1 + \cdots + v_m$ and $v = w_1 + \cdots + w_m$ are representations of v, then $\sum_{i=1}^{m} (v_i - w_i) = 0$, and so for each k, $v_k - w_k = \sum_{i \neq k} (w_i - v_i) \in W_k \cap \sum_{i \neq k} W_i = \{0\}$. Hence $v_k = w_k$ for all k, and v has a unique representation. The converse is obvious. ∎

6.1. Examples

1. Let V be a vector space over K, and let $\mathcal{B} = \{x_1, \ldots, x_n\}$ be a basis of V. Define $Kx_i = \{\alpha x_i \mid \alpha \in K\}$. Then Kx_i is a subspace of V. Note that $Kx_i = \langle x_i \rangle$. It is evident from the definition of basis that
$$V = Kx_1 \oplus \cdots \oplus Kx_n.$$

2. Let $I = (-a, a)$, $a > 0$ be an open interval in $\mathbb{R}$ and let $V = \mathbb{R}^I$, the space of all real valued functions defined on I. Let
$$V_e = \{f \in V \mid f(x) = f(-x) \text{ for all } x \in I\},$$
the set of all even functions on I, and let
$$V_o = \{f \in V \mid f(x) = -f(-x) \text{ for all } x \in I\},$$
the set of all odd functions on I. Then V_e and V_o are subspaces of V. Since for any $f \in V$, $f(x) = g(x) + h(x)$, $x \in I$, where $g(x) = [f(x) + f(-x)]/2 \in V_e$ and $h(x) = [f(x) - f(-x)]/2 \in V_o$, it follows that $V = V_e + V_o$. Also, if $f \in V_e \cap V_o$, then $f(x) = f(-x)$ and $f(x) = -f(-x)$, and so for all $x \in I$, $2f(x) = 0$, that is, $f \equiv 0$. Hence $V = V_e \oplus V_o$.

3. Let K be a field and Char $K \neq 2$. In particular, we can assume that $K = \mathbb{R}$ or $\mathbb{C}$. Then the vector space $K^{n \times n}$ has subspaces
$$\text{Sym} = \{A \in K^{n \times n} \mid A^t = A\}$$
and
$$\text{SSym} = \{A \in K^{n \times n} \mid A^t = -A\},$$
the space of symmetric and skew-symmetric matrices, respectively. It is easy to verify that $K^{n \times n} = \text{Sym} \oplus \text{SSym}$.

4. Let W be a nonzero proper subspace of vector space V. If X is a basis of W, then there is a subset Y of $V \setminus W$ such that $X \cup Y$ is a basis of V (Theorem 3.7). If $W' = \langle Y \rangle$, then clearly $V = W \oplus W'$. Such a subspace W' is called a **complementary subspace** of W in V. Note that, for a given subspace, its complementary subspace is not unique. Let x and y be fixed elements of X and Y, and let $Y_1 = \{x+y\} \cup Y \setminus \{y\}$. Then $X \cup Y_1$ is also a basis of V and $X \cap Y_1 = \emptyset$. Thus, if $W'' = \langle Y_1 \rangle$, then $V = W \oplus W''$. Now $W' \neq W''$. Otherwise $x + y \in W'$ which implies that $x + y$, and so x is a linear combination of elements of Y, a contradiction.

5. Let V and V' be finite dimensional vector spaces which are direct sum of vector spaces: $V = \oplus_{i=1}^m W_i$, $V' = \oplus_{i=1}^m W_i'$. For each i, let $T_i \in L(W_i, W_i')$. Then the mapping $T : V \to V'$ defined by $T(x) = \sum_{i=1}^m T_i(x_i)$, where $x = x_1 + \cdots + x_m$, is a linear transformation, called the direct sum of linear transformations $T_1, \ldots, T_m$. We denote this T by $\oplus_{i=1}^m T_i$.

   Clearly, $x \in \ker \oplus_{i=1}^m T_i$ if and only if for each i, $x_i \in \ker T_i$. Thus, $\ker \oplus_{i=1}^m T_i = \oplus_{i=1}^m \ker T_i$. Also, $\operatorname{im} \oplus_{i=1}^m T_i = \oplus_{i=1}^m \operatorname{im} T_i$.

6. Let $W_1, \ldots, W_m$ be subspaces of a vector space V. Then the mapping $T : V \to V/W_1 \oplus \cdots \oplus V/W_m$ defined by $T(v) = (v + W_1, \ldots, v + W_m)$ is a linear transformation, and its kernel is $\cap_{i=1}^m W_i$. Thus, $V/\cap_{i=1}^m W_i$ is isomorphic to a subspace of $V/W_1 \oplus \cdots \oplus V/W_m$.

   In particular, if each V/W_i is finite dimensional, then $\oplus_{i=1}^m V/W_i$ is also a finite dimensional vector space. Thus, $V/\cap_{i=1}^m W_i$ is also finite dimensional. Hence we have:

   > *if $W_1, \ldots, W_m$ are subspaces of a vector space V and if $\dim(V/W_i)$ is finite for $i = 1, \ldots, m$, then $\dim(V/\cap_{i=1}^m W_i)$ is also finite.* (See also Example 2, p.50).

6.2. Exercises

1. Let $V = \mathbb{R}^4$ and let $W = \left\langle \begin{bmatrix} 1 & 1 & 0 & 0 \end{bmatrix}^t, \begin{bmatrix} 1 & 0 & 1 & 0 \end{bmatrix}^t \right\rangle$. Find subspaces W_1 and W_2 of V such that $V = W \oplus W_1$ and $V = W \oplus W_2$ but $W_1 \neq W_2$.

2. Find nonzero subspaces W_1, W_2 and W_3 of $\mathbb{R}^3$ such that $\mathbb{R}^3 = W_1 + W_2 + W_3$, $W_i \cap W_j = \{0\}$ for $i \neq j$, but $\mathbb{R}^3 \neq W_1 \oplus W_2 \oplus W_3$.

3. Let S and T be linear operators on a finite dimensional vector space V. If $\operatorname{im} S = \operatorname{im} T$, $V = \operatorname{im} S \oplus \ker S$ and $V = \operatorname{im} T \oplus \ker T$, does it follow that $\ker S = \ker T$?

4. Consider the vector space $\mathbb{R}[x]$. Find a complementary subspace of W_m, $m \geq 1$ in $\mathbb{R}[x]$, where
$$W_m = \left\{ p(x) \in \mathbb{R}[x] \mid p^{(r)}(0) = 0,\ r = 0, 1, \ldots, m \right\}.$$

5. Let $W_1, \ldots, W_r$ be subspaces of a finite dimensional vector space V. Prove that the following statements are equivalent.
   (a) $V = W_1 \oplus \cdots \oplus W_r$;
   (b) If $\mathcal{B}_i$ is a basis of W_i, $i = 1, \ldots, r$, then $\mathcal{B}_i \cap \mathcal{B}_j = \emptyset$ for $i \neq j$ and $\mathcal{B}_1 \cup \cdots \cup \mathcal{B}_r$ is a basis of V;

(c) $V = W_1 + \cdots + W_r$ and $\dim V = \dim W_1 + \cdots + \dim W_r$.

6. Let S be a linear operator on $\mathbb{C}^{n\times n}$, $n \geq 1$, defined by $S(A) = A - A^*$. Show that $\mathbb{C}^{n\times n} = \operatorname{Im} S \oplus \ker S$.

7. Let V be a vector space over K and let $T \in L(V)$ such that $T^2 = \alpha T$ for some $\alpha \in K \setminus \{0\}$. Prove that $V = \operatorname{im} T \oplus \ker T$.

8. Let V be a vector space over K and let $V = W \oplus X$. Define $P : V \to V$ by $P(w+x) = w$ for $w \in W$, $x \in X$. Verify that $P \in L(V)$, called the **projection** on W along X. Show that if P is a projection on W along X, then
   (a) $I - P$ is a projection on X along W.
   (b) P is an idempotent, that is, $P^2 = P$.
   Therefore, by the above exercise, $V = \operatorname{im} P \oplus \ker P$. Verify that $\operatorname{im} P = W = \ker(I - P)$ and $\ker P = X = \operatorname{im}(I - P)$.

9. Prove that if T is an idempotent on a vector space V, then T is a projection on $\operatorname{im} T$ along $\ker T$.

   Thus from this and the above exercise, we can conclude that a linear operator is a projection if and only if it is an idempotent.

10. Let V be a vector space over K, $\operatorname{Char} K \neq 2$ and let $S, T \in L(V)$ be idempotents.
    (a) Show that if $S + T$ is an idempotent, then $S + T$ is a projection on $\operatorname{im} S \oplus \operatorname{im} T$ along $\ker S \cap \ker T$. Thus $V = \operatorname{im} S \oplus \operatorname{im} T \oplus \ker S \cap \ker T$.
    (b) Show that if $S - T$ is an idempotent, then $S - T$ is a projection on $\operatorname{im} S \cap \ker T$ along $\ker S \oplus \operatorname{im} T$. Thus $V = \operatorname{im} S \cap \ker T \oplus \ker S \oplus \operatorname{im} T$.

11. Prove that if $S, T \in L(V)$ are projections and $ST = TS$, then ST is a projection on $\operatorname{im} S \cap \operatorname{im} T$ along $\ker S \oplus \ker T$.

12. Let $P_1, \ldots, P_k \in L(V)$ be projections such that $P_i P_j = 0$ for $i \neq j$, and $P_1 + \cdots + P_k = I$. We refer this sum as a **resolution of the identity**.

    Show that if $P_1 + \cdots + P_k = I$ is a resolution of the identity, then $V = \operatorname{im} P_1 \oplus \cdots \oplus \operatorname{im} P_k$.

    Also, prove the converse of the above statement: if $V = W_1 \oplus \cdots \oplus W_k$ and P_i is a projection on W_i along $W_1 + \cdots + W_{i-1} + W_{i+1} + \cdots + W_k$, then $P_1 + \cdots + P_k = I$ is a resolution of the identity.

13. Let V be a vector space over K, $\operatorname{Char} K \neq 2$, and let $V = W \oplus X$. A **reflection** at W along X is a linear operator R on V defined by $R(w + x) = w - x$ for $w \in W$, $x \in X$. Prove that the following statements are equivalent:
    (a) R is a reflection;
    (b) R is an **involution**, that is, $R^2 = I$;
    (c) $V = \ker(R - I) \oplus \ker(R + I)$.

7. The matrix of a linear transformation

Let V be an n-dimensional vector space over a field K and let $\mathcal{B} = \{x_1, \ldots, x_n\}$ be a fixed ordered basis of V. Then for each vector v in V, there are uniquely determined scalars $\alpha_1, \dots, \alpha_n$ in K such that $v = \alpha_1 x_1 + \cdots + \alpha_n x_n$. Thus, we have a well defined mapping $[\,]_{\mathcal{B}} : V \to K^n$ given by

$$[v]_{\mathcal{B}} = \begin{bmatrix} \alpha_1 \\ \vdots \\ \alpha_n \end{bmatrix}, \text{ where } v = \alpha_1 x_1 + \cdots + \alpha_n x_n.$$

Clearly, $[\,]_{\mathcal{B}}$ is an isomorphism, that is, for u, $v \in V$ and $\alpha, \beta \in K$:

$$[\alpha u + \beta v]_{\mathcal{B}} = \alpha [u]_{\mathcal{B}} + \beta [v]_{\mathcal{B}}, \text{ and}$$
$$[v]_{\mathcal{B}} = 0 \ (\in K^n) \text{ if and only if } v = 0 \ (\in V).$$

Let V' be another vector space over K of finite dimension m, and let $\mathcal{B}' = \{y_1, \dots, y_m\}$ be its ordered basis. If $T \in L(V, V')$, then for each $j = 1, \dots, n$,

$$T(x_j) = \sum_{i=1}^{m} \alpha_{ij} y_i, \ \alpha_{ij} \in K,$$

and so

$$[T(x_j)]_{\mathcal{B}'} = \begin{bmatrix} \alpha_{1j} \\ \vdots \\ \alpha_{mj} \end{bmatrix}.$$

Now, if $v = \sum_{j=1}^{n} \beta_j x_j \in V$, $\beta_j \in K$, then $T(v) = \sum_{j=1}^{n} \beta_j T(x_j)$, and

$$[T(v)]_{\mathcal{B}'} = \sum_{j=1}^{n} \beta_j [T(x_j)]_{\mathcal{B}'} = \sum_{j=1}^{n} \beta_j \begin{bmatrix} \alpha_{1j} \\ \vdots \\ \alpha_{mj} \end{bmatrix} = \begin{bmatrix} \alpha_{11} & \cdots & \alpha_{1n} \\ \vdots & & \vdots \\ \alpha_{m1} & \cdots & \alpha_{mn} \end{bmatrix} \begin{bmatrix} \beta_1 \\ \vdots \\ \beta_n \end{bmatrix}.$$

Note that scalars $\alpha_{ij} \in K$ are uniquely determined by the choice of bases $\mathcal{B}$ and $\mathcal{B}'$. We define the **matrix of a linear transformation** T with respect to ordered bases $\mathcal{B}$–$\mathcal{B}'$ to be an $m \times n$ matrix:

$${}_{\mathcal{B}'}[T]_{\mathcal{B}} = \begin{bmatrix} \alpha_{11} & \cdots & \alpha_{1n} \\ \vdots & & \vdots \\ \alpha_{m1} & \cdots & \alpha_{mn} \end{bmatrix} = \begin{bmatrix} [T(x_1)]_{\mathcal{B}'} & \cdots & [T(x_n)]_{\mathcal{B}'} \end{bmatrix},$$

that is, the k–th column of ${}_{\mathcal{B}'}[T]_{\mathcal{B}}$ is $[T(x_k)]_{\mathcal{B}'}$.

Therefore, we have for all $v \in V$:

(5) $$[T(v)]_{\mathcal{B}'} = {}_{\mathcal{B}'}[T]_{\mathcal{B}} [v]_{\mathcal{B}}.$$

This derivation can be extended further. Let V, V' and V'' be vector spaces over K of dimensions n, m and l, respectively. Let $T \in L(V, V')$ and $S \in L(V', V'')$. If $\mathcal{B}$, $\mathcal{B}'$ and $\mathcal{B}''$ are ordered bases of V, V' and V'' then for $v \in V$:

$$\begin{aligned} {}_{\mathcal{B}''}[S \circ T]_{\mathcal{B}} [v]_{\mathcal{B}} &= [(S \circ T)(v)]_{\mathcal{B}''} = [S(T(v))]_{\mathcal{B}''} \\ &= {}_{\mathcal{B}''}[S]_{\mathcal{B}'} [T(v)]_{\mathcal{B}'} = {}_{\mathcal{B}''}[S]_{\mathcal{B}'} \, {}_{\mathcal{B}'}[T]_{\mathcal{B}} [v]_{\mathcal{B}}. \end{aligned}$$

Hence,

(6) $${}_{\mathcal{B}''}[S \circ T]_{\mathcal{B}} = {}_{\mathcal{B}''}[S]_{\mathcal{B}'} \, {}_{\mathcal{B}'}[T]_{\mathcal{B}}.$$

PROPOSITION 7.1. *Let V and V' be vector spaces over K of dimensions n and m and let $\mathcal{B}'$ and $\mathcal{B}''$ be ordered bases of V and V'. Then the mapping ${}_{\mathcal{B}'}[\,]_{\mathcal{B}}$: $L(V,V') \to K^{m\times n}$ given by $T \mapsto {}_{\mathcal{B}'}[T]_{\mathcal{B}}$ is an isomorphism.*

PROOF. Clearly, this mapping is a well defined linear transformation. Since $\dim L(V,V') = \dim K^{m\times n}$ (Proposition 4.5), it is enough to check that this mapping is injective (Proposition 5.10). Let $\mathcal{B} = \{x_1, \ldots, x_n\}$. Then ${}_{\mathcal{B}'}[T]_{\mathcal{B}} = 0$ if and only if $[T(x_j)]_{\mathcal{B}'} = 0$ for $j = 1, \ldots, n$, that is, $T(x_j) = 0$ for all j. Since T maps every basis vector to zero, $T \equiv 0$. Hence, ${}_{\mathcal{B}'}[\,]_{\mathcal{B}}$ is an isomorphism. ∎

We will write $[T]_{\mathcal{B}}$ for ${}_{\mathcal{B}}[T]_{\mathcal{B}}$, called the matrix of a linear operator with respect to an ordered basis $\mathcal{B}$. Clearly, if I is the identity linear operator on V, then $[I]_{\mathcal{B}} = I_n$, the $n\times n$ identity matrix, $n = \dim V$.

PROPOSITION 7.2. *Let $\mathcal{B}$ be an ordered basis of an n-dimensional vector space V over K. If S and T are linear operators on V, then $[S\circ T]_{\mathcal{B}} = [S]_{\mathcal{B}}\,[T]_{\mathcal{B}}$. In particular, T is a bijection if and only if $[T]_{\mathcal{B}}$ is an invertible matrix.*

PROOF. In eq.(6), let $V = V' = V''$ and $\mathcal{B} = \mathcal{B}' = \mathcal{B}''$. Then

$$(7) \qquad [S\circ T]_{\mathcal{B}} = [S]_{\mathcal{B}}\,[T]_{\mathcal{B}}\,.$$

If T is a bijection, there exists $S \in L(V)$ such that $S\circ T = I = T\circ S$, where I is the identity mapping on V. Hence, $I_n = [I]_{\mathcal{B}} = [S\circ T]_{\mathcal{B}} = [S]_{\mathcal{B}}\,[T]_{\mathcal{B}} = [T]_{\mathcal{B}}\,[S]_{\mathcal{B}}$, that is, $[T]_{\mathcal{B}}$ is an invertible matrix.

Conversely, let $T \in L(V)$ such that $[T]_{\mathcal{B}}$ is an invertible matrix. Since the mapping $T \mapsto [T]_{\mathcal{B}}$ is an isomorphism from $L(V)$ to $K^{n\times n}$, there exists a unique $S \in L(V)$ such that $[S]_{\mathcal{B}} = [T]_{\mathcal{B}}^{-1}$. Therefore, $[S\circ T]_{\mathcal{B}} = [S]_{\mathcal{B}}\,[T]_{\mathcal{B}} = I_n = [I]_{\mathcal{B}}$ and $[T\circ S]_{\mathcal{B}} = [I]_{\mathcal{B}}$. Hence, $S\circ T = I = T\circ S$, that is, T is bijective. ∎

PROPOSITION 7.3. ***Change of basis formula.*** *Let $\mathcal{B}$ and $\mathcal{B}'$ be ordered bases of a finite dimensional vector space V over K, and let $T \in L(V)$. Then there exists an invertible matrix P such that $[T]_{\mathcal{B}'} = P^{-1}\,[T]_{\mathcal{B}}\,P$.*

PROOF. Using eq.(6), we have

$$\begin{aligned}[T]_{\mathcal{B}'} &= [I\circ T]_{\mathcal{B}'} = {}_{\mathcal{B}'}[I]_{\mathcal{B}}\,[T]_{\mathcal{B}} \\ &= {}_{\mathcal{B}'}[I]_{\mathcal{B}}\,[T\circ I]_{\mathcal{B}} = {}_{\mathcal{B}'}[I]_{\mathcal{B}}\,[T]_{\mathcal{B}}\,{}_{\mathcal{B}}[I]_{\mathcal{B}'}\,,\end{aligned}$$

and ${}_{\mathcal{B}'}[I]_{\mathcal{B}}\,{}_{\mathcal{B}}[I]_{\mathcal{B}'} = [I]_{\mathcal{B}'} = I_n$, $n = \dim V$. Thus, if $P = {}_{\mathcal{B}}[I]_{\mathcal{B}'}$, then $P^{-1} = {}_{\mathcal{B}'}[I]_{\mathcal{B}}$. Hence, $[T]_{\mathcal{B}'} = P^{-1}\,[T]_{\mathcal{B}}\,P$. ∎

Matrices A and B in $K^{n\times n}$ are said to be **similar** if there is an invertible matrix P such that $B = P^{-1}AP$. The above proposition states that the matrices of a linear operator with respect to two different bases are similar. Linear operators S and T on a finite dimensional vector space V are said to be **similar** if their corresponding matrices with respect to a basis of V are similar. Equivalently, there are bases $\mathcal{B}$ and $\mathcal{B}'$ of V such that $[S]_{\mathcal{B}} = [T]_{\mathcal{B}'}$.

Let T be a linear operator on a finite dimensional vector space V over a field K. The **determinant of a linear operator** T is the determinant of the matrix $[T]_{\mathcal{B}}$, where $\mathcal{B}$ is an ordered basis of V; and the **trace of a linear operator** T is the trace of the matrix $[T]_{\mathcal{B}}$. We will denote the determinant of T by $\det T$ and the trace of T by $\operatorname{tr} T$. It follows from Proposition 7.3 that the determinant and the trace of a linear operator are independent of the choice of basis. Since, if $\mathcal{B}$ and

$\mathcal{B}'$ are ordered bases of V, then $[T]_{\mathcal{B}'} = P^{-1} [T]_{\mathcal{B}} P$ for some invertible matrix P, and so $\det [T]_{\mathcal{B}'} = \det [T]_{\mathcal{B}}$ and $\operatorname{tr} [T]_{\mathcal{B}'} = \operatorname{tr} [T]_{\mathcal{B}}$.

PROPOSITION 7.4. *Let V be a finite dimensional vector space over K. A linear operator T on V is invertible (that is, an isomorphism) if and only if* $\det T \neq 0$.

PROOF. Exercise. ∎

7.1. Examples

1. Consider the vector space $\mathbb{R}_3[x]$ of polynomials with real coefficients and of degree at most 3. The differential operator D is a linear operator on $\mathbb{R}_3[x]$. The matrix of D with respect to a basis $\mathcal{B} = \{1, x, x^2, x^3\}$ is:

$$[D]_{\mathcal{B}} = \begin{bmatrix} 0 & 1 & 0 & 0 \\ 0 & 0 & 2 & 0 \\ 0 & 0 & 0 & 3 \\ 0 & 0 & 0 & 0 \end{bmatrix}.$$

Verify that

$$\mathcal{B}_1 = \{p_1(x) = 1 + x, p_2(x) = x + x^2, p_3(x) = x^2 + x^3, p_4(x) = x + x^3\}$$

is also a basis of $\mathbb{R}_3[x]$. Now

$$\begin{aligned} D p_1(x) &= 1 = p_1(x) - \frac{1}{2} p_2(x) + \frac{1}{2} p_3(x) - \frac{1}{2} p_4(x); \\ D p_2(x) &= 1 + 2x = p_1(x) + \frac{1}{2} p_2(x) - \frac{1}{2} p_3(x) + \frac{1}{2} p_4(x); \\ D p_3(x) &= 2x + 3x^2 = \frac{5}{2} p_2(x) + \frac{1}{2} p_3(x) - \frac{1}{2} p_4(x); \\ D p_4(x) &= 1 + 3x^2 = p_1(x) + p_2(x) + 2p_3(x) - 2p_4(x). \end{aligned}$$

Thus, the matrix of D with respect to basis $\mathcal{B}_1$ is:

$$[D]_{\mathcal{B}_1} = \begin{bmatrix} 1 & 1 & 0 & 1 \\ -\frac{1}{2} & \frac{1}{2} & \frac{5}{2} & 1 \\ \frac{1}{2} & -\frac{1}{2} & \frac{1}{2} & 2 \\ -\frac{1}{2} & \frac{1}{2} & -\frac{1}{2} & -2 \end{bmatrix}.$$

If we denote $P = {}_{\mathcal{B}}[I]_{\mathcal{B}_1}$, then P is an invertible matrix given by:

$$P = \begin{bmatrix} 1 & 0 & 0 & 0 \\ 1 & 1 & 0 & 1 \\ 0 & 1 & 1 & 0 \\ 0 & 0 & 1 & 1 \end{bmatrix}$$

and according to Proposition 7.3, $[D]_{\mathcal{B}_1} = P^{-1} [D]_{\mathcal{B}} P$.

2. Let $T : \mathbb{R}^4 \to \mathbb{R}^3$ be a linear transformation given by:

$$T \begin{bmatrix} x_1 \\ x_2 \\ x_3 \\ x_4 \end{bmatrix} = \begin{bmatrix} x_1 - 3x_2 + x_3 + 2x_4 \\ x_1 - 2x_2 + x_3 - x_4 \\ 2x_1 - 5x_2 + x_4 \end{bmatrix}.$$

If $\mathcal{B}_0$ and $\mathcal{B}_0'$ are the standard bases of $\mathbb{R}^4$ and $\mathbb{R}^3$ respectively, then

$$_{\mathcal{B}_0'}[T]_{\mathcal{B}_0} = \begin{bmatrix} 1 & -3 & 1 & 2 \\ 1 & -2 & 1 & -1 \\ 2 & -5 & 0 & 1 \end{bmatrix}.$$

Note that the rank of this matrix is 3 $(= \dim \mathbb{R}^3)$ which implies that T is surjective. Let

$$\mathcal{B} = \left\{ v_1 = \begin{bmatrix} 7 \\ 3 \\ 1 \\ 1 \end{bmatrix}, v_2 = \begin{bmatrix} -4 \\ -2 \\ 0 \\ -1 \end{bmatrix}, v_3 = \begin{bmatrix} 5 \\ 2 \\ 0 \\ 1 \end{bmatrix}, v_4 = \begin{bmatrix} 7 \\ 3 \\ 0 \\ 1 \end{bmatrix} \right\}$$

and let

$$\mathcal{B}' = \left\{ w_1 = \begin{bmatrix} 1 \\ 1 \\ 0 \end{bmatrix}, w_2 = \begin{bmatrix} 0 \\ 1 \\ 1 \end{bmatrix}, w_3 = \begin{bmatrix} 1 \\ 0 \\ 1 \end{bmatrix} \right\}$$

be ordered bases of $\mathbb{R}^4$ and $\mathbb{R}^3$ respectively. Then $Tv_i = w_i$, $i = 1, 2, 3$, and $Tv_4 = 0$. Thus,

$$_{\mathcal{B}'}[T]_{\mathcal{B}} = \begin{bmatrix} 1 & 0 & 0 & 0 \\ 0 & 1 & 0 & 0 \\ 0 & 0 & 1 & 0 \end{bmatrix}.$$

Now to establish a relationship between the matrices $_{\mathcal{B}_0'}[T]_{\mathcal{B}_0}$ and $_{\mathcal{B}'}[T]_{\mathcal{B}}$, we use eq.(6):

$$\begin{aligned} _{\mathcal{B}'}[T]_{\mathcal{B}} &= {}_{\mathcal{B}'}[I \circ T]_{\mathcal{B}} = {}_{\mathcal{B}'}[I]_{\mathcal{B}_0'}\ {}_{\mathcal{B}_0'}[T]_{\mathcal{B}} \\ &= {}_{\mathcal{B}'}[I]_{\mathcal{B}_0'}\ {}_{\mathcal{B}_0'}[T \circ I]_{\mathcal{B}} = {}_{\mathcal{B}'}[I]_{\mathcal{B}_0'}\ {}_{\mathcal{B}_0'}[T]_{\mathcal{B}_0}\ {}_{\mathcal{B}_0}[I]_{\mathcal{B}}\ . \end{aligned}$$

Here, I appearing on the right of the matrix $_{\mathcal{B}_0'}[T]_{\mathcal{B}_0}$ is the identity operator on $\mathbb{R}^4$ and on the left is the identity operator on $\mathbb{R}^3$.

Clearly,

$$_{\mathcal{B}_0}[I]_{\mathcal{B}} = \begin{bmatrix} 7 & -4 & 5 & 7 \\ 3 & -2 & 2 & 3 \\ 1 & 0 & 0 & 0 \\ 1 & -1 & 1 & 1 \end{bmatrix},$$

and

$$_{\mathcal{B}'}[I]_{\mathcal{B}_0'} = {}_{\mathcal{B}_0'}[I]_{\mathcal{B}'}^{-1} = \begin{bmatrix} 1 & 0 & 1 \\ 1 & 1 & 0 \\ 0 & 1 & 1 \end{bmatrix}^{-1} = \frac{1}{2}\begin{bmatrix} 1 & 1 & -1 \\ -1 & 1 & 1 \\ 1 & -1 & 1 \end{bmatrix}.$$

Therefore,

$$_{\mathcal{B}'}[T]_{\mathcal{B}} = \frac{1}{2}\begin{bmatrix} 1 & 1 & -1 \\ -1 & 1 & 1 \\ 1 & -1 & 1 \end{bmatrix}\ {}_{\mathcal{B}_0'}[T]_{\mathcal{B}_0} \begin{bmatrix} 7 & -4 & 5 & 7 \\ 3 & -2 & 2 & 3 \\ 1 & 0 & 0 & 0 \\ 1 & -1 & 1 & 1 \end{bmatrix}.$$

3. Let V be a vector space over K and let T be a linear operator on V. A subspace W of V is said to be T-**invariant** if $T(W) \subseteq W$, that is, $T(w) \in W$ for all $w \in W$. Equivalently, if $\widehat{T}$ is the restriction of T to W, then $\widehat{T}$ is a linear operator on W. In this case, we say that T induces a linear operator on W, or $\widehat{T}$ is the linear operator on W induced by T.

   If W is a T–invariant subspace of V, then the mapping $T' : V/W \to V/W$ given by $T'(v+W) = T(v)+W$ is well defined. Since, if $x+W = y+W$, then $x-y \in W$, and so $T(x)-T(y) \in W$, that is, $T(x)+W = T(y)+W$. It is easy to verify that T' is a linear operator on V/W, called the linear operator induced by T on V/W.

   Let T be a linear operator on a finite dimensional vector space V and let W be a T-invariant subspace of V. Extend $\mathcal{B}_1 = \{x_1, \ldots, x_m\}$, an ordered basis of W to $\mathcal{B} = \{x_1, \ldots, x_m, x_{m+1}, \ldots, x_n\}$, an ordered basis of V. Then, $\mathcal{B}_2 = \{x_{m+1} + W, \ldots, x_n + W\}$ is an ordered basis of V/W. Let $\widehat{T}$ and T' be linear operators on W and V/W, respectively, induced by T. Also let $\left[\widehat{T}\right]_{\mathcal{B}_1} = A$ and let $[T']_{\mathcal{B}_2} = B$. Then by the definition

$$(*) \qquad \widehat{T}(x_j) = \sum_{i=1}^{m} [A]_{ij}\, x_i, \; j = 1, \ldots, m$$

and

$$(**) \qquad T'(x_j + W) = \sum_{i=m+1}^{n} [B]_{ij}\,(x_i + W), \; j = m+1, \ldots, n.$$

Thus,

$$T(x_j) - \sum_{i=m+1}^{n} [B]_{ij}\, x_i \in W, \; j = m+1, \ldots, n,$$

and so

$$T(x_j) - \sum_{i=m+1}^{n} [B]_{ij}\, x_i = \sum_{i=1}^{m} \gamma_{ij} x_i,$$

that is,

$$(***) \qquad T(x_j) = \sum_{i=1}^{m} \gamma_{ij} x_i + \sum_{i=m+1}^{n} [B]_{ij}\, x_i, \; j = m+1, \ldots, n.$$

Since $\widehat{T}(x_j) = T(x_j)$, from $(*)$ and $(***)$, we have

$$[T]_{\mathcal{B}} = \begin{bmatrix} A & C \\ 0 & B \end{bmatrix},$$

where $C \in K^{m \times (n-m)}$ and $[C]_{ij} = \gamma_{ij}$.

4. Let T be a linear operator on a finite dimensional vector space V over K. Let W_1 and W_2 be T-invariant subspaces of V of dimensions m_1 and m_2 such that $V = W_1 \oplus W_2$. Thus, if $\mathcal{B}_1$ and $\mathcal{B}_2$ are ordered bases of W_1 and W_2 then $\mathcal{B} = \mathcal{B}_1 \cup \mathcal{B}_2$ is a basis of V. Let $\widehat{T}_1$ and $\widehat{T}_2$ be linear operators on W_1 and W_2 induced by T and let $\left[\widehat{T}_1\right]_{\mathcal{B}_1} = A_1$ $(\in K^{m_1 \times m_1})$ and let

$\left[\widehat{T_2}\right]_{\mathcal{B}_2} = A_2 \ (\in K^{m_2 \times m_2})$. Since $T_i(x) \in W_i$ for $x \in W_i$, $i = 1, 2$, it is easy to verify that

$$[T]_{\mathcal{B}} = \begin{bmatrix} A_1 & 0 \\ 0 & A_2 \end{bmatrix}.$$

Similarly, we have the following.

Let $V = W_1 \oplus \cdots \oplus W_k$, a direct sum of T-invariant subspaces, $\dim W_i = m_i$, let $\mathcal{B}_i$ be an ordered basis of W_i for each i and let $\widehat{T_i}$ be the linear operator on W_i induced by T. If $\left[\widehat{T_i}\right]_{\mathcal{B}_i} = A_i$, $i = 1, \ldots, k$, then

$$[T]_{\mathcal{B}} = \begin{bmatrix} A_1 & 0 & \cdots & 0 \\ 0 & A_2 & \cdots & 0 \\ \vdots & & \ddots & \vdots \\ 0 & \cdots & 0 & A_k \end{bmatrix},$$

where $\mathcal{B} = \mathcal{B}_1 \cup \cdots \cup \mathcal{B}_k$.

5. We have seen in Example 1, p.42, that an $n \times n$ matrix A with entries from field K defines a linear operator τ_A on K^n. If $\mathcal{B} = \{e_1, \ldots, e_n\}$ is the standard basis of K^n, then $\tau_A(e_i) = Ae_i$, the i–th column of A. Therefore, $[\tau_A]_{\mathcal{B}} = A$.

   From now onwards, we will denote τ_A by A itself. More precisely, whenever we consider A as a linear operator, we mean the linear operator τ_A whose matrix representation with respect to standard basis is A.

6. Let V be a finite dimensional vector space over K of dimension n, and let $V = W_1 \oplus W_2$. The projection on W_1 along W_2 is a mapping $P : V \to V$ defined by $P(w_1 + w_2) = w_1$, $w_1 \in W_1$, $w_2 \in W_2$. The reflection at W_1 along W_2 is a mapping $R : V \to V$ defined by $R(w_1 + w_2) = w_1 - w_2$, $w_1 \in W_1$, $w_2 \in W_2$. Clearly, $P, R \in L(V)$. Also, these operators when restricted to W_1 are the identity operators on W_1. If $\mathcal{B}_1$ and $\mathcal{B}_2$ are ordered bases of W_1 and W_2 respectively, then $\mathcal{B} = \mathcal{B}_1 \cup \mathcal{B}_2$ is an ordered basis of V, and $[P]_{\mathcal{B}} = \begin{bmatrix} I_r & 0 \\ 0 & 0 \end{bmatrix}$ and $[R]_{\mathcal{B}} = \begin{bmatrix} I_r & 0 \\ 0 & -I_{n-r} \end{bmatrix}$, $r = \dim W_1$.

We now find the matrix representation of $f : \mathbb{R}^3 \to \mathbb{R}^3$ with respect to the standard basis of $\mathbb{R}^3$ where f is the projection on the plane $3x+y-2z = 0$ along the line passing through the origin and the point $\begin{bmatrix} 1 & 1 & 1 \end{bmatrix}^t$.

The plane $3x + y - 2z = 0$ defines a 2–dimensional subspace of $\mathbb{R}^3$. Any point on this plane is given by $\begin{bmatrix} 3x & 2z - 3x & -2z \end{bmatrix}^t$. Thus, if W_1 is this subspace, then

$$W_1 = \left\{ x \begin{bmatrix} 3 \\ -3 \\ 0 \end{bmatrix} + z \begin{bmatrix} 0 \\ 2 \\ -2 \end{bmatrix} \mid x, z \in \mathbb{R} \right\};$$

and if W_2 is the subspace determined by the line passing through the origin and $\begin{bmatrix} 1 & 1 & 1 \end{bmatrix}^t$, then

$$W_2 = \left\{ \alpha \begin{bmatrix} 1 \\ 1 \\ 1 \end{bmatrix} \mid \alpha \in \mathbb{R} \right\}.$$

Thus, $\left\{\begin{bmatrix}1\\-1\\0\end{bmatrix},\begin{bmatrix}0\\1\\-1\end{bmatrix}\right\}$ is a basis of W_1 and $\left\{\begin{bmatrix}1\\1\\1\end{bmatrix}\right\}$ is a basis of W_2. If we take $\mathcal{B}=\left\{\begin{bmatrix}1\\-1\\0\end{bmatrix},\begin{bmatrix}0\\1\\-1\end{bmatrix},\begin{bmatrix}1\\1\\1\end{bmatrix}\right\}$, a basis of $\mathbb{R}^3$, then

$$[f]_{\mathcal{B}}=\begin{bmatrix}1&0&0\\0&1&0\\0&0&0\end{bmatrix}.$$

If $\mathcal{B}_0$ is the standard basis of $\mathbb{R}^3$, then $[f]_{\mathcal{B}_0}={}_{\mathcal{B}_0}[I]_{\mathcal{B}}\,[f]_{\mathcal{B}}\,{}_{\mathcal{B}}[I]_{\mathcal{B}_0}$, where I is the identity operator on $\mathbb{R}^3$. Since ${}_{\mathcal{B}_0}[I]_{\mathcal{B}}=\begin{bmatrix}1&0&1\\-1&1&1\\0&-1&1\end{bmatrix}$, ${}_{\mathcal{B}}[I]_{\mathcal{B}_0}={}_{\mathcal{B}_0}[I]_{\mathcal{B}}^{-1}=\frac{1}{3}\begin{bmatrix}1&-1&-1\\2&1&-2\\1&1&1\end{bmatrix}$. Therefore, $[f]_{\mathcal{B}_0}=\frac{1}{3}\begin{bmatrix}1&-1&-1\\1&2&-1\\-2&-1&2\end{bmatrix}$.

7.2. Exercises

1. Let $T:\mathbb{R}^3\to\mathbb{R}^2$ given by $T\begin{bmatrix}x\\y\\z\end{bmatrix}=\begin{bmatrix}2x+4y-2z\\x-y+z\end{bmatrix}$. Write the matrix representation of T with respect to ordered bases

$$\left\{\begin{bmatrix}0&1&1\end{bmatrix}^t,\begin{bmatrix}1&0&1\end{bmatrix}^t,\begin{bmatrix}1&1&0\end{bmatrix}^t\right\}\text{ and }\left\{\begin{bmatrix}1&1\end{bmatrix}^t,\begin{bmatrix}1&-1\end{bmatrix}^t\right\}.$$

2. Let D be the differential operator on $\mathbb{R}_3[x]$. Write the matrix representation of T with respect to ordered basis $\{1-x,1+x^2,x-x^3,-x^2+x^3\}$.

3. Let T be a linear operator on $\mathbb{R}^3$, $T\begin{bmatrix}x\\y\\z\end{bmatrix}=\begin{bmatrix}2x+2y+z\\2y+z\\2x+3y+z\end{bmatrix}$. Write the matrix representation of T with respect to the basis

$$\mathcal{B}=\left\{\begin{bmatrix}1&1&0\end{bmatrix}^t,\begin{bmatrix}-1&1&0\end{bmatrix}^t,\begin{bmatrix}1&1&1\end{bmatrix}^t\right\}.$$

If $\mathcal{B}_0$ is the standard basis of $\mathbb{R}^3$, find an invertible matrix P such that $[T]_{\mathcal{B}_0}=P^{-1}[T]_{\mathcal{B}}P$.

4. Let $\mathcal{B}=\{u_1,\dots,u_n\}$ be a basis of a vector space V over K, and let T be an invertible linear operator on V. Then $\mathcal{B}'=\{Tu_1,\dots,Tu_n\}$ is also a basis of V. Show that for $S\in L(V)$, $[S]_{\mathcal{B}'}=[T]_{\mathcal{B}}^{-1}[S]_{\mathcal{B}}[T]_{\mathcal{B}}$.

5. Show that if T is a nilpotent linear operator on a finite dimensional vector space V, then $\det T=0$. Is the converse also true?

6. Let V be an n-dimensional vector space over K and let $T \in L(V)$, such that $(T - \lambda I)^n = 0$, $\lambda \in K$ and $(T - \lambda I)^{n-1} \neq 0$. Prove that there is a basis $\mathcal{B}$ of V such that

$$[T]_{\mathcal{B}} = \begin{bmatrix} \lambda & 1 & & 0 \\ & \ddots & \ddots & \\ & & \ddots & 1 \\ 0 & & & \lambda \end{bmatrix}$$

7. Let T be a linear operator on a finite dimensional vector space V over K. Prove that if the matrix representation of T with respect to all bases of V is the same, then $T = \alpha I$ for some $\alpha \in K$.

8. Prove that if $T \in L(V)$, $\dim V$ is finite and $\operatorname{rank} T = 1$, then $\det(I + T) = 1 + \operatorname{tr} T$.

9. Let $T \in L(V)$, $\dim V$ be finite and let $\dim \ker T = m$. Show that there exists an ordered basis $\mathcal{B}$ of V such that the first m columns of $[T]_{\mathcal{B}}$ are zero. Further, prove that if W is a complementary subspace of $\ker T$, which is also T-invariant, then $\mathcal{B}$ can be chosen such that the first m rows of $[T]_{\mathcal{B}}$ are also zero.

10. Let $T \in L(V, W)$, $\dim V = n$, $\dim W = m$ and let $\operatorname{rank} T = r$. Prove that there exists a basis $\mathcal{B}$ of V and a basis $\mathcal{B}'$ of W such that ${}_{\mathcal{B}'}[T]_{\mathcal{B}}$ is one of the following block matrix forms:

$$\begin{bmatrix} I_m & 0 \end{bmatrix} \text{ if } r = m, \text{ or } \begin{bmatrix} I_n \\ 0 \end{bmatrix} \text{ if } r = n, \text{ or } \begin{bmatrix} I_r & 0 \\ 0 & 0 \end{bmatrix} \text{ if } r < m \text{ and } r < n.$$

11. Let S and T be linear operators on a finite dimensional vector space V. Prove that the following statements are equivalent:
    (a) $S = P^{-1}TP$ for some invertible linear operator P;
    (b) $[S]_{\mathcal{B}}$ and $[T]_{\mathcal{B}}$ are similar matrices for any ordered basis $\mathcal{B}$ of V;
    (c) for a given ordered basis $\mathcal{B}$ of V there exists an ordered basis $\mathcal{B}'$ of V such that $[S]_{\mathcal{B}'} = [T]_{\mathcal{B}}$.

8. Duality

Let V be a vector space over a field K. A linear transformation from V to K is called a **linear form** on V. The vector space $L(V, K)$ of all linear forms on V is called the dual space of V, denoted by $V^\bullet$.

Let V be a finite dimensional vector space over K. It is evident from Proposition 4.5 and so by Theorem 4.7, V and $V^\bullet$ are isomorphic. Let $\mathcal{B} = \{x_1, \ldots, x_n\}$ be an ordered basis of V. For each i, define a linear form $x_i^\bullet$ on V whose action on $\mathcal{B}$ is given by $x_i^\bullet(x_j) = \delta_{ij}$. According to Proposition 4.4, each $x_i^\bullet$ is uniquely determined element of $V^\bullet$. The set $\mathcal{B}^\bullet = \{x_1^\bullet, \ldots, x_n^\bullet\}$, is now a basis of $V^\bullet$, called the **dual basis** of $\mathcal{B}$. To verify that $\mathcal{B}^\bullet$ is a basis, it is sufficient to show that $\mathcal{B}^\bullet$ is a linearly independent set (Corollary 3.8). Let $\alpha_1, \ldots, \alpha_n$ be scalars such that

$\alpha_1 x_1^\bullet + \cdots + \alpha_n x_n^\bullet = 0$. Then for each j,

$$0 = 0(x_j) = \left(\sum_{i=1}^{n} \alpha_i x_i^\bullet\right)(x_j) = \sum_{i=1}^{n} \alpha_i \delta_{ij} = \alpha_j.$$

Hence, $\alpha_1 = \cdots = \alpha_n = 0$, and $\mathcal{B}^\bullet$ is a linearly independent set. In fact, if $f \in V^\bullet$, then

$$f = f(x_1)\, x_1^\bullet + \cdots + f(x_n)\, x_n^\bullet.$$

Note that the above defined isomorphism from a finite dimensional vector space V to its dual $V^\bullet$ is not canonical, the isomorphism changes with a change of basis. If $\{x_1, \ldots, x_n\}$ is a basis of V with its dual basis $\{x_1^\bullet, \ldots, x_n^\bullet\}$, then we have an isomorphism $T_1 : V \to V^\bullet$ defined by $T_1(x_i) = x_i^\bullet$. Clearly,

$$\{y_1 = x_1 + x_2, y_2 = x_2, \ldots, y_n = x_n\}$$

is also a basis of V. If $\{y_1^\bullet, \ldots, y_n^\bullet\}$ is its dual basis and $T_2 : V \to V^\bullet$ is an isomorphism, $T_2(y_i) = y_i^\bullet$, then T_1 and T_2 are different isomorphisms, because $T_1(x_1 + x_2) \neq T_2(x_1 + x_2)$. Otherwise $x_1^\bullet + x_2^\bullet = y_1^\bullet$, and so $1 = y_1^\bullet(y_1) = (x_1^\bullet + x_2^\bullet)(x_1 + x_2) = 1 + 1$, a contradiction.

The vector space $V^{\bullet\bullet}\ (= (V^\bullet)^\bullet)$ is isomorphic to $V^\bullet$, and so it is isomorphic to V. Actually, there exists a canonical isomorphism from V to $V^{\bullet\bullet}$, an isomorphism which does not depend on the choice of basis. Let V be a vector space over K, and let $v \in V$. The **evaluation mapping**

$$e_v : V^\bullet \to K \text{ defined by } e_v(f) = f(v),$$

is a linear form on $V^\bullet$, that is, $e_v \in V^{\bullet\bullet}$. Indeed, for $\alpha, \beta \in K$ and $f, g \in V^\bullet$, $e_v(\alpha f + \beta g) = (\alpha f + \beta g)(v) = \alpha f(v) + \beta g(v) = \alpha e_v(f) + \beta e_v(g)$.

THEOREM 8.1. *Let V be a finite dimensional vector space over K. Then the mapping $e : V \to V^{\bullet\bullet}$ defined by $v \mapsto e_v$ is an isomorphism.*

PROOF. Since for $v \in V$, $e_v \in V^{\bullet\bullet}$, the mapping e is well defined. For $\alpha, \beta \in K$, $u, v \in V$ and $f \in V^\bullet$,

$$\begin{aligned} e_{\alpha u + \beta v}(f) &= f(\alpha u + \beta v) = \alpha f(u) + \beta f(v) \\ &= \alpha e_u(f) + \beta e_v(f) = (\alpha e_u + \beta e_v)(f). \end{aligned}$$

Therefore, $e_{\alpha u + \beta v} = \alpha e_u + \beta e_v$, and e is a linear transformation.

Since V is finite dimensional and $\dim V = \dim V^{\bullet\bullet}$, to prove that e is an isomorphism, it is sufficient to verify that e is injective (Proposition 5.10). Thus, we need to verify that $e_v = 0$ if and only if $v = 0$. If $v = 0$, then obviously $e_v = 0$. Conversely, if $v \neq 0$, then extend $x_1 = v$ to $\mathcal{B} = \{x_1, \ldots, x_n\}$, a basis of V. If $\{x_1^\bullet, \ldots, x_n^\bullet\}$ is the dual basis of $\mathcal{B}$, then $e_v(x_1^\bullet) = 1$, and so $e_v \neq 0$. This proves the theorem. ∎

Note that, according to the above theorem, for a finite dimensional vector space to each linear form F on $V^\bullet$, there is a unique vector v in V such that $F = e_v$. Thus, we can canonically identify V with $V^{\bullet\bullet}$.

Let S be a nonempty subset of a vector space V. Define

$$S^\circ = \{f \in V^\bullet \mid f(s) = 0 \text{ for all } s \in S\},$$

called the **annihilator** of S. It can be easily verified that S° is a subspace of $V^\bullet$. Since in $\langle S\rangle$, the elements are linear combination of elements of S, it is obvious that $\langle S\rangle^\circ = S^\circ$. Note that $\{0\}^\circ = V$ and $V^\circ = \{0\}$.

We have already pointed out that there is no canonical isomorphism from V and $V^\bullet$. However, there is a canonical bijection between the set of all k-dimensional subspaces of V, and $(n-k)$–dimensional subspaces of $V^\bullet$ ($n = \dim V$). If V is a vector space of dimension n (n finite) and W is a subspace of V of dimension k, then a basis $\{x_1, \ldots, x_k\}$ of W can be extended to $\{x_1, \ldots, x_k, \ldots, x_n\}$, a basis of V. If $\{x_1^\bullet, \ldots, x_n^\bullet\}$ is the dual basis of this basis, then $\{x_{k+1}^\bullet, \ldots, x_n^\bullet\}$ is a basis of W°. Indeed, if $f = \sum_{i=1}^n \alpha_i x_i^\bullet$, an element of $V^\bullet$ which is in W°, then $f(w) = 0$, for all $w \in W$, so $f(x_1) = \cdots = f(x_k) = 0$, and so $f = \sum_{i=k+1}^n \alpha_i x_i^\bullet \in \langle x_{k+1}^\bullet, \ldots, x_n^\bullet\rangle$. Thus, we have the following dimension formula for finite dimensional vector spaces:

$$\dim V = \dim W + \dim W^\circ,$$

for any subspace W of V.

PROPOSITION 8.2. *Let V be a finite dimensional vector space over K, and let X and Y be subspaces of V.*

(i) $(X+Y)^\circ = X^\circ \cap Y^\circ$.
(ii) $(X \cap Y)^\circ = X^\circ + Y^\circ$.
(iii) *Under the canonical identification of $V^{\bullet\bullet}$ with V, $(X^\circ)^\circ = X$.*

PROOF. (i) and (ii) are left as an exercise.

(iii) Since $e_x(f) = f(x) = 0$ for all $f \in X^\circ$ and for $x \in X$, it is clear that $X \subseteq (X^\circ)^\circ$. Also, by the above dimension formula: $\dim V^\bullet = \dim X^\circ + \dim (X^\circ)^\circ$, and so $\dim (X^\circ)^\circ = \dim V^\bullet - \dim X^\circ = \dim V - \dim X^\circ = \dim X$. Hence, $(X^\circ)^\circ = X$. ∎

Let V and W be vector spaces over K, and let $T : V \to W$ be a linear transformation. For each $f \in W^\bullet$, the mapping $f \circ T \in V^\bullet$. This defines a mapping:

$$T^\bullet : W^\bullet \to V^\bullet \text{ defined by } T^\bullet(f) = f \circ T.$$

The reader should verify that $T^\bullet$ is actually a linear transformation. This $T^\bullet$ is called the **dual transformation** of T. Note that under the usual identification of V with $V^{\bullet\bullet}$ and W with $W^{\bullet\bullet}$, we have $T^{\bullet\bullet} = T$.

The kernel and the image of T and of the dual transformation $T^\bullet$, for finite dimensional vector spaces, are related as follows.

PROPOSITION 8.3. *Let V and W be finite dimensional vector spaces over K, and let $T \in L(V, W)$. Then* $\ker T^\bullet = (\operatorname{im} T)^\circ$ *and* $\operatorname{im} T^\bullet = (\ker T)^\circ$.

PROOF. $T^\bullet(f) = 0$ if and only if $f(T(v)) = 0$ for all $v \in V$. Hence, $\ker T^\bullet = (\operatorname{im} T)^\circ$. Since $T^{\bullet\bullet} = T$, we have proved that $\ker T^\bullet = (\operatorname{im} T)^\circ$. Thus, $(\ker T)^\circ = ((\operatorname{im} T^\bullet)^\circ)^\circ = \operatorname{im} T^\bullet$. ∎

COROLLARY 8.4. *Let V and W be finite dimensional vector spaces over K, and let $T \in L(V, W)$. Then* $\operatorname{rank} T = \operatorname{rank} T^\bullet$.

PROOF. $\operatorname{rank} T^\bullet = \dim \operatorname{im} T^\bullet = \dim (\ker T)^\circ = \dim V - \dim \ker T = \dim \operatorname{im} T = \operatorname{rank} T$. ∎

Proposition 8.5. *Let V be a finite dimensional vector space over K and let T be a linear operator on V.*

(i) *If X is a T-invariant subspace of V, then X° is a $T^\bullet$-invariant subspace of $V^\bullet$.*

(ii) *If X and Y are T-invariant subspaces of V and $V = X \oplus Y$, then X° and Y° are $T^\bullet$-invariant subspaces of $V^\bullet$ and $V^\bullet = X^\circ \oplus Y^\circ$.*

Proof. (i) We need to check that for $f \in X^\circ$, $T^\bullet(f) \in X^\circ$. For $x \in X$, $T^\bullet(f)(x) = f(T(x)) = 0$, because $T(x) \in X$. Hence, $T^\bullet(f) \in X^\circ$.

(ii) By (i), X° and Y° are $T^\bullet$-invariant subspaces of $V^\bullet$. If $\mathcal{B}_1$ and $\mathcal{B}_2$ are bases of X and Y respectively, then $\mathcal{B} = \mathcal{B}_1 \cup \mathcal{B}_2$ is a basis of V. Let $\mathcal{B} = \{x_1, \ldots, x_k, \ldots, x_n\}$, where $\mathcal{B}_1 = \{x_1, \ldots, x_k\}$ and $\mathcal{B}_2 = \{x_{k+1}, \ldots, x_n\}$. Also, let $\mathcal{B}^\bullet$ be the dual basis of $\mathcal{B}$. Clearly, $\langle x^\bullet_{k+1}, \ldots, x^\bullet_n\rangle \subseteq X^\circ$. Also, $\dim X^\circ = \dim V - \dim X = \dim Y = n - k$. Thus, $X^\circ = \langle x^\bullet_{k+1}, \ldots, x^\bullet_n\rangle$. Similarly, $Y^\circ = \langle x^\bullet_1, \ldots, x^\bullet_k\rangle$. Hence, $V^\bullet = X^\circ \oplus Y^\circ$. ∎

8.1. Examples

1. If f is a linear form on K^n, then for any $x = \begin{bmatrix} x_1 & \cdots & x_n \end{bmatrix}^t$ in K^n,
$$f(x) = f(x_1 e_1 + \cdots + x_n e_n) = x_1 f(e_1) + \cdots + x_n f(e_n).$$
Thus, if $y = \begin{bmatrix} f(e_1) & \cdots & f(e_n) \end{bmatrix}^t$, then $f(x) = y^t x$ for all $x \in K^n$. Therefore, if for $y \in K^n$, $\widetilde{y} : K^n \to K$ defined by $\widetilde{y}(x) = y^t x$, then $(K^n)^\bullet = \{\widetilde{y} \mid y \in K^n\}$. Verify that the basis dual to the standard basis $\{e_1, \ldots, e_n\}$ is $\{\widetilde{e}_1, \ldots, \widetilde{e}_n\}$, $\widetilde{e}_i(x) = x_i$ for $x = \begin{bmatrix} x_1 & \cdots & x_n \end{bmatrix}^t$ in K^n.

2. From Example 3, p.43, it follows that
$$\left(K^{n\times n}\right)^\bullet = \left\{\mathrm{tr} \circ \lambda_A \mid A \in K^{n\times n}\right\},$$
where λ_A is the left matrix multiplication by A (see Example 4, p.43). Let $\mathcal{E}_{ij} : K^{n\times n} \to K$ defined by $\mathcal{E}_{ij}(A) = [A]_{ij}$. Then $\mathcal{E}_{ij}$ is a linear form on $K^{n\times n}$. The basis dual to the standard basis $\{E_{ij} \mid i, j = 1, \ldots, n\}$ is $\{\mathcal{E}_{ij} \mid i, j = 1, \ldots, n\}$. Verify that $\mathcal{E}_{ij} = \mathrm{tr} \circ \lambda_{E_{ji}}$.

3. The vector space $\mathbb{R}_n[x]$, $n \geq 1$, polynomials with coefficients from $\mathbb{R}$ and of degree at most n, has a basis $\mathcal{B} = \{1, x, \ldots, x^n\}$. For $p(x) \in \mathbb{R}_n[x]$, define $\mathcal{D}^k p(x) = p^{(k)}(0)$, $k = 0, 1, \ldots$. Clearly, $\mathcal{D}^0 p(x) = p(0)$. Then the dual basis of $\mathcal{B}$ is $\left\{\mathcal{D}^0, \frac{1}{1!}\mathcal{D}^1, \frac{1}{2!}\mathcal{D}^2, \ldots, \frac{1}{n!}\mathcal{D}^n\right\}$.

4. Let V and W be finite dimensional vector spaces over K. Let $T \in L(V, W)$ and let $\mathcal{B} = \{v_1, \ldots, v_n\}$ and $\mathcal{B}_1 = \{w_1, \ldots, w_m\}$ be ordered bases for V and W respectively. Then
$$Tv_j = \sum_{i=1}^{m} \alpha_{ij} w_i, \; j = 1, \ldots, n,$$
and so ${}_{\mathcal{B}_1}[T]_{\mathcal{B}}$ is an $m \times n$ matrix, whose (i,j)–th entry is α_{ij}. If $\mathcal{B}^\bullet = \{v^\bullet_1, \ldots, v^\bullet_n\}$ and $\mathcal{B}^\bullet_1 = \{w^\bullet_1, \ldots, w^\bullet_m\}$ are dual bases of $\mathcal{B}$ and $\mathcal{B}_1$ respectively, then
$$T^\bullet(w^\bullet_s) = \sum_{r=1}^{n} \beta_{rs} v^\bullet_r, \; s = 1, \ldots, m.$$

Now for $k = 1, \ldots, n$,

$$T^{\bullet}(w_s^{\bullet})(v_k) = \left(\sum_{r=1}^{n} \beta_{rs} v_r^{\bullet}\right)(v_k),$$

that is $w_s^{\bullet}(Tv_k) = \beta_{ks}$, then from the above equation

$$w_s^{\bullet}\left(\sum_{i=1}^{m} \alpha_{ik} w_i\right) = \beta_{ks}.$$

Thus $\beta_{ks} = \alpha_{sk}$, $s = 1, \ldots, m$, $k = 1, \ldots, n$. This shows that the transpose of the $m \times n$ matrix ${}_{\mathcal{B}_1}[T]_{\mathcal{B}}$ is ${}_{\mathcal{B}^{\bullet}}[T^{\bullet}]_{\mathcal{B}_1^{\bullet}}$.

8.2. Exercises

1. Let $x_1, \ldots, x_{n-1}$ be vectors in K^n and let $\phi : K^n \to K$ be a mapping given by $\phi(x) = \det\begin{bmatrix} x_1 & \cdots & x_{n-1} & x \end{bmatrix}$, the determinant of a matrix whose columns are $x_1, \ldots, x_{n-1}$ and x respectively. Prove that ϕ is a linear form on K^n and ϕ is nonzero if and only if $x_1, \ldots, x_{n-1}$ are linearly independent in K^n. Also prove that if $\phi \neq 0$, then $\{x_1, \ldots, x_{n-1}\}$ is a basis of $\ker\phi$.

2. Verify that $\mathcal{B} = \left\{\begin{bmatrix} 2 & 1 & 1 \end{bmatrix}^t, \begin{bmatrix} 3 & 4 & 1 \end{bmatrix}^t, \begin{bmatrix} 2 & 2 & 1 \end{bmatrix}^t\right\}$ is a basis of $\mathbb{R}^3$. Find a basis of $\left(\mathbb{R}^3\right)^{\bullet}$ dual to $\mathcal{B}$.

3. Let $f_1(x) = x_1 + 2x_2 + 3x_3$, $f_2(x) = 2x_1 + 3x_2 + 5x_3$, $f_3(x) = 3x_1 + 2x_2 + 4x_3$, $x = \begin{bmatrix} x_1 & x_2 & x_3 \end{bmatrix}^t$ be linear forms on $\mathbb{R}^3$. Prove that $\{f_1, f_2, f_3\}$ is a basis of $\left(\mathbb{R}^3\right)^{\bullet}$. Find vectors v_1, v_2, v_3 in $\mathbb{R}^3$ such that $f_i(v_j) = \delta_{ij}$, $i, j = 1, 2, 3$.

4. Show that nonzero linear forms f and g on a finite dimensional vector space are linearly dependent if and only if $\ker f = \ker g$.

5. Show that if $f_1, \ldots, f_m$ are linear forms on a finite dimensional vector space V, then $\dim \cap_{i=1}^{m} \ker f_i \geq n - m$ with equality if and only if $f_1, \ldots, f_m$ are linearly independent.

6. Let V be a finite dimensional vector space and let S be a subspace of $V^{\bullet}$ with a basis $\{f_1, \ldots, f_m\}$. Show that $S^{\circ} = \cap_{i=1}^{m} \ker f_i$ and $\dim S^{\circ} = \dim V - m$.

7. Let V be a vector space over K and let $f_1, \ldots, f_m$ be linearly independent elements of $V^{\bullet}$. Prove that $T : V \to K^m$ defined by $Tv = [f_1(v) \ \cdots \ f_m(v)]^t$ is a surjective linear transformation.

8. Let W be a subspace of a vector space V and let $v \in V \setminus W$. Prove that if f is a linear form on W then it can be extended to a linear form F on V such that $F(v) = 1$.

9. Prove that vectors $v_1, \ldots, v_m$ of a finite dimensional vector space V are linearly independent if and only if there are linear forms $f_1, \ldots, f_m$ on V such that $\det(f_i(v_j)) \neq 0$.

10. Let V and W be finite dimensional vector spaces over K, let $T \in L(V, W)$ and let $T^{\bullet} \in L(W^{\bullet}, V^{\bullet})$ be its dual mapping. Show that:
    (a) T is injective if and only if $T^{\bullet}$ is surjective;
    (b) T is surjective if and only if $T^{\bullet}$ is injective.

11. Let V be a finite dimensional vector space over K and let $T \in L(V)$. Show that if W is a T–invariant subspace of V, then for every projection P on W, $PTP = TP$. Conversely, if for a subspace W of V, $PTP = TP$ for any projection P on W, then W is T–invariant.

CHAPTER 3

Canonical Forms

1. Eigenvalues and eigenvectors

Let V be a vector space over a field K and let T be a linear operator on V. A scalar λ in K is called an **eigenvalue** of T if there exists a nonzero vector v in V such that $Tv = \lambda v$; each such nonzero vector is called an **eigenvector** of T corresponding to an eigenvalue λ.

Clearly, v is an eigenvector of T corresponding to an eigenvalue λ if and only if $(T - \lambda I)\, v = 0$, that is, $v \in \ker (T - \lambda I) \setminus \{0\}$. Thus, a nonzero linear combination of eigenvectors of T corresponding to λ is also an eigenvector. For this reason, the subspace $\ker (T - \lambda I)$ is also called an **eigenspace** of T corresponding to an eigenvalue λ.

For finite dimensional vector spaces, the following result characterizes the eigenvalues of a linear operator.

PROPOSITION 1.1. *Let V be a finite dimensional vector space over K and let T be a linear operator on V. Then a scalar λ in K is an eigenvalue of T if and only if* $\det (T - \lambda I) = 0$.

PROOF. $\lambda \in K$ is an eigenvalue of T if and only if $\ker (T - \lambda I) \neq \{0\}$, that is, $T - \lambda I$ is not an invertible linear operator on V. According to Proposition 2.7.4, this happens if and only if $\det (T - \lambda I) = 0$. ∎

Let T be a linear operator on an n–dimensional vector space V over K. If $\mathcal{B}$ is an ordered basis of V, then $\det (xI - T) = \det (xI_n - [T]_{\mathcal{B}})$ is a monic polynomial of degree n in $K[x]$. By the above proposition, λ is an eigenvalue of T if and only if λ is a root of this polynomial in K. The monic polynomial $\det (xI - T)$ in the variable x is called the **characteristic polynomial** of T, denoted by $c_T(x)$. Since the degree of characteristic polynomial of T is exactly equal to n, the dimension of V, T can not have more than n eigenvalues, counted with multiplicities.

If $A \in K^{n \times n}$, then by the characteristic polynomial of A, we mean the polynomial $c_A(x) = \det (xI_n - A)$. A scalar λ in K is an eigenvalue of A if there is a nonzero $v \in K^n$ such that $Av = \lambda v$. This nonzero vector is called an eigenvector of the matrix A corresponding to an eigenvalue λ. The eigenspace of A corresponding to an eigenvalue λ is $\ker (A - \lambda I_n)$, a subspace of K^n.

Now we give a general procedure of finding eigenvalues and corresponding eigenvectors of a linear operator T defined on a n–dimensional vector space V over K. Let $\mathcal{B} = \{v_1, \ldots, v_n\}$ be an ordered basis of V. A scalar λ in K is called an eigenvalue of T if and only if λ is a root of the characteristic polynomial $c_T(x) = \det (xI - T) = \det (xI_n - [T]_{\mathcal{B}})$. Thus, the linear operator T

and the matrix $A = [T]_{\mathcal{B}}$ have the same set of eigenvalues, counted with multiplicities. Let $a = \begin{bmatrix} a_1 & \cdots & a_n \end{bmatrix}^t \in K^n$ be an eigenvector of A corresponding to an eigenvalue λ. If we let $v = a_1 v_1 + \cdots + a_n v_n$, then $[v]_{\mathcal{B}} = a$, and $[Tv]_{\mathcal{B}} = [T]_{\mathcal{B}} [v]_{\mathcal{B}} = Aa = \lambda a = \lambda [v]_{\mathcal{B}}$. Since the mapping $V \to K^n$ given by $x \mapsto [x]_{\mathcal{B}}$ is an isomorphism, we have $Tv = \lambda v$. Therefore, we need to look for eigenvalues and eigenvectors of matrices only.

For computing the eigenvalues of $A \in K^{n \times n}$ and their corresponding eigenvectors, we perform the following steps.

1. The roots of the polynomial $c_A(x) = \det(xI_n - A)$ in K are the eigenvalues of A. If $c_A(x)$ has no roots in K, then A has no eigenvalues. For example, the 2×2 matrix $A = \begin{bmatrix} 0 & -1 \\ 1 & 0 \end{bmatrix}$ in $\mathbb{R}^{2 \times 2}$ has no eigenvalues because its characteristic polynomial $c_A(x) = x^2 + 1$ has no roots in $\mathbb{R}$.
2. If λ is an eigenvalue of A, then $a \in K^n$ is an eigenvector if and only if $a \in \ker(A - \lambda I_n) \setminus \{0\}$. Thus, a is a nonzero solution of the following homogeneous system of n linear equations in n unknowns $x_1, \ldots, x_n$:

$$\begin{aligned} (a_{11} - \lambda) x_1 + a_{12} x_2 + \cdots + a_{1n} x_n &= 0 \\ &\vdots \\ a_{n1} x_1 + a_{n2} x_2 + \cdots + (a_{nn} - \lambda) x_n &= 0 \end{aligned}$$

where $[A]_{ij} = a_{ij}$.

1.1. Examples

1. Let $A = \begin{bmatrix} 4 & 2 & 2 \\ 3 & 3 & 2 \\ -3 & -1 & 0 \end{bmatrix} \in \mathbb{R}^{3 \times 3}$. Then

$$c_A(x) = \det \begin{bmatrix} x-4 & -2 & -2 \\ -3 & x-3 & -2 \\ 3 & 1 & x \end{bmatrix} = (x-1)(x-2)(x-4),$$

and so 1, 2 and 4 are eigenvalues of A. Now we find eigenvectors corresponding to these eigenvalues.

If $y = \begin{bmatrix} y_1 \\ y_2 \\ y_3 \end{bmatrix} \in \mathbb{R}^3$ is an eigenvector corresponding to eigenvalue 1, then $(A - I_3) y = 0$, that is,

$$\begin{bmatrix} 3 & 2 & 2 \\ 3 & 2 & 2 \\ -3 & -1 & -1 \end{bmatrix} \begin{bmatrix} y_1 \\ y_2 \\ y_3 \end{bmatrix} = \begin{bmatrix} 0 \\ 0 \\ 0 \end{bmatrix}.$$

Subtracting the first row from the second, and adding the first row to the third, we get

$$\begin{bmatrix} 3 & 2 & 2 \\ 0 & 0 & 0 \\ 0 & 1 & 1 \end{bmatrix} \begin{bmatrix} y_1 \\ y_2 \\ y_3 \end{bmatrix} = \begin{bmatrix} 0 \\ 0 \\ 0 \end{bmatrix}.$$

Thus, $y_1 = 0$, and $y_2 + y_3 = 0$, and so

$$\ker(A - I_3) = \left\{ y_2 \begin{bmatrix} 0 \\ 1 \\ -1 \end{bmatrix} \mid y_2 \in \mathbb{R} \right\}.$$

In particular, $\begin{bmatrix} 0 \\ 1 \\ -1 \end{bmatrix}$ is an eigenvector of A corresponding to 1.

If $y = \begin{bmatrix} y_1 \\ y_2 \\ y_3 \end{bmatrix}$ is an eigenvector of A corresponding to eigenvalue 2, then $(A - 2I_3)\, y = 0$, that is,

$$\begin{bmatrix} 2 & 2 & 2 \\ 3 & 1 & 2 \\ -3 & -1 & -2 \end{bmatrix} \begin{bmatrix} y_1 \\ y_2 \\ y_3 \end{bmatrix} = \begin{bmatrix} 0 \\ 0 \\ 0 \end{bmatrix}.$$

Solving the system we have

$$\ker(A - 2I_3) = \left\{ y_1 \begin{bmatrix} 1 \\ 1 \\ -2 \end{bmatrix} \mid y_1 \in \mathbb{R} \right\}.$$

Thus, $\begin{bmatrix} 1 \\ 1 \\ -2 \end{bmatrix}$ is an eigenvector of A corresponding to eigenvalue 2.

Similarly, it can be verified that

$$\ker(A - 4I_3) = \left\{ y_3 \begin{bmatrix} 1 \\ 1 \\ -1 \end{bmatrix} \mid y_3 \in \mathbb{R} \right\}$$

and $\begin{bmatrix} 1 \\ 1 \\ -1 \end{bmatrix}$ is an eigenvector of A corresponding to eigenvalue 4.

2. Let $A = \begin{bmatrix} 0 & 0 & 1 \\ 1 & 0 & -1 \\ 0 & 1 & 1 \end{bmatrix} \in \mathbb{R}^{3\times 3}$. The characteristic polynomial of A is $(x-1)\left(x^2+1\right)$, and so A has only one eigenvalue 1 (in $\mathbb{R}$). The vector $\begin{bmatrix} 1 \\ 0 \\ 1 \end{bmatrix} \in \mathbb{R}^3$ is an eigenvector of A corresponding to eigenvalue 1.

Note that if $A \in \mathbb{R}^{n\times n}$, n is odd, then $c_A(x) \in \mathbb{R}[x]$, and so A has at least one eigenvalue.

If we consider the matrix A as an element of vector space $\mathbb{C}^{3\times 3}$ over $\mathbb{C}$, then A has eigenvalues 1, i and $-i$, and eigenvectors of A corresponding to these eigenvalues are $\begin{bmatrix} 1 \\ 0 \\ 1 \end{bmatrix}$, $\begin{bmatrix} 1 \\ -(1+i) \\ i \end{bmatrix}$ and $\begin{bmatrix} 1 \\ -(1-i) \\ -i \end{bmatrix}$, respectively.

3. The set of complex numbers $\mathbb{C}$ is a 2-dimensional vector space over $\mathbb{R}$, and $\mathcal{B} = \left\{1, i = \sqrt{-1}\right\}$ is a basis of $\mathbb{C}$. For any complex number $z = a + ib$,

$a, b \in \mathbb{R}$, the mapping $l_z : \mathbb{C} \to \mathbb{C}$ defined by $l_z(w) = zw$ is a linear operator on $\mathbb{C}$. It is easy to verify that

$$[l_z]_{\mathcal{B}} = \begin{bmatrix} a & -b \\ b & a \end{bmatrix},$$

and the characteristic polynomial of l_z is $(x-a)^2 + b^2$. This has no real roots, if $b \neq 0$. Thus, if $z \in \mathbb{C}$ and the imaginary part of z is nonzero, then l_z has no eigenvalues. If the imaginary part of z is zero, then $z = a \in \mathbb{R}$ and a is an eigenvalue of l_a and 1 and i are linearly independent eigenvectors of l_a.

It follows from the above that if $K = \mathbb{Q}$ or $\mathbb{R}$ and $A = \begin{bmatrix} a & -b \\ b & a \end{bmatrix} \in K^{2\times 2}$, $b \neq 0$, then A has no eigenvalues. Find an eigenvector of A, if $b = 0$.

However, if K is a field of characteristic 5 and $A = \begin{bmatrix} 1 & -2 \\ 2 & 1 \end{bmatrix}$, then $c_A(x) = x(x-2)$, and so 0 and 2 are eigenvalues of A. The reader can verify that $\begin{bmatrix} 1 \\ -2 \end{bmatrix}$ and $\begin{bmatrix} -2 \\ 1 \end{bmatrix}$ are eigenvectors of A corresponding to eigenvalues 0 and 2 respectively.

4. Let K be a field and let $p(x) = x^n + a_{n-1}x^{n-1} + \cdots + a_0$ be a monic polynomial of degree n. Let A be an $n \times n$ matrix given by:

$$A = \begin{bmatrix} 0 & \cdots & 0 & -a_0 \\ 1 & \cdots & 0 & -a_1 \\ \vdots & \ddots & \vdots & \vdots \\ 0 & \cdots & 1 & -a_{n-1} \end{bmatrix}.$$

The characteristic polynomial of A is $p(x)$. The matrix A is called the **companion matrix** associated with $p(x)$.

Consider

$$\det(xI_n - A) = \det \begin{bmatrix} x & 0 & \cdots & 0 & a_0 \\ -1 & x & \cdots & 0 & a_1 \\ \vdots & \ddots & & \vdots & \vdots \\ \vdots & & \ddots & \vdots & \vdots \\ 0 & \cdots & \cdots & -1 & x + a_{n-1} \end{bmatrix}.$$

Multiply the last row by x and add it to the $(n-1)$–th row, and then in the resulting matrix, multiply the $(n-1)$–th row by x and add it to the $(n-2)$–th row, and continuing similarly we have

$$\det(xI_n - A) = \det \begin{bmatrix} 0 & \cdots & 0 & p(x) \\ -1 & \cdots & 0 & * \\ \vdots & \ddots & \vdots & \vdots \\ 0 & \cdots & -1 & * \end{bmatrix},$$

where $*$ denotes those entries which are of no interest to us. Hence, $c_A(x) = p(x)$.

5. The differential operator D on $\mathbb{R}_3[x]$ has matrix representation with respect to a basis $\{1, x, x^2, x^3\}$:

$$\begin{bmatrix} 0 & 1 & 0 & 0 \\ 0 & 0 & 2 & 0 \\ 0 & 0 & 0 & 3 \\ 0 & 0 & 0 & 0 \end{bmatrix}.$$

The characteristic polynomial of D is x^4. Thus, 0 is the only eigenvalue of D and a constant polynomial is an eigenvector of D corresponding to 0.

6. Let $A \in K^{n\times n}$ be such that the sum of entries of each row is equal to λ, that is, $\sum_{j=1}^{n} [A]_{ij} = \lambda$ for all $i = 1, \ldots, n$. Then λ is an eigenvalue of A and $\begin{bmatrix} 1 \\ \vdots \\ 1 \end{bmatrix} \in K^n$ is corresponding eigenvector of A.

7. *If A and B are similar $n \times n$ matrices with entries from a field K, then A and B have the same characteristic polynomial.*

Let $P \in K^{n\times n}$ be an invertible matrix such that $B = P^{-1}AP$. Then $xI_n - B = P^{-1}(xI_n - A)P$, and so

$$\begin{aligned} c_B(x) &= \det(xI_n - B) = \det\left(P^{-1}(xI_n - A)P\right) \\ &= (\det P)^{-1} c_A(x) \det P = c_A(x). \end{aligned}$$

Note that two nonsimilar matrices can also have the same characteristic polynomial. For example, $A = \begin{bmatrix} 1 & 0 \\ 0 & 1 \end{bmatrix}$ and $B = \begin{bmatrix} 1 & 1 \\ 0 & 1 \end{bmatrix}$.

8. Let K be a field and let A be a block triangular matrix:

$$A = \begin{bmatrix} B & C \\ 0 & D \end{bmatrix} \in K^{n\times n}, B \in K^{m\times n}, m < n.$$

Then the characteristic polynomial of A is given by:

$$\begin{aligned} c_A(x) &= \det(xI_n - A) = \det\begin{bmatrix} xI_m - B & -C \\ 0 & xI_{n-m} - D \end{bmatrix} \\ &= \det(xI_m - B)\det(xI_{n-m} - D) = c_B(x)\, c_D(x). \end{aligned}$$

Combining this with Example 3, p.60 and Example 4, p.60, we have the following results:

Let T be a linear operator on a finite dimensional vector space V over a field K.

(i) *If W is a T–invariant subspace of V, then $c_T(x) = c_{\widehat{T}}(x)c_{T'}(x)$, where $\widehat{T}$ and T' are linear operators induced by T on W and V/W, respectively.*

(ii) *If $W_1, \ldots, W_k$ are T–invariant subspaces of V such that $V = W_1 \oplus \cdots \oplus W_k$, then $c_T(x) = c_{\widehat{T}_1}(x) \cdots c_{\widehat{T}_k}(x)$, where each $\widehat{T}_i$ is the linear operator on W_i induced by T.*

9. *Let $A, B \in K^{n\times n}$. If $AB = BA$, then AB and BA have the same characteristic polynomial.*

Let $P = \begin{bmatrix} I_n & A \\ 0 & I_n \end{bmatrix} \in K^{2n\times 2n}$. Then $P^{-1} = \begin{bmatrix} I_n & -A \\ 0 & I_n \end{bmatrix}$. Now let $R = \begin{bmatrix} AB & 0 \\ B & 0 \end{bmatrix} \in K^{2n\times 2n}$. Then $P^{-1}RP = \begin{bmatrix} 0 & 0 \\ B & BA \end{bmatrix}$. Thus, if $S = \begin{bmatrix} 0 & 0 \\ B & BA \end{bmatrix}$, then R and S are similar matrices, and so by Example 7, $c_R(x) = c_S(x)$. By Example 8, $c_R(x) = x^n c_{AB}(x)$ and $c_S(x) = x^n c_{BA}(x)$. Hence, $c_{AB}(x) = c_{BA}(x)$.

10. Let $A \in K^{n\times n}$. The left multiplication by A defines a linear operator

$$\lambda_A : K^{n\times m} \to K^{n\times m} \text{ such that } \lambda_A(B) = AB.$$

A scalar α is an eigenvalue of λ_A if and only if α is an eigenvalue of A.

If α is an eigenvalue of A, then $Ab = \alpha b$ for some $b \in K^n \setminus \{0\}$. Let B be an $n\times m$ matrix whose first column is b and all other columns zero. Then B is a nonzero matrix and $\lambda_A(B) = AB = \alpha B$, that is, α is an eigenvalue of λ_A. Conversely, if α is an eigenvalue of λ_A, then there is a nonzero matrix $B \in K^{n\times m}$ such that $\lambda_A(B) = \alpha B$. Let b be one of the nonzero columns of B. Then $Ab = \alpha b$, and so α is an eigenvalue of A.

11. Consider an infinite dimensional vector space over $\mathbb{R}$: $V = C^\infty(\mathbb{R})$, the space of all real valued functions of real variable which can be differentiated any number of times. Let D be the differential operator on V. Then for any $\lambda \in \mathbb{R}$, the function $f_\lambda(x) = \exp \lambda x$ is such that $Df_\lambda = \lambda f_\lambda$. Hence every real number is an eigenvalue of D. On the other hand the integral operator J defined by $Jf(x) = \int_0^x f(t)\,dt$ is a linear operator on V which has no eigenvalues.

Also, note that if W is a subspace of V consisting of polynomials functions only, then W is a D-invariant subspace of V, then considering D as a linear operator in W, D has only one eigenvalue 0. Since if $\lambda \in \mathbb{R}$ and $p(x) = a_0 + a_1x + \cdots + a_nx^n \in W$ such that $Dp(x) = \lambda p(x)$. Then $(\lambda a_0 - a_1) + (\lambda a_1 - 2a_2)x + \cdots + (\lambda a_{n-1} - na_n)x^{n-1} + \lambda a_n x^n = 0$. This implies that $p(x) = a_0$ and $\lambda = 0$.

1.2. Exercises. In the following exercises, V is a finite dimensional vector space over K and $T \in L(V)$.

1. Let v and w be eigenvectors of T corresponding to two distinct eigenvalues λ and μ, respectively. Show that for nonzero scalars α and β, $\alpha v + \beta w$ is not an eigenvector of T.
2. Prove that T is invertible if and only if 0 is not an eigenvalue of T.
3. Prove that if every nonzero vector is an eigenvector of T, then T is a scalar multiple of the identity operator.
4. Show that if λ is an eigenvalue of T and $p(x) \in K[x]$, then $p(\lambda)$ is an eigenvalue of $p(T)$.
5. Find eigenvalues and eigenvectors for the following matrices over $\mathbb{C}$:

$$\begin{bmatrix} 6 & -1 & 2 \\ 4 & 1 & 2 \\ -10 & 0 & 3 \end{bmatrix}, \begin{bmatrix} 7 & -5 & -4 \\ 2 & -1 & -3 \\ 2 & -2 & -2 \end{bmatrix}, \begin{bmatrix} 3 & -2 & 2 \\ 10 & -9 & 10 \\ 6 & -6 & 7 \end{bmatrix},$$

$$\begin{bmatrix} 2 & 1 & 0 & 0 \\ 1 & 2 & 1 & 0 \\ 0 & 1 & 2 & 1 \\ 0 & 0 & 1 & 2 \end{bmatrix}, \begin{bmatrix} 0 & 1 & 2 & 3 \\ 1 & 0 & 3 & 2 \\ 2 & 3 & 0 & 1 \\ 3 & 2 & 1 & 0 \end{bmatrix}, \begin{bmatrix} 1 & 1 & 0 & 3 \\ 0 & 1 & 0 & 0 \\ 0 & 5 & 1 & -14 \\ 0 & 0 & 0 & 1 \end{bmatrix}.$$

6. Prove that if $\operatorname{Char} K \neq 2$, then T can be expressed as the sum of two invertible linear operators.

7. Let $A \in K^{n\times n}$. Show that A and A^t have the same characteristic polynomials.

8. Verify that if $a, b \in K^n$ and $A = ab^t$, then $c_A(x) = x^{n-1}(x - a^t b)$.

9. Let $A = \begin{bmatrix} a & b \\ c & d \end{bmatrix} \in K^{2\times 2}$. If $b \neq 0$ and $\mu \in K$ is a root of the polynomial $bx^2 + (a-d)x + c = 0$, then $a + b\mu$ is an eigenvalue of A with eigenvector $\begin{bmatrix} 1 \\ \mu \end{bmatrix}$. If $b = 0$, then d is an eigenvalue of A with eigenvector $\begin{bmatrix} 0 \\ 1 \end{bmatrix}$.

10. If $A \in \mathbb{C}^{n\times n}$, then the eigenvalues of A^* are complex conjugates of the eigenvalues of A, counting with multiplicities.

11. Let $p(x) = x^n + a_{n-1}x^{n-1} + \cdots + a_1 x + a_0 \in K[x]$ and let A be the companion matrix of $p(x)$. For $\lambda \in K$, define

$$p_i(\lambda) = \left.\frac{p_{i-1}(x) - p_{i-1}(0)}{x}\right|_{x=\lambda},$$

$i = 1, \ldots, n$ and $p_0(x) = p(x)$. Show that if λ is a root of $p(x)$ in K, then $\begin{bmatrix} p_1(\lambda) & \cdots & p_n(\lambda) \end{bmatrix}^t$ is an eigenvector of A corresponding to λ.

12. Let $A \in \mathbb{C}^{n\times n}$ and let A_i be the submatrix of A obtained by deleting the i–th row and the i–th column. Prove that

$$\frac{d}{dx} c_A(x) = \sum_{i=1}^{n} c_{A_i}(x).$$

Conclude that if λ is an eigenvalue of A with multiplicity 1, then

$$\operatorname{rank}(A - \lambda I_n) = n - 1.$$

13. An $n \times n$ **primary permutation matrix** is:

$$P = \begin{bmatrix} 0 & 1 & 0 & \cdots & 0 \\ 0 & 0 & 1 & \cdots & 0 \\ \vdots & \vdots & \vdots & \ddots & \vdots \\ 0 & 0 & 0 & \cdots & 1 \\ 1 & 0 & 0 & \cdots & 0 \end{bmatrix} \in \mathbb{C}^{n\times n}.$$

Show that the eigenvalues of P are $1, \xi, \ldots, \xi^{n-1}$, where ξ is a primitive n–th root of unity, and that $u_k = \begin{bmatrix} 1 & \xi^k & \cdots & \xi^{(n-1)k} \end{bmatrix}^t$ is an eigenvector of P corresponding to ξ^k, $k = 0, 1, \ldots, n-1$.

14. An $n \times n$ **circulant matrix** is:

$$A = \begin{bmatrix} a_0 & a_1 & a_2 & \cdots & a_{n-1} \\ a_{n-1} & a_0 & a_1 & \cdots & a_{n-2} \\ a_{n-2} & a_{n-1} & a_0 & \cdots & a_{n-3} \\ \vdots & \vdots & \vdots & \ddots & \vdots \\ a_1 & a_2 & a_3 & \cdots & a_0 \end{bmatrix} \in \mathbb{C}^{n\times n}.$$

Verify that $A = a_0 I_n + a_1 P + \cdots + a_{n-1}P^{n-1}$, where P is the permutation matrix. Hence, compute the eigenvalues and corresponding eigenvectors of A.

15. Let $A \in \mathbb{C}^{n\times n}$ and let $M = \max\left\{\sum_{i=1}^{n}\left|[A]_{ij}\right| \mid j = 1, \ldots, n\right\}$. Prove that if λ is an eigenvalue of A, $|\lambda| \le M$.

2. The minimal polynomial

Let V be a finite dimensional vector space over K of dimension n. Then $L(V)$ is an n^2–dimensional vector space over K. Thus, if T is a linear operator on V, then $I = T^0, T, \ldots, T^{n^2}$, are linearly dependent, that is, there are scalars $\alpha_0, \alpha_1, \ldots, \alpha_{n^2}$ not all of which are zero such that

$$\alpha_0 I + \alpha_1 T + \cdots + \alpha_{n^2} T^{n^2} = 0 \ (\in L(V)).$$

Therefore, if $f(x) = \alpha_0 + \alpha_1 x + \cdots + \alpha_{n^2} x^{n^2}$, then $f(T) = 0$, and the set $S = \{p(x) \in K[x] \mid p(T) = 0\}$ has a nonzero nonconstant polynomial. It is immediate that if $p(x)$ is in S and α is the leading coefficient of $p(x)$, then the monic polynomial $\alpha^{-1}p(x)$ is also in S. Let $m(x)$ be the monic polynomial of the least degree in S. Obviously, $m(x)$ is unique. Also, for any nonzero $p(x)$ in S, $m(x)$ divides $p(x)$. Otherwise, by the division algorithm $p(x) = q(x)m(x) + r(x)$ for $q(x), r(x) \in K[x]$ with $0 < \deg r(x) < \deg m(x)$, and so $r(T) = q(T)m(T) - p(T) = 0$, a contradiction to the choice of $m(x)$. The polynomial $m(x)$ is called the **minimal polynomial** of T. We will denote the minimal polynomial of T by $m_T(x)$.

Also, for a fixed vector y in V, the vectors $y, Ty, \ldots, T^n y$ are linearly dependent, and so there are scalars $\beta_0, , \ldots, \beta_n$ in K not all of which are zero such that $\beta_0 y + \beta_1 Ty \cdots + \beta_n T^n y = 0$. Thus, if $p(x) = \beta_0 + \beta_1 x + \cdots + \beta_n x^n$, then $p(T)y = 0$. Let $m_T^y(x)$ be the monic polynomial of the least degree such that $m_T^y(T)y = 0$, called the **annihilator** of y with respect to the linear operator T. Similar arguments as in the previous paragraph show that if $q(x) \in K[x]$ and $q(T)y = 0$, then $m_T^y(x)$ divides $q(x)$. In particular, $m_T^y(x)$ divides the minimal polynomial $m_T(x)$.

Clearly, for any $y \in V$ and $T \in L(V)$, $\deg m_T^y(x) \le \dim V$. Our next result implies that the same is true for minimal polynomials.

PROPOSITION 2.1. *Let V be a finite dimensional vector space over K and let T be a linear operator on V. Then there exists a vector y in V such that $m_T^y(x) = m_T(x)$.*

PROOF. For any $v \in V$, the polynomial $m_T^v(x)$ divides $m_T(x)$. Thus, each $m_T^v(x)$ is a factor of $m_T(x)$ in $K[x]$. Therefore, as v varies over the whole of V, we get only finite number of polynomials. Let these be $p_1(x), \ldots, p_k(x)$ corresponding to vectors $y_1, \ldots, y_k$, that is, $m_T^{y_i}(x) = p_i(x)$, $i = 1, \ldots, k$. Let $V_i = \ker p_i(T)$.

Then $V \subseteq \cup_{i=1}^{k} V_i$, and so $V = V_r$ for some r (Exercise 11, p.31). Therefore, $p_r(T)v = 0$ for all $v \in V$. Hence, $p_r(T) = m_T^{y_r}(x) = m_T(x)$. ∎

COROLLARY 2.2. *Let V be a finite dimensional vector space over K of dimension n, and let T be a linear operator on V. The minimal polynomial of T is of degree at most n.*

PROOF. Since $m_T(x) = m_T^v(x)$ for some $v \in V$, $\deg m_T(x) = \deg m_T^v(x) \leq n$. ∎

THEOREM 2.3. ***Primary decomposition theorem.*** *Let V be a finite dimensional vector space over K and let T be a linear operator on V. Let the minimal polynomial of T in $K[x]$ be*

$$m_T(x) = p_1(x)^{r_1} \cdots p_k(x)^{r_k},$$

where $p_1(x), \ldots, p_k(x)$ are monic irreducible polynomials and $r_1, \ldots, r_k$ are positive integers. Then

$$V = \ker p_1(T)^{r_1} \oplus \cdots \oplus \ker p_k(T)^{r_k},$$

a direct sum of T-invariant subspaces of V. If for each i, T_i is the linear operator on $\ker p_i(T)^{r_i}$ induced by T, then the minimal polynomial of T_i is $p_i(x)^{r_i}$.

PROOF. For each i, let $q_i(x) = p_1(x)^{r_1} \cdots p_{i-1}(x)^{r_{i-1}} p_{i+1}(x)^{r_{i+1}} \cdots p_k(x)^{r_k} = m_T(x)/p_i(x)^{r_i}$. Then $\gcd(q_1(x), \ldots, q_k(x)) = 1$, and so there are polynomials $s_1(x), \ldots, s_k(x)$ such that

$$s_1(x)q_1(x) + \cdots + s_k(x)q_k(x) = 1.$$

Thus, $s_1(T)q_1(T) + \cdots + s_k(T)q_k(T) = I$, and for any v in V,

$$v = s_1(T)q_1(T)v + \cdots + s_k(T)q_k(T)v.$$

Since $p_i(T)^{r_i} s_i(T)q_i(T)v = s_i(T)m_T(T)v = 0$, so $s_i(T)q_i(T)v \in \ker p_i(T)^{r_i}$. Therefore,

$$V = \ker p_1(T)^{r_1} + \cdots + \ker p_k(T)^{r_k}.$$

Next, to show that this sum is actually direct, assume that $v_1 + \cdots + v_k = 0$, $v_i \in \ker p_i(T)^{r_i}$. Clearly, for $i \neq j$, $q_i(T)v_j = 0$. Now for any i,

$$v_i = -(v_1 + \cdots + v_{i-1} + v_{i+1} + \cdots + v_k),$$

and so $q_i(T)v_i = 0$. Also $p_i(T)^{r_i} v_i = 0$. Since $\gcd(q_i(x), p_i(x)^{r_i}) = 1$, there are $a_i(x), b_i(x) \in K[x]$ such that $a_i(x)q_i(x) + b_i(x)p_i(x)^{r_i} = 1$. Thus, $v_i = a_i(T)q_i(T)v_i + b_i(T)p_i(T)^{r_i} v_i = 0$. Hence, the sum is direct.

For the second part, note that $p_i(T)^{r_i} v = 0$ for all v in V_i. Thus, if $m_i(x)$ is the minimal polynomial of T_i, then $m_i(x)$ divides $p_i(x)^{r_i}$. Since $p_i(x)$ is monic and irreducible, $m_i(x) = p_i(x)^s$ for some $s \leq r_i$. Now the polynomial $f(x) = q_i(x)p_i(x)^s$ is such that $f(T) = 0$. Therefore, $m_T(x)$ divides $f(x)$. Hence $s = r_i$ and $m_i(x) = p_i(x)^{r_i}$. ∎

LEMMA 2.4. *Let V be a finite dimensional vector space over K of dimension n and let T be a linear operator on V. If $m_T(x) = p(x)^r$, where $p(x)$ is a monic irreducible polynomial of degree m, then m divides n.*

PROOF. Let $y \in V \setminus \{0\}$ such that $m_T^y(x) = m_T(x)$ (Proposition 2.1). Then $y, Ty, \dots, T^{mr-1}y$ are linearly independent and the subspace W generated by these vectors is T–invariant and of dimension mr. If $V = W$, then the result is proved.

If $V \neq W$, then $\dim V/W = n - mr < n$. Suppose that the statement is true for all vector spaces of dimension at most $n-1$. Let T' be the linear operator on V/W induced by T. Since the minimal polynomial of T' divides $m_T(x)$, so the minimal polynomial of T' is $p(x)^t$ for some $t < r$. By the induction hypothesis, m divides $n - mr$ and so m divides n. ∎

THEOREM 2.5. ***Cayley Hamilton theorem.*** *Let V be a finite dimensional vector space over K of dimension n. Let the minimal polynomial of T be $m_T(x) = p_1(x)^{r_1} \cdots p_k(x)^{r_k}$, where each $p_i(x)$ is monic irreducible polynomial of degree m_i, and let $\dim \ker p_i(T)^{r_i} = n_i$. Then m_i divides n_i and the characteristic polynomial $c_T(x) = p_1(x)^{n_1/m_1} \cdots p_k(x)^{n_k/m_k}$. In particular, $m_T(x)$ divides $c_T(x)$.*

PROOF. First we consider the case when $k = 1$, that is, $m_T(x) = p(x)^r$, $\deg p(x) = m$. Then we require to show that $c_T(x) = p(x)^{n/m}$. We prove it by induction on n. Note by Lemma 2.4, it follows that n/m is a positive integer.

If $n = 1$, then $V = Kv$ for any $v \in V \setminus \{0\}$, and so $Tv = \alpha v$, $\alpha \in K$. Therefore $m_T(x) = x - \alpha = c_T(x)$. Assume that the hypothesis is true for all vector spaces of dimension at most $n-1$. Let $\dim V = n$ and let $y \in V$ such that $m_T^y(x) = m_T(x)$ (Proposition 2.1). Then $y, Ty, \dots, T^{mr-1}y$ are linearly independent and $W = \langle y, Ty, \dots, T^{mr-1}y \rangle$ is a nonzero T–invariant subspace of dimension mr.

If $V = W$, then $n = mr$. Let $\mathcal{B} = \{y_1 = y,\ y_2 = Ty, \dots,\ y_{mr} = T^{mr-1}y\}$ be an ordered basis of V and let $(p(x))^r = \alpha_0 + \alpha_1 x + \cdots + \alpha_{mr-1}x^{mr-1} + x^{mr}$. Then

$$
\begin{aligned}
Ty_1 &= y_2, \\
Ty_2 &= y_3, \\
&\ \ \vdots \\
Ty_{mr-1} &= y_{mr}, \\
Ty_{mr} = T^{mr}y &= (-\alpha_0)\,y_1 + (-\alpha_1)\,y_2 + \cdots + (-\alpha_{mr-1})\,y_{mr},
\end{aligned}
$$

and so

$$
[T]_{\mathcal{B}} = \begin{bmatrix} 0 & 0 & \cdots & 0 & -\alpha_0 \\ 1 & 0 & \cdots & 0 & -\alpha_1 \\ 0 & 1 & \cdots & 0 & -\alpha_2 \\ \vdots & \vdots & \ddots & \vdots & \vdots \\ 0 & 0 & \cdots & 1 & -\alpha_{mr} \end{bmatrix},
$$

the companion matrix of $p(x)^r$. Therefore, $c_T(x) = p(x)^r = p(x)^{n/m}$.

If $W \neq V$, and if T induces linear operators $\widehat{T}$ and T' on W and V/W respectively, then $c_T(x) = c_{\widehat{T}}(x)\, c_{T'}(x)$. Since $\dim W = mr < n$, by the induction hypothesis, $c_{\widehat{T}}(x) = p(x)^{mr/m}$. Also as $\dim V/W = n - mr < n$, so $c_{T'}(x) = p(x)^{(n-mr)/m}$. Hence $c_T(x) = p(x)^{n/m}$.

Now consider the general case, $m_T(x) = p_1(x)^{r_1} \cdots p_k(x)^{r_k}$, where each $p_i(x)$ is monic irreducible polynomial and each r_i is a positive integer. The subspace $V_i = \ker p_i(T)^{r_i}$ is a T–invariant subspace of V. Let T_i be the linear operator on

V_i induced by T. Then by Theorem 2.3, it follows that $V = V_1 \oplus \cdots \oplus V_k$ and $m_{T_i}(x) = p_i(x)^{r_i}$, $i = 1, \ldots, k$. Also by Example 8, p.73, we have that $c_T(x) = c_{T_1}(x) \cdots c_{T_k}(x)$. By the above paragraph, we have $c_{T_i}(x) = p_i(x)^{n_i/m_i}$, where $n_i = \dim V_i$ and $m_i = \deg p_i(x)$. Therefore, $c_T(x) = p_1(x)^{n_1/m_1} \cdots p_k(x)^{n_k/m_k}$.

Also $m_{T_i}(x) = p_i(x)^{r_i}$ in V_i, and so by Proposition 2.1, there exists $y_i \in V_i$ such that $m_{T_i}^{y_i}(x) = m_{T_i}(x)$. Hence, $y_i, Ty_i, \ldots, T^{m_i r_i - 1} y_i$ are linearly independent in V_i. Thus $m_i r_i \leq n_i$, and $r_i \leq n_i/m_i$. Hence, $m_T(x)$ divides $c_T(x)$. ∎

From the above result, if T is a linear operator on a finite dimensional vector space V and $c_T(x) = p_1(x)^{m_1} \cdots p_k(x)^{m_k}$, where $p_1(x), \ldots, p_k(x)$ are monic irreducible polynomials and $m_1, \ldots, m_k$ are positive integers. Then $c_T(T) = 0$ and $m_T(x) = p_1(x)^{r_1} \cdots p_k(x)^{r_k}$, where each r_i is a positive integer and $r_i \leq m_i$. For example, if V is a vector space over $\mathbb{R}$ of dimension 7 and for a linear operator T on V, $c_T(x) = (x^2+1)^2 (x-1)^3$. Then $m_T(x)$ is one of the following polynomials: $(x^2+1)(x-1)$, $(x^2+1)(x-1)^2$, $(x^2+1)(x-1)^3$, $(x^2+1)^2(x-1)$, $(x^2+1)^2(x-1)^2$or $(x^2+1)^2(x-1)^3$. In particular, λ is an eigenvalue of T if and only if λ is a root of its minimal polynomial.

2.1. Examples

1. *If A and B are similar matrices with entries from K, then $m_A(x) = m_B(x)$.*

   If P is an invertible matrix and $B = P^{-1}AP$, then
   $$m_A(B) = m_A(P^{-1}AP) = P^{-1} m_A(A) P = 0.$$
   Thus, $m_B(x)$ divides $m_A(x)$. Similarly, $A = PBP^{-1}$ implies $m_A(x)$ divides $m_B(x)$. Hence, $m_A(x) = m_B(x)$.

   However, two non similar matrices of the same size can have the same minimal polynomial. Let
   $$A = \begin{bmatrix} 0 & 1 & 0 & 0 \\ 0 & 0 & 0 & 0 \\ 0 & 0 & 0 & 0 \\ 0 & 0 & 0 & 0 \end{bmatrix} \text{ and } B = \begin{bmatrix} 0 & 1 & 0 & 0 \\ 0 & 0 & 0 & 0 \\ 0 & 0 & 1 & 0 \\ 0 & 0 & 0 & 0 \end{bmatrix}.$$
   Then $A^2 = 0 = B^2$, and so $m_A(x) = x^2 = m_B(x)$. Matrices A and B are not similar. Otherwise $AP = PB$ for some invertible matrix P. In particular, $APe_3 = PBe_3 = Pe_3$. This means that either $Pe_3 = 0$ or is an eigenvector of A corresponding to eigenvalue 1. Since 0 is the only eigenvalue of A, $Pe_3 = 0$. But then the third column of P is zero, and P is not invertible.

2. Let V be a finite dimensional vector space over K of dimension n and let T be a linear operator on V. Then $c_T(x) = m_T(x)$ if and only if $\deg m_T(x) = n$. Also, by Proposition 2.1, there is a vector y in V such that $m_T(x) = m_T^y(x)$. Therefore the characteristic polynomial and minimal polynomial of a linear operator T coincide if and only if for a certain vector y in V, vectors $y, Ty, \ldots, T^{n-1}y$ are linearly independent, that is, the set $\{y, Ty, \ldots, T^{n-1}y\}$ is a basis of V.

   Hence, if $T \in L(V)$ and $c_T(x) = a_0 + a_1 x + \cdots + a_{n-1}x^{n-1} + x^n$ which is also the minimal polynomial of T, then there exists a vector y such that

$\mathcal{B}=\{y,Ty,\ldots,T^{n-1}y\}$ is a basis of V. Clearly, then

$$[T]_{\mathcal{B}}=\begin{bmatrix}0&0&\cdots&0&-a_0\\1&0&\cdots&0&-a_1\\0&1&\cdots&0&-a_2\\\vdots&\vdots&\ddots&\vdots&\vdots\\0&0&\cdots&1&-a_{n-1}\end{bmatrix},$$

the companion matrix associated with $c_T(x)$. The matrix analogue of the above is: for an $n\times n$ matrix A over K, $c_A(x)=m_A(x)$ if and only if A is similar to the companion matrix associated with $c_A(x)$.

For an $n\times n$ companion matrix A, observe that $Ae_1=e_2,\ldots,Ae_{n-1}=e_n$, and so $e_1,Ae_1,A^2e_1,\ldots,A^{n-1}e_1$ are linearly independent vectors of K^n. Hence, $c_A(x)=m_A(x)=m_A^{e_1}(x)$.

3. The $n\times n$ matrix A over K:

$$A=\begin{bmatrix}0&1&&\\&\ddots&\ddots&\\&&\ddots&1\\&&&0\end{bmatrix}$$

has the same minimal and characteristic polynomial, x^n. Indeed, if $x=\begin{bmatrix}1&\cdots&1\end{bmatrix}^t\in K^n$, then $x,Ax,\ldots,A^{n-1}x$ are linearly independent in K. A direct verification shows that $A^k\neq 0$ for $k\leq n-1$ and $A^n=0$. Also, it can be verified that

$$\begin{bmatrix}&&1\\&\cdot&\\1&&\end{bmatrix}^{-1}\begin{bmatrix}0&1&&\\&\ddots&\ddots&\\&&\ddots&1\\&&&0\end{bmatrix}\begin{bmatrix}&&1\\&\cdot&\\1&&\end{bmatrix}=\begin{bmatrix}0&&&\\1&\ddots&&\\&\ddots&\ddots&\\&&1&0\end{bmatrix},$$

that is, A is similar to a matrix which is precisely the companion matrix of the polynomial x^n. This also shows that $m_A(x)=x^n$.

The $n\times n$ matrix

$$J_n(\lambda)=\begin{bmatrix}\lambda&1&&\\&\ddots&\ddots&\\&&\ddots&1\\&&&\lambda\end{bmatrix}\in K^{n\times n}$$

is called the **Jordan block** of order n corresponding to the eigenvalue λ. It is now immediate that the minimal polynomial of $J_n(\lambda)$ is $(x-\lambda)^n$, the characteristic polynomial of $J_n(\lambda)$.

4. If T is an invertible linear operator on a finite dimensional vector space over a field K, then the minimal polynomial of T^{-1} is $m_T(0)^{-1}x^r m_T\left(\frac{1}{x}\right)$, where $r=\deg m_T(x)$.

First observe that T is invertible if and only if $m_T(0)\neq 0$. If $f(x)=a_0+a_1x+\cdots+a_kx^k\in K[x]$, $a_k\neq 0$, then $x^kf\left(\frac{1}{x}\right)=a_k+a_{k-1}x+\cdots+a_0x^k$.

Thus, for $S = T^{-1}$ and for $g(x) = x^k f\left(\frac{1}{x}\right)$, we have

$$\begin{aligned} g(S) &= a_k I + a_{k-1} S + \cdots + a_0 S^k \\ &= S^k \left(a_k T^k + a_{k-1} T^{k-1} + \cdots + a_0\right) \\ &= S^k f(T). \end{aligned}$$

Hence, $f(T) = 0$ if and only if $g(S) = 0$. In particular, $\deg m_S(x) = \deg m_T(x)$. The polynomial $p(x) = x^r m_T\left(\frac{1}{x}\right)$ is of degree r and $p(S) = 0$. Therefore, the minimal polynomial of T^{-1} is $m_T(0)^{-1} p(x)$, that is, $m_T(0)^{-1} x^r m_T\left(\frac{1}{x}\right)$.

5. Let $A = \begin{bmatrix} R & * \\ 0 & S \end{bmatrix}$, where $R \in K^{r\times r}$, $S \in K^{s\times s}$ and $*$ denotes those entries of the matrix which are of no interest to us. If $p(x)$ is any polynomial in $K[x]$, then $p(A) = \begin{bmatrix} p(R) & * \\ 0 & p(S) \end{bmatrix}$. Thus, if $p(A) = 0$, then $p(R) = 0$ $(\in K^{r\times r})$, and $p(S) = 0$ $(\in K^{s\times s})$. Therefore, $m_R(x)$ divides $m_A(x)$ and also $m_S(x)$ divides $m_A(x)$. Hence, if $m(x)$ is the lcm (least common multiple) of $m_R(x)$ and $m_S(x)$, then $m(x)$ divides $m_A(x)$.

Next,

$$m_R(A) = \begin{bmatrix} m_R(R) & * \\ 0 & m_R(S) \end{bmatrix} = \begin{bmatrix} 0 & * \\ 0 & m_R(S) \end{bmatrix},$$

and

$$m_S(A) = \begin{bmatrix} m_S(R) & * \\ 0 & m_S(S) \end{bmatrix} = \begin{bmatrix} m_S(R) & * \\ 0 & 0 \end{bmatrix}.$$

Therefore, $m_R(A)\, m_S(A) = 0$. Hence, $m_A(x)$ divides $m_R(x)\, m_S(x)$. Thus, we have the monic polynomials $m(x)$, $m_A(x)$ and $m_R(x)\, m_S(x)$ in $K[x]$ such that $m(x)$ divides $m_A(x)$ and $m_A(x)$ divides $m_R(x)\, m_S(x)$.

In the case of characteristic polynomials, we have $c_A(x) = c_R(x)\, c_S(x)$. However the corresponding statement for minimal polynomials is not true. We give examples to show that $m_A(x)$ can be $m(x)$, or $m_R(x)\, m_S(x)$, or even can be different from these two.

If $R = \begin{bmatrix} 1 & 1 & 0 \\ 0 & 1 & 0 \\ 0 & 0 & 2 \end{bmatrix}$, $S = \begin{bmatrix} 2 & 1 & 0 \\ 0 & 2 & 0 \\ 0 & 0 & 1 \end{bmatrix}$ and $A = \begin{bmatrix} R & 0 \\ 0 & S \end{bmatrix}$, then $m_R(x) = (x-1)^2 (x-2)$, $m_S(x) = (x-1)(x-2)^2$ and

$$m_A(x) = (x-1)^2 (x-2)^2 = m(x).$$

If $R = \begin{bmatrix} 1 & 1 \\ 0 & 1 \end{bmatrix}$, $S = \begin{bmatrix} 1 & 0 \\ 0 & 2 \end{bmatrix}$ and

$$A = \begin{bmatrix} 1 & 1 & 0 & 0 \\ 0 & 1 & 1 & 0 \\ 0 & 0 & 1 & 0 \\ 0 & 0 & 0 & 2 \end{bmatrix},$$

then $m_R(x) = (x-1)^2$, $m_S(x) = (x-1)(x-2)$ and

$$m_A(x) = (x-1)^3 (x-2) = m_R(x)\, m_S(x).$$

If $R = \begin{bmatrix} 1 & 1 \\ 0 & 1 \end{bmatrix}$, $S = \begin{bmatrix} 1 & 1 & 0 \\ 0 & 1 & 0 \\ 0 & 0 & 2 \end{bmatrix}$ and

$$A = \begin{bmatrix} 1 & 1 & 1 & 0 & 0 \\ 0 & 1 & 0 & 1 & 0 \\ 0 & 0 & 1 & 1 & 0 \\ 0 & 0 & 0 & 1 & 0 \\ 0 & 0 & 0 & 0 & 2 \end{bmatrix},$$

then $m_R(x) = (x-1)^2$, $m_S(x) = (x-1)^2(x-2)$ and

$$m_A(x) = (x-1)^3(x-2)$$

which is neither the lcm of $m_R(x)$ and $m_S(x)$ nor the product $m_R(x)\,m_S(x)$.

6. This example illustrates a method of finding the minimal polynomial of a linear operator. This is based on the observation that if T is a linear operator on an n–dimensional vector space V over K and if $\mathcal{B} = \{v_1, \dots, v_n\}$ is a basis of V, then the minimal polynomial of T is the lcm of the annihilator of the elements of $\mathcal{B}$ with respect to T, that is, the lcm of the polynomials $m_T^{v_1}(x), \dots, m_T^{v_n}(x)$ (Verify!). If A is the matrix of T with respect to $\mathcal{B}$ then for each i, $m_T^{v_i}(x) = m_A^{e_i}(x)$, where $\{e_1, \dots, e_n\}$ is the standard basis of K^n. The degree of each $m_A^{e_i}(x)$ is the largest positive integer k_i such that $e_i, Ae_i, \dots, A^{k_i-1}e_i$ are linearly independent. Obviously, k_i can not exceed n and $k_i = 1$ if and only if e_i is an eigenvalue of A. Also, for any $r \geq k_i$, $A^r e_i \in \langle e_i, Ae_i, \dots, A^{k_i-1}e_i \rangle$. Therefore for each i, $\deg m_A^{e_i}(x)$ is the rank of an $n \times n$ matrix whose columns are the i–th columns of $I_n, A, \dots, A^{n-1}$, respectively; and $\deg m_A^{e_i}(x) = k_i$ if and only if first k_i columns of this matrix are linearly independent.

Consider

$$A = \begin{bmatrix} 1 & 1 & -2 & -2 \\ 0 & 1 & 1 & 1 \\ 1 & 2 & 1 & 0 \\ -1 & -2 & 0 & 1 \end{bmatrix} \in \mathbb{R}^{4\times 4}.$$

Then

$$A^2 = \begin{bmatrix} 1 & 2 & -3 & -3 \\ 0 & 1 & 2 & 2 \\ 2 & 5 & 1 & 0 \\ -2 & -5 & 0 & 1 \end{bmatrix}, \quad A^3 = \begin{bmatrix} 1 & 3 & -3 & -3 \\ 0 & 1 & 3 & 3 \\ 3 & 9 & 2 & 1 \\ -3 & -9 & -1 & 0 \end{bmatrix}.$$

Since the matrix formed by first columns of I, A, A^2, A^3 is

$$\begin{bmatrix} 1 & 1 & 1 & 1 \\ 0 & 0 & 0 & 0 \\ 0 & 1 & 2 & 3 \\ 0 & -1 & -2 & -3 \end{bmatrix},$$

which is of rank 2, so $\deg m_A^{e_1}(x) = 2$. Thus there are $\alpha_0, \alpha_1 \in \mathbb{R}$ such that $A^2e_1 = \alpha_0 e_1 + \alpha_1 Ae_1$, that is,

$$\begin{bmatrix} 1 \\ 0 \\ 2 \\ -2 \end{bmatrix} = \alpha_0 \begin{bmatrix} 1 \\ 0 \\ 0 \\ 0 \end{bmatrix} + \alpha_1 \begin{bmatrix} 1 \\ 0 \\ 1 \\ -1 \end{bmatrix}.$$

On solving the system, we have $\alpha_0 = -1$, $\alpha_1 = 2$, and so $A^2e_1 = -e_1 + 2Ae_1$, that is, $(A^2 - 2A + I)e_1 = 0$. Therefore, $m_A^{e_1}(x) = x^2 - 2x + 1 = (x-1)^2$.

The matrix formed by second columns of I, A, A^2, A^3 is

$$\begin{bmatrix} 0 & 1 & 2 & 3 \\ 1 & 1 & 1 & 1 \\ 0 & 2 & 5 & 9 \\ 0 & -2 & -5 & -9 \end{bmatrix},$$

and is of rank 3. Thus there are $\alpha_0, \alpha_1, \alpha_2 \in \mathbb{R}$ such that $A^3e_2 = \alpha_0 e_2 + \alpha_1 Ae_2 + \alpha_2 A^2 e_2$, that is,

$$\begin{bmatrix} 3 \\ 1 \\ 9 \\ -9 \end{bmatrix} = \alpha_0 \begin{bmatrix} 0 \\ 1 \\ 0 \\ 0 \end{bmatrix} + \alpha_1 \begin{bmatrix} 1 \\ 1 \\ 2 \\ -2 \end{bmatrix} + \alpha_2 \begin{bmatrix} 2 \\ 1 \\ 5 \\ -5 \end{bmatrix}.$$

On solving the system, we have $\alpha_0 = 1$, $\alpha_1 = -3$, $\alpha_2 = 3$. Therefore, $m_A^{e_2}(x) = (x-1)^3$.

Similarly, it can be verified that $m_A^{e_3}(x)$ is of degree 4 and $A^4e_3 = -e_3 + 4Ae_3 - 6A^2e_3 + 4A^3e_3$. Thus, $m_A^{e_3}(x) = (x-1)^4$. Since $\deg m_A^{e_3}(x) = 4$, $m_A^{e_3}(x) = m_A(x)$ which is also the characteristic polynomial of A.

We consider another matrix $B \in \mathbb{R}^{4\times 4}$:

$$B = \begin{bmatrix} 1 & 0 & 0 & 2 \\ 1 & 2 & 0 & -2 \\ 0 & 0 & -1 & 1 \\ 0 & 0 & 0 & -1 \end{bmatrix}.$$

Then

$$B^2 = \begin{bmatrix} 1 & 0 & 0 & 0 \\ 3 & 4 & 0 & 0 \\ 0 & 0 & 1 & -2 \\ 0 & 0 & 0 & 1 \end{bmatrix} \text{ and } B^3 = \begin{bmatrix} 1 & 0 & 0 & 2 \\ 7 & 8 & 0 & -2 \\ 0 & 0 & -1 & 3 \\ 0 & 0 & 0 & -1 \end{bmatrix}.$$

Clearly, $Be_2 = 2e_2$ and $Be_3 = -e_3$. Thus, $m_B^{e_2}(x) = x-2$ and $m_B^{e_3}(x) = x+1$. Since the matrix $\begin{bmatrix} e_1 & Be_1 & B^2e_1 & B^3e_1 \end{bmatrix}$ is of rank 2, $\deg m_B^{e_2}(x) = 2$. Verify that $B^2e_1 = -2e_1 + 3Be_1$, and so $m_B^{e_1}(x) = (x-1)(x-2)$. Similarly, $m_B^{e_4}(x) = (x+1)^2(x-1)$. Hence,

$$\begin{aligned} m_B(x) &= \operatorname{lcm}\left\{(x-1)(x-2), x-2, x+1, (x+1)^2(x-1)\right\} \\ &= (x+1)^2(x-1)(x-2). \end{aligned}$$

Observe that $m_B(x) = m_B^v(x)$ for $v = e_2 + e_4$.

2.2. Exercises

1. Verify that the following matrices have the same characteristic polynomial, but different minimal polynomials.

$$\begin{bmatrix} 1 & 1 & 0 & 0 \\ 0 & 1 & 1 & 0 \\ 0 & 0 & 1 & 1 \\ 0 & 0 & 0 & 1 \end{bmatrix}, \begin{bmatrix} 1 & 0 & 1 & 0 \\ 0 & 1 & 0 & 1 \\ 0 & 0 & 1 & 0 \\ 0 & 0 & 0 & 1 \end{bmatrix}, \begin{bmatrix} 1 & 0 & 0 & 1 \\ 0 & 1 & 0 & 0 \\ 0 & 0 & 1 & 0 \\ 0 & 0 & 0 & 1 \end{bmatrix}.$$

2. Obtain the minimal polynomials for the following matrices. Also for each of the following matrices find a vector such that its annihilator is equal to the minimal polynomial.

$$\begin{bmatrix} 3 & -2 & 2 \\ 10 & -9 & 10 \\ 6 & -6 & 7 \end{bmatrix}, \begin{bmatrix} 15 & 9 & -12 \\ -10 & -4 & 10 \\ 8 & 6 & -5 \end{bmatrix}, \begin{bmatrix} 6 & -1 & 2 \\ 4 & 1 & 2 \\ -10 & 0 & -3 \end{bmatrix}, \begin{bmatrix} 1 & -1 & -1 \\ 1 & -1 & -1 \\ 0 & 0 & -1 \end{bmatrix},$$

$$\begin{bmatrix} 0 & 1 & 1 & 1 \\ 1 & 0 & 1 & 1 \\ 1 & 1 & 0 & 1 \\ 1 & 1 & 1 & 0 \end{bmatrix}, \begin{bmatrix} 1 & -3 & 0 & 3 \\ -2 & -6 & 0 & 13 \\ 0 & -3 & 1 & 3 \\ -1 & -4 & 0 & 8 \end{bmatrix}, \begin{bmatrix} 1 & 1 & 0 & -3 \\ 0 & 1 & 0 & 0 \\ 0 & 5 & 1 & -14 \\ 0 & 0 & 0 & 1 \end{bmatrix}, \begin{bmatrix} 1 & 2 & 3 & 4 \\ 0 & 1 & 2 & 3 \\ 0 & 0 & 1 & 2 \\ 0 & 0 & 0 & 1 \end{bmatrix}.$$

3. Show that if D is a diagonal matrix, then $m_D(x) = \prod_{i=1}^{k}(x - d_i)$, where $d_1, \ldots, d_k$ are distinct diagonal entries of D.

4. Let T be a linear operator on a finite dimensional vector space over K and let $T^\bullet$ be its dual. Prove that T and $T^\bullet$ have the same minimal polynomial.

5. Give an example of a linear operator on $\mathbb{R}^4$ whose characteristic and minimal polynomials are $(x-1)^2(x-2)(x-3)$ and $(x-1)(x-2)(x-3)$ respectively.

6. Give an example of a linear operator on $\mathbb{R}^4$ whose minimal polynomial is $(x^2-1)(x+1)$ and the polynomial is $(x^2-1)^2(x+1)$.

7. Give an example of a linear operator on $\mathbb{R}^4$ whose minimal polynomial is $(x+1)^3$. What will be its characteristic polynomial?

8. Let $A \in \mathbb{Q}^{n \times n}$ and let p be a prime, $p > n+1$. Show that if $A^p = I_n$, then $A = I_n$.
Hint: use irreducibility of $1 + x + x^2 + \cdots + x^{p-1}$.

9. Let T be a linear operator on an n–dimensional vector space V over K. Show that if T is invertible, then there is a polynomial $q(x) \in K[x]$ of degree at most $n-1$ such that $T^{-1} = q(T)$.

10. Using the above result find the inverse of the matrix

$$A = \begin{bmatrix} 5 & 2 & -3 \\ 1 & 3 & -1 \\ 2 & 2 & -1 \end{bmatrix}.$$

11. Compute A^{25}, if

$$A = \begin{bmatrix} 1 & 1 & 0 \\ -1 & 1 & 2 \\ 2 & 0 & -1 \end{bmatrix}.$$

12. Let V be a 3-dimensional vector space over $\mathbb{R}$ and let $T \in L(V)$ whose action on a basis $\{v_1, v_2, v_3\}$ is given by:

$$\begin{aligned} Tv_1 &= 6v_1 + 4v_2 - 10v_3, \\ Tv_2 &= -v_1 + v_2, \\ Tv_3 &= 2v_1 + 2v_2 - 3v_3. \end{aligned}$$

Using primary decomposition theorem, show that there is a basis $\mathcal{B}$ of V such that

$$[T]_{\mathcal{B}} = \begin{bmatrix} 2 & 0 & 0 \\ 0 & 1 & 2 \\ 0 & -2 & 1 \end{bmatrix}.$$

13. This exercise is another proof of Cayley Hamilton theorem. Let $A \in K^{n\times n}$. Verify that the entries of $\operatorname{adj}(xI_n - A)$ are polynomials in $K[x]$, each of degree at most $n-1$. Now write $\operatorname{adj}(xI_n - A) = B_0 + B_1x + \cdots + B_{n-1}x^{n-1}$, each $B_i \in K^{n\times n}$, and let $c_A(x) = a_0 + a_1x + \cdots + a_{n-1}x^{n-1} + x^n$. Then using the identity

$$(xI_n - A)\operatorname{adj}(xI_n - A) = c_A(x)\, I_n,$$

and equating the coefficients of the corresponding powers of x, obtain the following equations:

$$\begin{aligned} -AB_0 &= I_n, \\ B_0 - AB_1 &= a_1 I_n, \\ &\vdots \\ B_{n-2} - AB_{n-1} &= a_{n-1} I_n, \\ B_{n-1} &= I_n. \end{aligned}$$

Multiply the i–th equation by A^{i-1} on the left and add all these equations to get $c_A(A) = 0$.

3. Diagonalizable and triangulable operators

In this section, V will always denote a vector space over a field K of finite dimension n.

A linear operator T on V is said to be **diagonalizable** if V has a basis consisting of eigenvectors of T only. Equivalently, T has n linearly independent eigenvectors. Clearly, a scalar multiple of identity operator is diagonalizable.

Let T be diagonalizable linear operator on V. If $\mathcal{B} = \{x_1, \ldots, x_n\}$ is an ordered basis of V and each x_i is an eigenvector of T, that is, $Tx_i = \lambda_i x_i$, $i = 1, \ldots, n$, where $\lambda_1, \ldots, \lambda_n$ are in K, need not be distinct, then $[T]_{\mathcal{B}} = \operatorname{diag}(\lambda_1, \ldots, \lambda_n)$. Thus, a linear operator T on V is diagonalizable if and only if the matrix of T with respect to a certain basis of V is diagonal. The corresponding matrix analogue is:

an $n \times n$ matrix over K is diagonalizable if it is similar to a diagonal matrix over K, that is, there exists an invertible matrix P with entries from K such that $P^{-1}AP$ is a diagonal matrix.

PROPOSITION 3.1. *Eigenvectors corresponding to distinct eigenvalues of a linear operator are linearly independent.*

PROOF. Let $T \in L(V)$ and let $x_1, \ldots, x_m$ be eigenvectors of T corresponding to distinct eigenvalues $\lambda_1, \ldots, \lambda_m$. For $j = 1, \ldots, m$, let

$$S_j = \prod_{i=1,\, i \neq j}^{m} (\lambda_i - \lambda_j)^{-1} (T - \lambda_i I).$$

Then $S_j \in L(V)$ and $S_j(x_k) = \delta_{jk} x_k$. Now if $\alpha_1, \ldots, \alpha_m$ are scalars such that $\alpha_1 x_1 + \cdots + \alpha_m x_m = 0$, then operating by S_j, we get

$$0 = S_j\left(\sum_{k=1}^{m} \alpha_k x_k\right) = \sum_{k=1}^{m} \alpha_k S_j(x_k) = \alpha_j x_j.$$

Since $x_j \neq 0$, $\alpha_j = 0$. Hence, $x_1, \ldots, x_m$ are linearly independent. ∎

COROLLARY 3.2. *Let T be a linear operator on an n–dimensional vector space V over a field K. If T has n distinct eigenvalues, then T is diagonalizable.*

COROLLARY 3.3. *Let $\lambda_1, \ldots, \lambda_k$ be distinct eigenvalues of a linear operator T. If for each i, the vectors $x_{i1}, \ldots, x_{in_i}$ are linearly independent eigenvectors of T corresponding to eigenvalue λ_i, then $x_{11}, \ldots, x_{1n_1}, \ldots, x_{k1}, \ldots, x_{kn_k}$ are linearly independent.*

PROOF. Let S_j be the linear operator as defined in Proposition 3.1. If

$$\sum_{r=1}^{n_k} \sum_{i=1}^{k} \alpha_{ir} x_{ir} = 0,$$

$\alpha_{ir} \in K$, then operating by S_j on both sides, we have $\sum_{r=1}^{n_j} \alpha_{jr} x_{jr} = 0$. Thus, $\alpha_{j1} = \cdots = \alpha_{jn_j} = 0$. ∎

Let T be a linear operator on V and let λ be an eigenvalue of T. Recall that the subspace $E(\lambda) = \ker(T - \lambda I)$ is called the eigenspace of T corresponding to λ. The dimension of $E(\lambda)$ is called the **geometric multiplicity** of λ, and the multiplicity of λ as a root of the characteristic polynomial is called the **algebraic multiplicity** of λ.

PROPOSITION 3.4. *The geometric multiplicity of an eigenvalue of a linear operator can not exceed its algebraic multiplicity.*

PROOF. Let $T \in L(V)$ and let λ be an eigenvalue of T with the geometric multiplicity r. Then there are r linearly independent eigenvectors of T corresponding to λ. Let these be $x_1, \ldots, x_r$. If $r = n$, the dimension of V, then $\mathcal{B} = \{x_1, \ldots, x_r\}$ is a basis of V and $[T]_{\mathcal{B}} = \lambda I_r$. Thus, $c_T(x) = (x - \lambda)^r$, and in this case the algebraic multiplicity of λ is r.

If $r < n$, then extend $\{x_1, \ldots, x_r\}$ to $\mathcal{B} = \{x_1, \ldots, x_r, \ldots, x_n\}$ so that it forms an ordered basis of V. The matrix representation of T with respect to $\mathcal{B}$ is of the form

$$[T]_{\mathcal{B}} = \begin{bmatrix} \lambda I_r & C \\ 0 & B \end{bmatrix},$$

and so $c_T(x) = (x-\lambda)^r \det(\lambda I_{n-r} - B)$. Hence the algebraic multiplicity of λ is at least r. ∎

Now we have the following characterization for diagonalizable linear operators. First, recall that a polynomial $p(x)$ in $K[x]$ is said to split over K if

$$p(x) = a_0 (x - a_1)^{r_1} \cdots (x - a_k)^{r_k},$$

where $a_0, a_1, \ldots, a_k \in K$ and $r_1, \ldots, r_k$ are positive integers.

THEOREM 3.5. *Let T be a linear operator on an n–dimensional vector space V over a field K and let $\lambda_1, \ldots, \lambda_k$ be all the distinct eigenvalues of T. Then the following statements are equivalent:*

(i) *T is diagonalizable;*

(ii) *$V = E(\lambda_1) \oplus \cdots \oplus E(\lambda_k)$, $E(\lambda_i) = \ker(T - \lambda_i I)$;*

(iii) *the characteristic polynomial of T splits over K and the algebraic multiplicity of each eigenvalue equals its geometric multiplicity;*

(iv) *the minimal polynomial of T is $(x - \lambda_1) \cdots (x - \lambda_k)$.*

PROOF. *(i)* ⇒ *(ii)* If T is diagonalizable, then V has an ordered basis

$$\mathcal{B} = \{x_{11}, \ldots, x_{1n_1}, \ldots, x_{k1}, \ldots, x_{kn_k}\}$$

such that each x_{ir} is an eigenvector of T corresponding to eigenvalue λ_i. Clearly, $n_1 + \cdots + n_k = n$. Thus, it is enough to show that for each i, $\{x_{i1}, \ldots, x_{in_i}\}$ is a basis of $E(\lambda_i)$. Suppose on the contrary that for some $i = 1, \ldots, k$, $\{x_{i1}, \ldots, x_{in_i}\}$ is not a basis of $E(\lambda_i)$. Then there exists $y_i \in E(\lambda_i)$ such that $\{x_{i1}, \ldots, x_{in_i}, y_i\}$ is a linearly independent subset of $E(\lambda_i)$. But then by Corollary 3.3, $\mathcal{B} \cup \{y_i\}$ is a linearly independent subset of V, a contradiction.

(ii) ⇒ *(i)* Obvious.

(i) ⇒ *(iii)* If T is diagonalizable, then there is an ordered basis $\mathcal{B}$ of V such that $[T]_{\mathcal{B}}$ is diagonal matrix with entries from K, and so $c_T(x)$ splits over K. Since $V = E(\lambda_1) \oplus \cdots \oplus E(\lambda_k)$, where each $E(\lambda_i)$ is T–invariant, so if $\mathcal{B}_i$ is a basis of $E(\lambda_i)$, then $\mathcal{B} = \cup_{i=1}^k \mathcal{B}_i$ is a basis of V and $[T]_{\mathcal{B}} = \operatorname{diag}\left([T_1]_{\mathcal{B}_1}, \ldots, [T_k]_{\mathcal{B}_k}\right)$, where T_i is the linear operator on $E(\lambda_i)$ induced by T (see Example 4, p.60). Let $\dim E(\lambda_i) = n_i$. Since $T_i v = Tv = \lambda_i v$ for all $v \in E(\lambda_i)$, $[T_i]_{\mathcal{B}_i} = \lambda_i I_{n_i}$. Hence, $c_T(x) = (x - \lambda_1)^{n_1} \cdots (x - \lambda_k)^{n_k}$, and the algebraic multiplicity of each λ_i equals its geometric multiplicity.

(iii) ⇒ *(iv)* Since $c_T(x)$ splits over K, $c_T(x) = (x - \lambda_1)^{n_1} \cdots (x - \lambda_k)^{n_k}$. Also, the geometric and the algebraic multiplicities of each eigenvalue are equal, so for each λ_i, there are n_i linearly independent eigenvectors of T: $x_{i1}, \ldots, x_{in_i}$. Therefore, $\mathcal{B} = \{x_{11}, \ldots, x_{1n_1}, \ldots, x_{k1}, \ldots, x_{kn_k}\}$ is a basis of V (Corollary 3.3), and the linear operator $\prod_{i=1}^k (T - \lambda_i I)$ maps all the basis elements of $\mathcal{B}$ to the zero vector. Hence, $(T - \lambda_1 I) \cdots (T - \lambda_k I) = 0$, and $m_T(x) = (x - \lambda_1) \cdots (x - \lambda_k)$.

(iv) ⇒ *(i)* A direct consequence of Primary decomposition theorem (Theorem 2.3). ∎

COROLLARY 3.6. *A nonzero nilpotent operator on V is not diagonalizable.*

PROOF. Let $T \in L(V)$ be a nilpotent operator. Then $c_T(x) = x^n$, $n = \dim V$. If T is diagonalizable, then by the above theorem, $m_T(x) = x$, and so by Cayley Hamilton theorem, $T = 0$. ∎

If T is diagonalizable linear operator on a finite dimensional vector space V over a field K, then its characteristic polynomial must split. Corollary 3.6 shows that the converse of this statement need not be true. For example, the differential operator D on $\mathbb{R}_m[x]$, the space of all polynomials over $\mathbb{R}$ of degree at most m, $m \geq 1$, is nilpotent.

A linear operator T on a finite dimensional vector space V over K is said to be **triangulable** if there is an ordered basis of V with respect to which T is represented by an upper triangular matrix. The matrix analogue is: an $n \times n$ matrix A is triangulable over K if there is an invertible matrix P with entries from K such that $P^{-1}AP$ is an upper triangular matrix.

Our next result is a characterization of those linear operators whose characteristic polynomials admit splitting.

THEOREM 3.7. *Let V be a finite dimensional vector space over K of dimension n and let T be a linear operator on V. Then the following statements are equivalent:*

(i) *the characteristic polynomial of T splits over K;*
(ii) *every nonzero T–invariant subspace of V contains an eigenvector of T;*
(iii) *T is triangulable.*

PROOF. *(i)* ⇒ *(ii)* Let W be a nonzero proper T–invariant subspace of V. If T_1 is a linear operator on W induced by T, then $c_{T_1}(x)$ divides $c_T(x)$, and so $c_{T_1}(x)$ splits over K. If $\lambda \in K$ is a root of $c_{T_1}(x)$, then λ is an eigenvalue of T_1, and so there is $w \in W \setminus \{0\}$ such that $T_1 w = \lambda w$. Hence, $Tw = T_1 w = \lambda w$, and w is an eigenvector of T in W.

(ii) ⇒ *(i)* Let $q(x)$ be monic irreducible factor of $c_T(x)$ in $K[x]$. Then we need to show that $q(x)$ is linear. The subspace $W = \ker q(T)$ is a T–invariant subspace of V and $W \neq \{0\}$. Therefore, W has an eigenvector of T. Let w be an eigenvector of T in W corresponding to eigenvalue λ. If T_1 is the linear operator on W induced by T, then $T_1 w = \lambda w$ and $q(T_1) = 0$. Since $q(x)$ is monic irreducible, the minimal polynomial of T_1 is $q(x)$, and λ is a root of $q(x)$. Hence, $q(x) = x - \lambda$ and $c_T(x)$ splits over K.

(i) ⇒ *(iii)* By induction on $n = \dim V$. If $n = 1$, then the statement is obviously true. Assume that the result is true for vector spaces over K of dimension at most $n - 1$. Let V be an n–dimensional vector space over K and $T \in L(V)$ such that $c_T(x)$ splits over K. Thus, T has an eigenvalue λ in K. Let x_1 be an eigenvector of T corresponding to λ, and let $W = \langle x_1 \rangle$. Then W is a T–invariant subspace of V and so T induces a linear operator $\widehat{T}$ on V/W. Since $c_{\widehat{T}}(x)$ divides $c_T(x)$, $c_{\widehat{T}}(x)$ also splits over K. Now $\dim V/W < n$, so by the induction hypothesis, V/W has an ordered basis $\overline{B} = \{\overline{x}_2, \ldots, \overline{x}_n\}$, $\overline{x}_i = x_i + W$, such that

$$\left[\widehat{T}\right]_{\overline{B}} = \begin{bmatrix} \alpha_{22} & \cdots & \cdots & \alpha_{2n} \\ 0 & \ddots & & \vdots \\ \vdots & \ddots & \ddots & \vdots \\ 0 & \cdots & 0 & \alpha_{nn} \end{bmatrix},$$

that is, for $j = 2, \ldots, n$,

$$\widehat{T}(\overline{x}_j) = \sum_{i=2}^{j} \alpha_{ij}\overline{x}_i,$$

or

$$T(x_j) = \sum_{i=2}^{j} \alpha_{ij}x_i + \alpha_{1j}x_j,\ \alpha_{1j} \in K.$$

Since $\mathcal{B} = \{x_1, x_2, \ldots, x_n\}$ is a basis of V, so if $\alpha_{11} = \lambda$, then the matrix of T with respect to $\mathcal{B}$ is

$$[T]_{\mathcal{B}} = \begin{bmatrix} \alpha_{11} & \alpha_{12} & \cdots & \cdots & \alpha_{1n} \\ 0 & \alpha_{22} & \cdots & \cdots & \alpha_{2n} \\ \vdots & \ddots & \ddots & & \vdots \\ \vdots & & \ddots & \ddots & \vdots \\ 0 & \cdots & \cdots & 0 & \alpha_{nn} \end{bmatrix}.$$

Hence, T is triangulable.

$(iii) \Rightarrow (i)$ Obvious. ∎

3.1. Examples

1. If $A \in K^{n\times n}$ is diagonalizable, then there exists a basis $\mathcal{B}$ of K^n so that the matrix of A with respect to $\mathcal{B}$ is a diagonal matrix, that is, $P^{-1}AP$ is a diagonal matrix for some invertible matrix P. We now give a method of finding this matrix P.

   Let $\lambda_1, \ldots, \lambda_n$ be eigenvalues of A, counted with multiplicities, and let $u_1, \ldots, u_n$ be corresponding eigenvectors, that is, $Au_i = \lambda_i u_i$. Since A is diagonalizable, we can choose $u_1, \ldots, u_n$ so that they are linearly independent. Let $P = \begin{bmatrix} u_1 & \cdots & u_n \end{bmatrix}$ be an $n \times n$ matrix whose i–th column is u_i. Then P is an invertible matrix and $P^{-1}AP = \operatorname{diag}(\lambda_1, \ldots, \lambda_n)$. Since for each i, the i–th column of $P^{-1}AP$ is

$$\begin{aligned} \left(P^{-1}AP\right)e_i &= P^{-1}A(Pe_i) = P^{-1}Au_i \\ &= P^{-1}(\lambda_i u_i) = \lambda_i P^{-1}(Pe_i) = \lambda_i e_i, \end{aligned}$$

   which is the i–th column of $\operatorname{diag}(\lambda_1, \ldots, \lambda_n)$.

   Consider the real 3×3 matrix

$$A = \begin{bmatrix} 4 & 2 & 2 \\ 3 & 3 & 2 \\ -3 & -1 & 0 \end{bmatrix}.$$

   This has the characteristic polynomial $(x-1)(x-2)(x-4)$. Since eigenvalues of A are distinct, A is diagonalizable, that is, A is similar to a diagonal matrix. Vectors $u_1 = \begin{bmatrix} 0 \\ 1 \\ -1 \end{bmatrix}$, $u_2 = \begin{bmatrix} 1 \\ 1 \\ -2 \end{bmatrix}$, $u_3 = \begin{bmatrix} 1 \\ 1 \\ -1 \end{bmatrix}$ are eigenvectors of A corresponding to eigenvalues 1, 2 and 4 respectively. Thus, if $P = \begin{bmatrix} u_1 & u_2 & u_3 \end{bmatrix}$, then $P^{-1}AP = \operatorname{diag}(1, 2, 4)$.

The matrix

$$B = \begin{bmatrix} 0 & 0 & 1 \\ 1 & 0 & -1 \\ 0 & 1 & 1 \end{bmatrix} \in \mathbb{R}^{3\times 3}$$

is not diagonalizable as its characteristic polynomial is $(x-1)\left(x^2+1\right)$. In fact, it is not triangulable. However, if B is considered as a matrix in $\mathbb{C}^{3\times 3}$, it is diagonalizable, and $P^{-1}AP = \operatorname{diag}(1, i, -i)$ for

$$P = \begin{bmatrix} 1 & i & -i \\ 0 & 1-i & 1+i \\ 1 & -1 & -1 \end{bmatrix}.$$

2. Linear operators S and T on V are said to be **simultaneously diagonalizable** if they have the same set of eigenvectors which form a basis of V. Equivalently, there is a basis $\mathcal{B}$ of V such that $[S]_{\mathcal{B}}$ and $[T]_{\mathcal{B}}$ are diagonal matrices.

Two diagonalizable linear operators S and T on V are simultaneously diagonalizable if and only if they commute, that is, $ST = TS$.

If S and T are simultaneously diagonalizable, then $[S]_{\mathcal{B}}$ and $[T]_{\mathcal{B}}$ are diagonal matrices with respect to some ordered basis $\mathcal{B}$ of V. Since any two diagonal matrices of the same size commute, $[ST]_{\mathcal{B}} = [S]_{\mathcal{B}}[T]_{\mathcal{B}} = [T]_{\mathcal{B}}[S]_{\mathcal{B}} = [TS]_{\mathcal{B}}$. Hence, $ST = TS$.

Conversely, assume that S and T are diagonalizable and $ST = TS$. Let $\lambda_1, \dots, \lambda_k$ be distinct eigenvalues of S. Then there is an ordered basis $\mathcal{B} = \mathcal{B}_1 \cup \cdots \cup \mathcal{B}_k$ of V such that $[S]_{\mathcal{B}} = \operatorname{diag}(\lambda_1 I_{n_1}, \dots, \lambda_k I_{n_k})$, where $|\mathcal{B}_i| = n_i$, the number of linearly independent eigenvectors of S corresponding to λ_i. Write

$$[T]_{\mathcal{B}} = \begin{bmatrix} A_{11} & \cdots & A_{1k} \\ \vdots & & \vdots \\ A_{k1} & \cdots & A_{kk} \end{bmatrix}$$

in the block form conformal with the blocks of $[S]_{\mathcal{B}}$. Since $ST = TS$, $[S]_{\mathcal{B}}[T]_{\mathcal{B}} = [T]_{\mathcal{B}}[S]_{\mathcal{B}}$, and this on multiplication, we have

$$\begin{bmatrix} \lambda_1 A_{11} & \cdots & \lambda_1 A_{1k} \\ \vdots & & \vdots \\ \lambda_k A_{k1} & \cdots & \lambda_k A_{kk} \end{bmatrix} = \begin{bmatrix} \lambda_1 A_{11} & \cdots & \lambda_k A_{1k} \\ \vdots & & \vdots \\ \lambda_1 A_{k1} & \cdots & \lambda_k A_{kk} \end{bmatrix}.$$

Thus, for $i \neq j$, $\lambda_i A_{ij} = \lambda_j A_{ij}$, that is, $A_{ij} = 0$. Hence,

$$[T]_{\mathcal{B}} = \operatorname{diag}(A_{11}, \dots, A_{kk}).$$

Now if $V_i = \langle \mathcal{B}_i \rangle$, then $V = V_1 \oplus \cdots \oplus V_k$ and the matrix representation of T with respect to $\mathcal{B}$ shows that each V_i is T–invariant. Let T_i be the linear operator induced by T on V_i. Since $m_T(T_i) = 0$, $m_{T_i}(x)$, the minimal polynomial of T_i, divides $m_T(x)$, and so each T_i is a diagonalizable linear operator on V_i. Let $\mathcal{B}_i'$ be a basis of V_i consisting of eigenvectors of T_i. Then each $[T_i]_{\mathcal{B}_i'}$ is a diagonal matrix, and so $\mathcal{B}' = \mathcal{B}_1' \cup \cdots \cup \mathcal{B}_k'$ is a basis of V and $[T]_{\mathcal{B}'} = \operatorname{diag}\left([T_1]_{\mathcal{B}_1'}, \dots, [T_k]_{\mathcal{B}_k'}\right)$, a diagonal matrix. Also, for each $x \in V_i$,

$Sx = \lambda_i x$. Therefore, $[S]_{\mathcal{B}'} = [S]_{\mathcal{B}}$. Hence, S and T are simultaneously diagonalizable.

3. Let

$$A = \begin{bmatrix} 3 & -1 & -1 \\ 1 & 1 & -1 \\ 1 & -1 & 1 \end{bmatrix} \text{ and } B = \begin{bmatrix} 3 & -2 & -2 \\ 1 & 0 & -2 \\ 3 & -3 & -1 \end{bmatrix}.$$

Since $m_A(x) = (x-1)(x-2)$ and $m_B(x) = (x^2-1)(x-2)$, A and B are diagonalizable matrices. Also $AB = BA$. The above example shows that there exists an invertible matrix P such that $P^{-1}AP$ and $P^{-1}BP$ are diagonal matrices. We now find this matrix P.

The matrix A has linearly independent eigenvectors $\begin{bmatrix} 1 \\ 1 \\ 0 \end{bmatrix}, \begin{bmatrix} 2 \\ 1 \\ 1 \end{bmatrix}$ and $\begin{bmatrix} 1 \\ 1 \\ 1 \end{bmatrix}$ corresponding to eigenvalues 2, 2 and 1. Therefore, for

$$P_1 = \begin{bmatrix} 1 & 2 & 1 \\ 1 & 1 & 1 \\ 0 & 1 & 1 \end{bmatrix},$$

$P_1^{-1}AP_1 = \operatorname{diag}(2,2,1) = \operatorname{diag}(2I_2, 1)$, and $P_1^{-1}BP_1 = \operatorname{diag}(B_1, -1)$, where $B_1 = \begin{bmatrix} 3 & 2 \\ -1 & 0 \end{bmatrix}$.

Note that both these matrices are in block diagonal form and having blocks of the same sizes respectively. Now if we consider B_1, then its eigenvalues are 1 and 2 and $\begin{bmatrix} 1 \\ -1 \end{bmatrix}, \begin{bmatrix} 2 \\ -1 \end{bmatrix}$ are corresponding eigenvectors. Thus, if $Q_2 = \begin{bmatrix} 1 & 2 \\ -1 & -1 \end{bmatrix}$, then $Q_2^{-1}B_1Q_2 = \operatorname{diag}(1,2)$, and so for $P_2 = \operatorname{diag}(Q_2, 1)$:

$$\begin{aligned} P_2^{-1}\left(P_1^{-1}BP_1\right)P_2 &= \operatorname{diag}\left(Q_2^{-1}, 1\right)\operatorname{diag}(B_1, -1)\operatorname{diag}(Q_2, 1) \\ &= \operatorname{diag}\left(Q_2^{-1}B_1Q_2, -1\right) = \operatorname{diag}(1, 2, -1), \end{aligned}$$

and

$$\begin{aligned} P_2^{-1}\left(P_1^{-1}AP_1\right)P_2 &= \operatorname{diag}\left(Q_2^{-1}, 1\right)\operatorname{diag}(2I_2, 1)\operatorname{diag}(Q_2, 1) \\ &= \operatorname{diag}(2I_2, 1). \end{aligned}$$

Hence, if $P = P_1P_2$, then $P^{-1}AP$ and $P^{-1}BP$ are diagonal matrices.

4. Let S and T be diagonalizable linear operators on V and let $ST = TS$. Assume that all the eigenvalues of T are distinct. Let $\mathcal{B}$ be a basis of V consisting of eigenvectors of T corresponding to distinct eigenvalues $\lambda_1, \ldots, \lambda_n$. Then $[T]_{\mathcal{B}} = \operatorname{diag}(\lambda_1, \ldots, \lambda_n)$. We have seen in Example 3 that $[S]_{\mathcal{B}}$ will also have the same block diagonal form, so $[S]_{\mathcal{B}}$ is also a diagonal matrix. Let $[S]_{\mathcal{B}} = \operatorname{diag}(\mu_1, \ldots, \mu_n)$. The following system of linear equations in

unknowns $x_1, \ldots, x_n$:

$$\begin{aligned} x_1 + \lambda_1 x_2 + \cdots + \lambda_1^{n-1} x_n &= \mu_1 \\ &\vdots \\ x_1 + \lambda_n x_2 + \cdots + \lambda_n^{n-1} x_n &= \mu_n \end{aligned}$$

will have a unique solution because its coefficient matrix is the Vandermonde matrix

$$\begin{bmatrix} 1 & \lambda_1 & \cdots & \lambda_1^{n-1} \\ \vdots & \vdots & & \vdots \\ 1 & \lambda_n & \cdots & \lambda_n^{n-1} \end{bmatrix}$$

and $\lambda_1, \ldots, \lambda_n$ are all distinct. Let $x_1 = a_0, \ldots, x_n = a_{n-1}$ be the solution of this system and let $p(x) = a_0 + a_1 x + \cdots + a_{n-1} x^{n-1} \in K[x]$. Then $p(\lambda_i) = \mu_i$ for $i = 1, \ldots, n$, and so $[S]_{\mathcal{B}} = p([T]_{\mathcal{B}}) = [p(T)]_{\mathcal{B}}$. Hence, $S = p(T)$. This proves the following:

if S and T are commuting diagonalizable linear operators on an n-dimensional vector space V and all eigenvalues of T are distinct, then there is a polynomial $p(x)$ in $K[x]$ of degree at most $n-1$ such that $S = p(T)$.

In the above example, the matrix

$$B = \begin{bmatrix} 3 & -2 & -2 \\ 1 & 0 & -2 \\ 3 & -3 & -1 \end{bmatrix}$$

has distinct eigenvalues 1, 2, -1. If

$$P = \begin{bmatrix} 1 & 0 & 1 \\ 1 & 1 & 1 \\ 0 & -1 & 1 \end{bmatrix},$$

whose columns are eigenvectors corresponding to eigenvalues 1, 2, -1, respectively, then $P^{-1}BP = \operatorname{diag}(1, 2, -1)$. Since $AB = BA$ and A is diagonalizable, $P^{-1}AP$ will also be a diagonal matrix.

Now, as above, on solving the following system

$$\begin{aligned} x_1 + 1x_2 + 1^2 x_3 &= 2, \\ x_1 + 2x_2 + 2^2 x_3 &= 2, \\ x_1 + (-1)\, x_2 + (-1)^2\, x_3 &= 1, \end{aligned}$$

we have $x_1 = 5/3$, $x_2 = 1/2$, $x_3 = -1/6$. Therefore

$$A = \frac{5}{3} I_3 + \frac{1}{2} B - \frac{1}{6} B^2.$$

5. The result proved in Example 2 can be extended to finite number of commuting diagonalizable linear operators on V. In fact, it is actually true for an arbitrary class of commuting linear operators on V. The proof of this is by induction on dimension of V.

   A commuting family $\mathcal{F}$ of diagonalizable linear operators on V is simultaneously diagonalizable, that is, there is a basis $\mathcal{B}$ of V such that $[T]_{\mathcal{B}}$ is a diagonal matrix for every T in $\mathcal{F}$.

The result is obviously true if $\dim V = 1$. Assume that the statement is true for all vector spaces of dimension less than n. Choose any $T \in \mathcal{F}$ which is not a scalar multiple of the identity operator. Let $\lambda_1, \ldots, \lambda_k$ be distinct eigenvalues of T. Then $V = E(\lambda_1) \oplus \cdots \oplus E(\lambda_k)$, where each $E(\lambda_i) = \ker(T - \lambda_i I)$. Since for any $S \in \mathcal{F}$, $ST = TS$, each $E(\lambda_i)$ is also invariant under S. Let $\mathcal{F}_i$ be the family of linear operators on $E(\lambda_i)$ obtained by restricting the operators in $\mathcal{F}$ to $E(\lambda_i)$. Clearly, $\mathcal{F}_i$ is a commuting family, and for each $S_i \in \mathcal{F}_i$, since the minimal polynomial of S_i divides the minimal polynomial of S, where $S_i = S|_{E(\lambda_i)}$, each S_i is diagonalizable. Now $\dim E(\lambda_i) < \dim V$, so the operators in $\mathcal{F}_i$ can be diagonalized simultaneously.. Thus, there is a basis $\mathcal{B}_i$ of $E(\lambda_i)$ such that for every $S_i \in \mathcal{F}_i$, $[S_i]_{\mathcal{B}_i}$ is a diagonal matrix. Let $\mathcal{B} = \mathcal{B}_1 \cup \cdots \cup \mathcal{B}_k$. Then $[S]_{\mathcal{B}}$ is clearly a diagonal matrix for $S \in \mathcal{F}$.

6. Let $A \in K^{n \times n}$, λ_1 an eigenvalue of A and u_1 the corresponding eigenvector. Let $\{u_1, \ldots, u_n\}$ be any ordered basis of K^n with first element u_1. Then the matrix $P_1 = \begin{bmatrix} u_1 & \cdots & u_n \end{bmatrix}$ is invertible and the first column of $P_1^{-1}AP_1$ is:

$$\left(P_1^{-1}AP_1\right)e_1 = P_1^{-1}Au_1 = \lambda_1 P_1^{-1}u_1 = \lambda_1 P_1^{-1}(P_1 e_1) = \lambda_1 e_1.$$

Thus the matrix $P_1^{-1}AP_1$ is of the form

$$P_1^{-1}AP_1 = \begin{bmatrix} \lambda_1 & v_1^t \\ 0 & A_2 \end{bmatrix}, \; v_1 \in K^{n-1}, \; A_2 \in K^{(n-1)\times(n-1)}.$$

If A_2 is an upper triangular matrix, then this shows that A is similar to an upper triangular matrix. Otherwise, if A_2 has an eigenvalue λ_2, then similar arguments give an invertible matrix Q_2 such that

$$Q_2^{-1}A_2Q_2 = \begin{bmatrix} \lambda_2 & v_2^t \\ 0 & A_3 \end{bmatrix}, \; v_2 \in K^{n-2}, \; A_3 \in K^{(n-2)\times(n-2)}.$$

Now let $P_2 = \begin{bmatrix} 1 & 0 \\ 0 & Q_2 \end{bmatrix}$. Then

$$\begin{aligned}(P_1P_2)^{-1}A(P_1P_2) &= \begin{bmatrix} 1 & 0 \\ 0 & Q_2^{-1} \end{bmatrix}\begin{bmatrix} \lambda_1 & v_1^t \\ 0 & A_2 \end{bmatrix}\begin{bmatrix} 1 & 0 \\ 0 & Q_2 \end{bmatrix} \\ &= \begin{bmatrix} \lambda_1 & v_1^tQ_2 \\ 0 & Q_2^{-1}A_2Q_2 \end{bmatrix}.\end{aligned}$$

This process can be continued further if A_3 has an eigenvalue and A_3 is not an upper triangular matrix. In particular, if A is triangulable, by this method we can obtain an invertible matrix P such that $P^{-1}AP$ is an upper triangular matrix.

We illustrate this procedure by considering the following 4×4 matrix with entries from $\mathbb{R}$:

$$A = \begin{bmatrix} 1 & 0 & 0 & 0 \\ 0 & 1 & -1 & 1 \\ -1 & 0 & 2 & 0 \\ 0 & 0 & 0 & 2 \end{bmatrix}.$$

Clearly, $Ae_2 = e_2$, that is, e_2 is an eigenvector of A corresponding to eigenvalue 1. We extend e_2 to form a basis of $\mathbb{R}^4$. One such ordered basis is

$\{e_2, e_1, e_3, e_4\}$. Thus for $P_1 = \begin{bmatrix} e_2 & e_1 & e_3 & e_4 \end{bmatrix}$, $P_1^{-1} = P_1$ and

$$P_1^{-1} A P_1 = \begin{bmatrix} 1 & 1 & -1 & 1 \\ 0 & 1 & 0 & 0 \\ 0 & -1 & 2 & 0 \\ 0 & 0 & 0 & 2 \end{bmatrix}.$$

Now if $A_2 = \begin{bmatrix} 1 & 0 & 0 \\ -1 & 2 & 0 \\ 0 & 0 & 2 \end{bmatrix}$, then A_2 has an eigenvalue 2 with eigenvector e_2 ($\in \mathbb{R}^3$). Thus we can take $Q_2 = \begin{bmatrix} e_2 & e_1 & e_3 \end{bmatrix} \in \mathbb{R}^{3\times 3}$, and so

$$Q_2^{-1} A_2 Q_2 = \begin{bmatrix} 2 & -1 & 0 \\ 0 & 1 & 0 \\ 0 & 0 & 2 \end{bmatrix}.$$

Now for $P_2 = \operatorname{diag}(1, Q_2) = \begin{bmatrix} 1 & 0 & 0 & 0 \\ 0 & 0 & 1 & 0 \\ 0 & 1 & 0 & 0 \\ 0 & 0 & 0 & 1 \end{bmatrix}$, we have

$$(P_1 P_2)^{-1} A (P_1 P_2) = P_2^{-1} \left(P_1^{-1} A P_1\right) P_2 = \begin{bmatrix} 1 & -1 & 1 & 1 \\ 0 & 2 & -1 & 0 \\ 0 & 0 & 1 & 0 \\ 0 & 0 & 0 & 2 \end{bmatrix}.$$

Since this matrix is an upper triangular matrix, we stop here. The matrix $P = P_1 P_2 = \begin{bmatrix} e_2 & e_3 & e_1 & e_4 \end{bmatrix}$ triangularizes A.

7. Linear operators S and T are said to be **simultaneously triangulable** if there is an ordered basis $\mathcal{B}$ of V such that $[S]_{\mathcal{B}}$ and $[T]_{\mathcal{B}}$ are upper triangular matrices.

   If S and T are triangulable linear operators on V and they commute, then S and T can be triangularized simultaneously. Two upper triangular matrices of the same size may not commute. Thus, two simultaneously triangulable linear operators may not be commuting.

   Let λ be an eigenvalue of T and let $U = \ker(T - \lambda I)$. Then, for $u \in U$, $(T - \lambda I) Su = S(T - \lambda I) u = 0$. Thus U is S-invariant subspace of V. Let x_1 be an eigenvector of S in U. Then for some $\mu \in K$, $Sx_1 = \mu x_1$, and also $Tx_1 = \lambda x_1$. If $W = \langle x_1 \rangle$, then W is invariant under S as well as under T. Let S' and T' be linear operators on V/W induced by S and T respectively. Then S' and T' are triangulable and $S'T' = T'S'$. Since $\dim V/W < \dim V$, by induction there is an ordered basis $\mathcal{B}'$ of V/W such that $[T']_{\mathcal{B}'}$ and $[S']_{\mathcal{B}'}$ are upper triangular matrices. Let $\{x_2, \ldots, x_n\}$ be an ordered subset of V such that $\mathcal{B}' = \{x_2 + W, \ldots, x_n + W\}$. Then for $\mathcal{B} = \{x_1, \ldots, x_n\}$, $[T]_{\mathcal{B}}$ and $[S]_{\mathcal{B}}$ are upper triangular matrices.

   An obvious modification in the above discussion will actually prove the following result.

   A commuting family of triangulable linear operators can be triangularized simultaneously.

8. Let T be a linear operator on V and let $p(x) = \prod_{i=1}^{k} (x - \lambda_i)^{m_i}$, $\lambda_i \in K$, $m_i > 0$ be the minimal polynomial of T. Then using partial fractions we can write

$$\frac{1}{p(x)} = \sum_{i=1}^{k} \sum_{j=1}^{m_i} \frac{a_{ij}}{(x - \lambda_i)^j}, \ a_{ij} \in K.$$

Thus, if $p_j(x) = \prod_{i=1, i \neq j}^{k} (x - \lambda_i)^{m_i}$, then we have

$$1 = \sum_{i=1}^{k} \left(\sum_{j=1}^{m_i} a_{ij} (x - \lambda_i)^{m_i - j} \right) p_i(x).$$

Let $q_i(x) = \sum_{j=1}^{m_i} a_{ij} (x - \lambda_i)^{m_i - j}$, then

(*) $$\sum_{i=1}^{k} p_i(x) q_i(x) = 1.$$

Now for each i, define $E_i = p_i(T) q_i(T)$. Then $E_i \in L(V)$ and satisfies the following:

$$E_1 + \cdots + E_k = I,$$
$$E_i E_j = 0 \text{ for } i \neq j,$$
$$E_i^2 = E_i,$$

that is, $E_1 + \cdots + E_k = I$ is a resolution of identity (Exercise 12, p.55).

It follows directly from (*) that $E_1 + \cdots + E_k = I$. For $i \neq j$, $p_i(x) p_j(x)$ is a multiple of $p(x)$, and so $p_i(T) p_j(T) = 0$. Hence,

$$\begin{aligned} E_i E_j &= (p_i(T) q_i(T)) (p_j(T) q_j(T)) \\ &= (p_i(T) p_j(T)) (q_i(T) q_j(T)) = 0. \end{aligned}$$

Now $E_i = E_i I = E_i (E_1 + \cdots + E_k) = E_i^2$.

Since for each i, $E_i^2 = E_i$, the minimal polynomial of all these E_i's is $x(x-1)$. Thus, $\{E_1, \ldots, E_k\}$ is a commuting set of diagonalizable linear operators on V. Therefore there is an ordered basis $\mathcal{B}$ of V such that for each i:

$$[E_i]_{\mathcal{B}} = \text{diag}\left(0_{n_1}, \ldots, 0_{n_{i-1}}, I_{n_i}, 0_{n_{i+1}}, \ldots, 0_{n_k}\right),$$

where 0_{n_r} is the $n_r \times n_r$ zero matrix and for $j = 1, \ldots, k$, n_j is the geometric multiplicity of eigenvalue 1 of E_j.

Let $D = \sum_{i=1}^{k} \lambda_i E_i$. Since $[D]_{\mathcal{B}} = \text{diag}(\lambda_1 I_{n_1}, \ldots, \lambda_{n_k} I_{n_k})$, D is a diagonalizable linear operator on V. Let $N = T - D$. Then $NE_i = TE_i - DE_i = (T - \lambda_i I) E_i$, and so $(NE_i)^{m_i} = (T - \lambda_i I)^{m_i} E_i = (T - \lambda_i I)^{m_i} p_i(T) q_i(T) = p(T) q_i(T) = 0$. Thus, if $r = \max\{m_1, \ldots, m_k\}$, then

$$N^r = N^r (E_1 + \cdots + E_k) = (NE_1)^r + \cdots + (NE_k)^r = 0,$$

that is, N is a nilpotent linear operator on V. Also as N and D are polynomials in T, $ND = DN$. This proves the following statement.

> *If T is a triangulable linear operator on V, then $T = D + N$, where D is diagonalizable and N is a nilpotent linear operator on V, and $ND = DN$. Moreover, D and N are uniquely determined by these properties.*

For the uniqueness part, suppose that we also have $T = D' + N'$, where D' is diagonalizable and N' is nilpotent, and $N'D' = D'N'$. Then $D'T = D'(D' + N') = TD'$. Since D' commutes with T and D is a polynomial in T, $DD' = D'D$. Therefore, D and D' are simultaneously diagonalizable and $D - D'$ is also a diagonalizable linear operator on V. Now $NN' = (T - D)(T' - D') = N'N$. Since N and N' are nilpotent and $NN' = N'N$, so $N' - N$ is nilpotent. But $T = D + N = D' + N'$ implies that $D - D' = N' - N$, that is, $D - D'$ is diagonalizable and nilpotent. Hence, $D - D' = 0$ or $D = D'$ and $N = N'$.

If we take

$$A = \begin{bmatrix} 1 & 0 & 0 & 2 \\ 1 & 2 & 0 & -2 \\ 0 & 0 & -1 & 1 \\ 0 & 0 & 0 & -1 \end{bmatrix},$$

then $m_A(x) = (x+1)^2(x-1)(x-2)$. For $\lambda_1 = -1$, $\lambda_2 = 1$, $\lambda_3 = 2$, we have

$$E_1 = \begin{bmatrix} 0 & 0 & 0 & -1 \\ 0 & 0 & 0 & 1 \\ 0 & 0 & 1 & 0 \\ 0 & 0 & 0 & 1 \end{bmatrix}, E_2 = \begin{bmatrix} 1 & 0 & 0 & 1 \\ -1 & 0 & 0 & -1 \\ 0 & 0 & 0 & 0 \\ 0 & 0 & 0 & 0 \end{bmatrix}, E_3 = \begin{bmatrix} 0 & 0 & 0 & 0 \\ 1 & 1 & 0 & 0 \\ 0 & 0 & 0 & 0 \\ 0 & 0 & 0 & 0 \end{bmatrix}.$$

The matrix $D = -E_1 + E_2 + 2E_3 = \begin{bmatrix} 1 & 0 & 0 & 2 \\ 1 & 2 & 0 & -2 \\ 0 & 0 & -1 & 0 \\ 0 & 0 & 0 & -1 \end{bmatrix}$ is diagonalizable and $N = A - D = \begin{bmatrix} 0 & 0 & 0 & 0 \\ 0 & 0 & 0 & 0 \\ 0 & 0 & 0 & 1 \\ 0 & 0 & 0 & 0 \end{bmatrix}$ is nilpotent, and $A = D + N$.

3.2. Exercises. In the following exercises, V is a finite dimensional vector space over K.

1. Verify that for each of the following matrices, 1 is an eigenvalue, and find its geometric and algebraic multiplicities.

$$\begin{bmatrix} 3 & -2 & 2 \\ 10 & -9 & 10 \\ 6 & -6 & 7 \end{bmatrix}, \quad \begin{bmatrix} 1 & 2 & 2 \\ 2 & 1 & -2 \\ 2 & -2 & 1 \end{bmatrix},$$

$$\begin{bmatrix} 3 & 0 & 0 & -1 \\ -2 & 1 & 1 & 0 \\ 2 & 1 & 1 & -1 \\ 6 & 0 & -1 & 1 \end{bmatrix}, \begin{bmatrix} 4 & -9 & -18 & -9 \\ -2 & 8 & 14 & 7 \\ 3 & -10 & -19 & -10 \\ -3 & 10 & 20 & 11 \end{bmatrix}.$$

2. Determine which of the matrices given in Exercise 2, p.84, are diagonalizable.
3. Give an example of a 3×3 matrix which is not triangulable over $\mathbb{R}$, but is diagonalizable over $\mathbb{C}$.

4. Prove that for $a, b \in K^n$, the matrix ab^t is diagonalizable.

5. Prove that $T \in L(V)$ is nilpotent if and only if 0 is the only eigenvalue of T.

6. Deduce from Exercise 9, p.75, that a 2×2 symmetric matrix is diagonalizable.

7. Let

$$A = \begin{bmatrix} a & b & \cdots & b \\ b & a & \ddots & \vdots \\ \vdots & \ddots & \ddots & b \\ b & \cdots & b & a \end{bmatrix} \in K^{n\times n}.$$

Prove that A is diagonalizable. Obtain an invertible matrix P such that $P^{-1}AP$ is diagonal.

8. For each of the following matrices A, find a matrix P such that $P^{-1}AP$ is upper triangular, or diagonal if A is diagonalizable.

$$\begin{bmatrix} 2 & -1 & 2 \\ 3 & -2 & -4 \\ 1 & -1 & 0 \end{bmatrix}, \begin{bmatrix} 0 & 1 & 2 \\ -1 & 0 & 1 \\ -2 & 1 & 0 \end{bmatrix}, \begin{bmatrix} 8 & 7 & -2 \\ -10 & -9 & 3 \\ -9 & -9 & 4 \end{bmatrix}, \begin{bmatrix} -6 & 1 & 3 \\ 11 & -1 & -5 \\ -24 & 3 & 11 \end{bmatrix},$$

$$\begin{bmatrix} -1 & -3 & 0 & 3 \\ -2 & -6 & 0 & 13 \\ 0 & -3 & 1 & 3 \\ -1 & -4 & 0 & 8 \end{bmatrix}, \begin{bmatrix} 1 & 1 & 0 & 3 \\ 0 & 1 & 0 & 0 \\ 0 & 5 & 1 & -14 \\ 0 & 0 & 0 & 1 \end{bmatrix}.$$

9. Find two matrices B and C with $C \neq -B$ such that $C^2 = A = B^2$, where

$$A = \begin{bmatrix} -3 & 1 & 0 \\ 6 & -3 & 2 \\ 8 & -6 & 5 \end{bmatrix}.$$

Hint: A is diagonalizable.

10. Give an example to show that the sum of two triangulable linear operators is not always a triangulable linear operator.

11. $S, T \in L(V)$ are called similar if there are bases $\mathcal{B}$ and $\mathcal{B}'$ of V such that $[S]_{\mathcal{B}'} = [T]_{\mathcal{B}}$. Prove that two diagonalizable linear operators are similar if and only if they have the same characteristic polynomial.

12. Give an example of two diagonalizable linear operators which have the same minimal polynomial and are not similar.

13. Prove that if $T \in L(V)$ is diagonalizable, then $V = \ker T \oplus \operatorname{im} T$.

14. Let λ be an eigenvalue of $T \in L(V)$ with algebraic multiplicity m. Prove that T is not diagonalizable if $\operatorname{rank}(T - \lambda I) > n - m$, where $n = \dim V$.

15. Prove that if $S, T \in L(V)$ and are simultaneously triangulable, then every eigenvalue of $ST - TS$ must be 0.

16. Prove that projection and refection operators on V are diagonalizable.

17. Let $T \in L(V)$. Prove that T is diagonalizable if and only if $T = \lambda_1 P_1 + \cdots + \lambda_k P_k$, for some $\lambda_1, \ldots, \lambda_k$ distinct elements of K and $P_1 + \cdots + P_k = I$ is a resolution of identity.
Hint: if $\mathcal{B}$ is a basis of V such that $[T]_{\mathcal{B}} = \operatorname{diag}(\lambda_1 I_{n_1}, \ldots, \lambda_k I_{n_k})$, $n_1 + \cdots + n_k = \dim V$, then define $P_i \in L(V)$ such that

$$[P_i]_{\mathcal{B}} = \operatorname{diag}\left(0_{n_1}, \ldots, 0_{n_{i-1}}, I_{n_i}, 0_{n_i}, \ldots, 0_{n_k}\right),$$

0_r is $r \times r$ zero matrix.

18. Let T be a triangulable linear operator on V. Using primary decomposition theorem, show that there is a basis $\mathcal{B}$ of V such that $[T]_{\mathcal{B}} = \operatorname{diag}(A_1, \ldots, A_k)$, where each A_i is an $n_i \times n_i$ upper triangular matrix with λ_i on the diagonal; $\lambda_1, \ldots, \lambda_k$ are distinct eigenvalues of T.

19. Let

$$A = \begin{bmatrix} \mu_1 & & * \\ & \ddots & \\ 0 & & \mu_k \end{bmatrix} \in K^{n \times n}$$

be an upper triangular matrix. For $\alpha \in K$, compute $Q^{-1}AQ$ where $Q = I_n + \alpha E_{ij}$. Conclude that if $[A]_{ij} \neq 0$ and $\mu_i \neq \mu_j$, then we can choose α so that the (i,j)–th entry of $Q^{-1}AQ$ is 0. Now conclude the matrix if $A \in K^{n \times n}$ is triangulable with $\lambda_1, \ldots, \lambda_k$ distinct eigenvalues, then A is similar to the block diagonal matrix $\operatorname{diag}(A_1, \ldots, A_k)$, where each A_i is an $n_i \times n_i$ upper triangular matrix with λ_i on the diagonal and $n_1 + \cdots + n_k = n$.

20. Let $A, B \in K^{n \times n}$ be triangulable matrices. The matrix equation $AX - XB = C$ ($C \in K^{n \times n}$) has a unique solution if and only if A and B have no common eigenvalues.
Hint: $AX - XB = C$ has a unique solution if and only if $\lambda_A - \rho_B$ is invertible operator on $K^{n \times n}$ (see Example 4, p.43). If α and β are eigenvalues of A and B and $x, y \in K^{n \times n}$ are eigenvectors of A and B^t corresponding to α and β, then xy^t is an eigenvector of $\lambda_A - \rho_B$ corresponding to $\alpha - \beta$.

21. Let $A, B \in \mathbb{C}^{n \times n}$. Show that if A and B have no common eigenvalues then matrices $\begin{bmatrix} A & C \\ 0 & B \end{bmatrix}$ and $\begin{bmatrix} A & 0 \\ 0 & B \end{bmatrix}$, $C \in \mathbb{C}^{n \times n}$, are similar.
Hint: use Exercise 20.

22. Let σ be a bijective mapping on $\{1, \ldots, n\}$ and let $A \in K^{n \times n}$. Define $A(\sigma) \in K^{n \times n}$ such that $[A(\sigma)]_{ij} = [A]_{\sigma(i)\sigma(j)}$, $i, j = 1, \ldots, n$. Prove that A and $A(\sigma)$ are similar matrices. Obtain an invertible matrix P such that $A(\sigma) = P^{-1}AP$.

4. The Jordan form

In the previous section, we have shown that if the characteristic polynomial of a linear operator defined on a finite dimensional vector space splits then its matrix representation with respect to a certain basis is an upper triangular matrix. In this section we show that in fact a basis can be found so that the representation is almost a diagonal matrix.

Let T be a linear operator on a finite dimensional vector space V over a field K, λ be an eigenvalue of T and for each positive integer r, let $E_r(\lambda) = \ker(T - \lambda I)^r$. Since V is finite dimensional, there is a positive integer p such that

$$\{0\} = E_0(\lambda) \subset E_1(\lambda) \subset \cdots \subset E_p(\lambda) = E_{p+1}(\lambda) = \cdots .$$

Clearly, $E_1(\lambda)$ is the eigenspace of T associated with λ. The subspace $E_r(\lambda)$, $1 \le r \le p$, in the above chain is referred to as a **generalized eigenspace** of order r associated with λ. A vector $x \in E_r(\lambda) \setminus E_{r-1}(\lambda)$ is called a **generalized eigenvector** of T of order r corresponding to eigenvalue λ. In particular, a generalized eigenvector of order 1 is a usual eigenvector. Note that if x is a generalized eigenvector of T corresponding to λ of order r, $r > 1$, then $(T - \lambda I)x$ is a generalized eigenvector of T of order $r - 1$, and $(T - \lambda I)^{r-1} x$ is an eigenvector of T corresponding to λ.

A sequence of nonzero vectors $x_1, \dots, x_k$, $k \ge 1$, is called a **Jordan chain** for T of length k associated with eigenvalue λ if

$$\begin{aligned} Tx_1 &= \lambda x_1, \\ Tx_2 &= \lambda x_2 + x_1, \\ &\ \vdots \\ Tx_k &= \lambda x_k + x_{k-1}. \end{aligned}$$

An eigenvector is itself a Jordan chain of length 1. Note that if $x_1, \dots, x_k$ is a Jordan chain for T associated with eigenvalue λ, then each x_r $(1 \le r \le k)$ is a generalized eigenvector of T of order r.

PROPOSITION 4.1. *A Jordan chain consists of linearly independent vectors.*

PROOF. Let $x_1, \dots, x_k$ be a Jordan chain for T associated with eigenvalue λ. Let $\alpha_1 x_1 + \cdots + \alpha_k x_k = 0$, $\alpha_i \in K$, and let r be the largest index such that $\alpha_r \neq 0$. Then $r > 1$ and $x_r = \sum_{i=1}^{r-1} \left(-\alpha_r^{-1}\alpha_i\right) x_i$. Now operating by the linear operator $(T - \lambda I)^{r-1}$ on both sides, we have $x_1 = (T - \lambda I)^{r-1} x_r = \sum_{i=1}^{r-1} \left(-\alpha_r^{-1}\alpha_i\right)(T - \lambda I)^{r-1} x_i = 0$, a contradiction. ∎

It is obvious from the above proposition that in a finite dimensional vector space V the length of any Jordan chain can not exceed the dimension of V. If $\mathcal{B}' = \{x_1, \dots, x_k\}$ is a Jordan chain for T associated with eigenvalue λ, then $J = \langle x_1, \dots, x_k \rangle$ is a k-dimensional T-invariant subspace of V, called a **Jordan subspace** of V. If T' is the restriction of T to J, then T' is a linear operator on J, and it is easy to verify that

$$[T']_{\mathcal{B}'} = \begin{bmatrix} \lambda & 1 & & \\ & \ddots & \ddots & \\ & & \ddots & 1 \\ & & & \lambda \end{bmatrix},$$

a $k \times k$ matrix with λ on the diagonal, 1 on the superdiagonal and all other entries zero. This matrix is called a Jordan block of order k with eigenvalue λ, denoted by $J_k(\lambda)$. Clearly, $\lambda I_k - J_k(\lambda)$ is a nilpotent matrix of index k (Example 3, p.80).

Therefore, if V has a basis which is a disjoint union of Jordan chains for T, then the matrix representation of T with respect to this basis is a block diagonal matrix with Jordan blocks on the diagonal. We will call such a basis of V as a **Jordan basis** for T and the corresponding matrix representation as a **Jordan canonical form** for linear operator T.

THEOREM 4.2. ***Existence of Jordan canonical form.*** *Let T be a linear operator on an n-dimensional vector space V over a field K. If the characteristic polynomial of T splits over K, then V has a basis which is a disjoint union of Jordan chains for T, that is, V has a Jordan basis for T.*

PROOF. Assume first that T is nilpotent. We prove the result for nilpotent operators by induction on $n = \dim V$. If $T = 0$, in particular, $\dim V = 1$, then any basis of V is a Jordan basis for T. Assume that $T \neq 0$ and the statement is true for all vector spaces of dimension less than n.

Since T is a nonzero nilpotent operator on V, $\operatorname{im} T$ is a nonzero proper T-invariant subspace of V. Therefore, by the induction hypothesis, $\operatorname{im} T$ has a basis $\mathcal{B}'$ which is a disjoint union of Jordan chains for T', where T' is the restriction of T to $\operatorname{im} T$. Let $\mathcal{B}' = \mathcal{B}'_1 \cup \cdots \cup \mathcal{B}'_k$, and for each i, $\mathcal{B}'_i = \{x_{i1}, \ldots, x_{i\,n_i}\}$, a Jordan chain for T' of length n_i, that is, $T'x_{i1} = Tx_{i1} = 0$ and $T'x_{ij} = Tx_{ij} = x_{i\,j-1}$, $j = 2, \ldots, n_i$, $i = 1, \ldots, k$.

Now $x_{11}, \ldots, x_{k1}$ are linearly independent elements of $\ker T$. We extend it to form a basis of $\ker T$: $\{x_{11}, \ldots, x_{k1}, y_1, \ldots, y_q\}$. Also as each $x_{i\,n_i} \in \operatorname{Im} T$, so there exists $x_{i\,n_i+1} \in V$ such that $Tx_{i\,n_i+1} = x_{i\,n_i}$. Let $\mathcal{B} = \mathcal{B}_1 \cup \cdots \cup \mathcal{B}_k \cup \mathcal{B}_{k+1} \cup \cdots \cup \mathcal{B}_{k+q}$, where for $i = 1, \ldots, k$, $\mathcal{B}_i = \mathcal{B}'_i \cup \{x_{i\,n_i+1}\}$, and for $i = 1, \ldots, q$, $\mathcal{B}_{k+i} = \{y_i\}$. Then $\mathcal{B}$ is a disjoint union of Jordan chains for T. We now show that $\mathcal{B}$ is a basis of V, and then this will complete the proof for nilpotent operators.

Clearly, $|\mathcal{B}| = |\mathcal{B}'| + k + q = \dim \operatorname{im} T + \dim \ker T = \dim V$. Suppose that

$$\sum_{i=1}^{k} \sum_{j=1}^{n_i+1} \alpha_{ij} x_{ij} + \sum_{r=1}^{q} \beta_r y_r = 0, \ \alpha_{ij}, \beta_r \in K.$$

Then by operating T on both sides, we have

$$\sum_{i=1}^{k} \sum_{j=2}^{n_i+1} \alpha_{ij} x_{i\,j-1} = 0.$$

Since the vectors involved in the above linear combinations are the elements of $\mathcal{B}'$, so $\alpha_{ij} = 0$, $j = 1, \ldots, n_i$, $i = 1, \ldots, k$. Thus,

$$\sum_{i=1}^{k} \alpha_{i1} x_{i1} + \sum_{r=1}^{q} \beta_r y_r = 0.$$

Since $\{x_{11}, \ldots, x_{k1}, y_1, \ldots, y_q\}$ is a basis of $\ker T$, so $\alpha_{11} = \cdots = \alpha_{k1} = \beta_1 = \cdots = \beta_r = 0$. Hence, $\mathcal{B}$ is a basis of V.

Finally, if T is an arbitrary linear operator on V whose characteristic polynomial splits over K, then its minimal polynomial is of the form $(x - \lambda_1)^{m_1} \cdots (x - \lambda_k)^{m_k}$. By the primary decomposition theorem $V = V_1 \oplus \cdots \oplus V_k$, where $V_i = \ker(T - \lambda_i I)^{m_i}$, a T-invariant subspace of V. If T_i is the linear operator induced by T on V_i, then $S_i = T_i - \lambda_i I$ is a nilpotent linear operator on V_i. Therefore, by the first part of the proof, V_i has a basis $\mathcal{B}_i$ consisting of disjoint union of Jordan chains for S_i, and so a disjoint union of Jordan chains for T associated with eigenvalue λ_i. Hence $\mathcal{B} = \mathcal{B}_1 \cup \cdots \cup \mathcal{B}_k$ is a basis of V which is a disjoint union of Jordan chains for T. ∎

We now show that Jordan canonical form of a linear operator is essentially unique, that is, it is independent of the choice of the Jordan basis, up to the order of Jordan chains.

LEMMA 4.3. *Let T be a linear operator on a finite dimensional vector space V over K. Assume that V has a basis $\mathcal{B}$ which is a disjoint union of Jordan chains for T. Then the number of generalized eigenvectors of T in $\mathcal{B}$ corresponding to eigenvalue λ and of order up to s is* $\dim\ker(T-\lambda I)^s$.

PROOF. Let $\mathcal{B}$ be a basis which is a disjoint union of Jordan chains for T:

$$\begin{gathered} x_{11},\ldots,x_{1m_1},\\ x_{21},\ldots,x_{2m_2},\\ \vdots\\ x_{r1},\ldots,x_{rm_r},\end{gathered}$$

where the i–th chain $x_{i1},\ldots,x_{im_i}$ has length m_i and is associated with eigenvalue λ_i. Assume further that $\lambda_1=\cdots=\lambda_d=\lambda$ and $\lambda_i\neq\lambda$ for $i>d$.

For a positive integer p, define

$$\widehat{x}_{ip}=\begin{cases} x_{ip} & \text{if } p\le m_i\\ 0 & \text{if } p>m_i.\end{cases}$$

We prove by induction on s that $\ker(T-\lambda I)^s$ has a basis consisting of nonzero elements of the set $\{\widehat{x}_{ij}\mid i=1,\ldots,d;\ j=1,\ldots,s\}$. This will clearly prove the lemma.

Since $(T-\lambda I)\,x_{i1}=0$ for $i=1,\ldots,d$, $\{x_{11},\ldots,x_{d1}\}$ is a linearly independent subset of $\ker(T-\lambda I)$. Thus to prove the induction hypothesis for $s=1$, we need to verify that this set is a basis of $\ker(T-\lambda I)$.

If $v=\sum_{i=1}^{r}\sum_{j=1}^{m_i}\alpha_{ij}x_{ij}\in\ker(T-\lambda I)$, then

$$\begin{aligned} 0 &= (T-\lambda I)\,v\\ &= \sum_{i=1}^{d}\sum_{j=2}^{m_i}\alpha_{ij}x_{i\,j-1}+\sum_{i=d+1}^{r}\sum_{j=1}^{m_i}\alpha_{ij}\,(\lambda_i-\lambda)\,x_{ij}+\sum_{i=d+1}^{r}\sum_{j=2}^{m_i}\alpha_{ij}x_{i\,j-1}\\ &= \sum_{i=1}^{d}\sum_{j=2}^{m_i-1}\alpha_{i\,j+1}x_{ij}\\ &\quad+\sum_{i=d+1}^{r}\left(\sum_{j=1}^{m_i-1}\left((\lambda_i-\lambda)\,\alpha_{ij}+\alpha_{i\,j+1}\right)x_{ij}+(\lambda_i-\lambda)\,\alpha_{i\,m_i}x_{i\,m_i}\right).\end{aligned}$$

Therefore, $\alpha_{ij}=0$, for $j=2,\ldots,m_i$, $i=1,\ldots,d$, and for $j=1,\ldots,m_i$, $i=d+1,\ldots,r$. Hence, $v=\sum_{i=1}^{d}\alpha_{i1}x_{i1}$, and $\{x_{11},\ldots,x_{d1}\}$ is a basis of $\ker(T-\lambda I)$.

Now assume that the induction hypothesis is true for s. We then show that $\ker(T-\lambda I)^{s+1}$ has a basis consisting of nonzero elements of the set

$$\{\widehat{x}_{ij}\mid i=1,\ldots,d;\ j=1,\ldots,s+1\}.$$

Since the nonzero elements of this set are in $\mathcal{B}$, we only need to verify that every element of $\ker(T-\lambda I)^{s+1}$ is a linear combination of elements of this set.

Let $v = \sum_{i=1}^{r} \sum_{j=1}^{m_i} \alpha_{ij} x_{ij} \in \ker (T - \lambda I)^{s+1}$. Then

$$(T - \lambda I)\, v \in \ker (T - \lambda I)^{s} = \langle \widehat{x}_{ij} \mid i = 1, \ldots, d;\ j = 1, \ldots, s \rangle .$$

Since

$$\begin{aligned}(T - \lambda I)\, v &= \sum_{i=1}^{d} \sum_{j=1}^{m_i - 1} \alpha_{i\, j+1} x_{ij} \\ &\quad + \sum_{i=d+1}^{r} \left(\sum_{j=1}^{m_i - 1} \left((\lambda_i - \lambda)\, \alpha_{ij} + \alpha_{i\, j+1} \right) x_{ij} + (\lambda_i - \lambda)\, \alpha_{i\, m_i} x_{i\, m_i} \right),\end{aligned}$$

we have $\alpha_{ij} = 0$, for $j = 1, \ldots, m_i$, $i = d+1, \ldots, r$, and also $\alpha_{ij} = 0$ for $i = 1, \ldots, d$, $j = s+2, \ldots, m_i$, whenever $m_i > s+1$. Therefore, we have $v = \sum_{i=1}^{d} \sum_{j=1}^{s+1} \alpha_{ij} \widehat{x}_{ij}$. This proves the lemma. ∎

THEOREM 4.4. *Let V be a finite dimensional vector space over K and let T be a linear operator on V. Assume that V has a Jordan basis which is a disjoint union of Jordan chains for T. Then the number of Jordan chains for T of length m associated with eigenvalue λ is*

$$2 \dim \ker (T - \lambda I)^{m} - \dim \ker (T - \lambda I)^{m+1} - \dim \ker (T - \lambda I)^{m-1} .$$

PROOF. The number of Jordan chains for T of length at least k associated with eigenvalue λ is exactly equal to the number of generalized eigenvectors of T of order k corresponding to λ which appear in a Jordan basis. By Lemma 4.3, it is equal to $\dim \ker (T - \lambda I)^{k} - \dim \ker (T - \lambda I)^{k-1}$. Therefore, the number of Jordan chains for T of length m associated with λ is

$$\begin{aligned}&\left(\dim \ker (T - \lambda I)^{m} - \dim \ker (T - \lambda I)^{m-1} \right) \\ &- \left(\dim \ker (T - \lambda I)^{m+1} - \dim \ker (T - \lambda I)^{m} \right) \\ =\ & 2 \dim \ker (T - \lambda I)^{m} - \dim \ker (T - \lambda I)^{m+1} - \dim \ker (T - \lambda I)^{m-1} .\end{aligned}$$

∎

According to the above theorem, the number of Jordan chains for a linear operator T associated with eigenvalue λ does not depend on the choice of Jordan basis for T, it depends on T only. Therefore, in the matrix representation of T with respect to any Jordan basis, the number of Jordan blocks of a given order associated with an eigenvalue λ is the same.

Let T be a linear operator on an n-dimensional vector space V over K. Let the characteristic polynomial of T be $c_T(x) = (x - \lambda_1)^{n_1} \cdots (x - \lambda_k)^{n_k}$, where $\lambda_1, \ldots, \lambda_k$ are distinct roots in K. According to Theorem 4.2, the matrix representation of T with respect to a Jordan basis for V is given by

$$J = \operatorname{diag}(J_1, \ldots, J_k),$$

where for $i = 1, \ldots, k$,

$$J_i = \operatorname{diag}\left(J_{m(i,1)}(\lambda_i), \ldots, J_{m(i,r_i)}(\lambda_i) \right).$$

We also order a Jordan basis so that the Jordan blocks for each eigenvalue λ_i are in descending order, that is, $m(i,1) \geq \cdots \geq m(i, r_i)$. We will call such a matrix

J a **Jordan canonical form**, or simply a **Jordan matrix**, of a linear operator T. Further, for each eigenvalue λ_i, the number r_i and $m(i,1), \ldots, m(i,r_i)$, that is, the number of Jordan blocks and their sizes respectively, are uniquely determined by T only. For each eigenvalue λ_i, $m(i,1) + \cdots + m(i,r_i) = n_i$, the algebraic multiplicity of λ_i, and r_i is the geometric multiplicity of λ_i. Also, for $J_i = \operatorname{diag}\left(J_{m(i,1)}(\lambda_i), \ldots, J_{m(i,r_i)}(\lambda_i)\right)$, it is easy to verify that $(J_i - \lambda_i I_{n_i})$ is a nilpotent matrix of index $m(i,1)$. Therefore, the minimal polynomial of T is given by $m_T(x) = (x - \lambda_1)^{m(1,1)} \cdots (x - \lambda_k)^{m(k,r_1)}$.

4.1. Examples

1. Let $A \in K^{n \times n}$ and let $m_A(x) = \prod_{i=1}^{k} (x - \lambda_i)^{m_i} \in K[x]$, where $m_i > 0$ and $\lambda_1, \ldots, \lambda_k$ are distinct. By primary decomposition theorem,

$$K^n = V_1 \oplus \cdots \oplus V_k,$$

where $V_i = \ker(A - \lambda_i I_n)^{m_i}$, and each V_i is A–invariant.

Now a set of all linearly independent solutions of $(A - \lambda_i I_n)^{m_i} x = 0$ will give a basis of V_i. Clearly $\dim V_i = n_i$, the algebraic multiplicity of eigenvalue λ_i. Let $\mathcal{B}_i = \{x_{i1}, \ldots, x_{i\,n_i}\}$ be an ordered basis of V_i. Then $\mathcal{B} = \mathcal{B}_1 \cup \cdots \cup \mathcal{B}_k$ is an ordered basis of K^n. Let

$$P = \begin{bmatrix} x_{11} & \cdots & x_{1\,n_1} & \cdots & x_{k1} & \cdots & x_{k\,n_k} \end{bmatrix}$$

be an $n \times n$ matrix whose columns are from the elements of $\mathcal{B}$ respectively. Then $P^{-1}AP = \operatorname{diag}(A_1, \ldots, A_k)$, each A_i is an $n_i \times n_i$ matrix whose characteristic polynomial is $(x - \lambda_i)^{n_i}$ and the minimal polynomial is $(x - \lambda_i)^{m_i}$.

If Q_i is an invertible matrix such that $Q_i^{-1} A_i Q_i = J_i$, the Jordan form of A_i, then for $Q = \operatorname{diag}(Q_1, \ldots, Q_k)$,

$$(PQ)^{-1} A (PQ) = Q^{-1} \left(P^{-1}AP\right) Q = \operatorname{diag}(J_1, \ldots, J_k),$$

which is the Jordan form of A and the columns of PQ form an ordered basis of K^n which is a Jordan basis.

Thus, for Jordan form of A, we need to know a method of computing matrices Q_i. We now give one such method. Actually for the above procedure, it is not necessary to know the minimal polynomial of A. If $c_A(x) = \prod_{i=1}^{k} (x - \lambda_i)^{n_i}$, then as $\ker(A - \lambda_i I_n)^{m_i} = \ker(A - \lambda_i I_n)^{r}$ for all $r \geq m_i$, all linearly independent solutions of equation $(A - \lambda_i I_n)^{n_i} x = 0$ will give a basis $\mathcal{B}_i$. In fact, if λ_i is an eigenvalue of A, then m_i, the multiplicity of λ_i as a root of $m_A(x)$, is the smallest positive integer such that $\operatorname{rank}(A - \lambda_i I_n)^{m_i} = \operatorname{rank}(A - \lambda_i I_n)^{m_i + 1}$. This gives another method of computing the minimal polynomial of a triangulable linear operator, if we know all its eigenvalues.

We now determine the Jordan form of A in $\mathbb{R}^{4 \times 4}$ given by

$$A = \begin{bmatrix} -2 & 5 & 1 & 0 \\ -2 & 4 & 1 & 0 \\ -1 & 2 & 1 & 0 \\ -1 & 2 & 0 & 1 \end{bmatrix}.$$

The characteristic polynomial of A is $(x-1)^4$. The rank of the matrix

$$A - I_4 = \begin{bmatrix} -3 & 5 & 1 & 0 \\ -2 & 3 & 1 & 0 \\ -1 & 2 & 0 & 0 \\ -1 & 2 & 0 & 0 \end{bmatrix}$$

is 2. Therefore, the geometric multiplicity of eigenvalue 1 is $4-2=2$. Hence, there are two Jordan blocks in the Jordan form of A. Thus, it is either $\operatorname{diag}(J_2(1), J_2(1))$ or $\operatorname{diag}(J_3(1), J_1(1))$. The matrix

$$(A - I_4)^2 = \begin{bmatrix} -2 & 2 & 2 & 0 \\ -1 & 1 & 1 & 0 \\ -1 & 1 & 1 & 0 \\ -1 & 1 & 1 & 0 \end{bmatrix},$$

which is nonzero, and so the minimal polynomial of A can not be $(x-1)^2$. This implies that $\operatorname{diag}(J_2(1), J_2(1))$ can not be the Jordan form of A. Thus, the Jordan form of A is given by $\operatorname{diag}(J_3(1), J_1(1))$, and $Q^{-1}AQ = J$ for some invertible matrix Q. We now compute this Q.

Let $Q = \begin{bmatrix} q_1 & q_2 & q_3 & q_4 \end{bmatrix}$. Then $AQ = QJ$, and so equating the corresponding columns, we have

$$\begin{aligned} Aq_1 &= (AQ)e_1 = (QJ)e_1 = Qe_1 = q_1, \\ Aq_2 &= (AQ)e_2 = (QJ)e_2 = Q(e_1+e_2) = q_1 + q_2, \\ Aq_3 &= (AQ)e_3 = (QJ)e_3 = Q(e_2+e_3) = q_2 + q_3, \\ Aq_4 &= (AQ)e_4 = (QJ)e_4 = Qe_4 = q_4. \end{aligned}$$

The vectors q_1 and q_4 are eigenvectors of A, and q_2 and q_3 are generalized eigenvectors of orders 2 and 3 respectively. Clearly, q_4 can be any eigenvector of A. We let $q_4 = e_4$. Since $\ker(A-I_4) \subsetneq \ker(A-I_4)^2 \subsetneq \ker(A-I_4)^3 = \mathbb{R}^4$, q_3 is a generalized eigenvector of order 3, that is, $q_3 \in \mathbb{R}^4 \setminus \ker(A-I_4)^2$. Now to find a basis of $\ker(A-I_4)^2$, we solve the equation $(A-I_4)^2 x = 0$, that is,

$$\begin{bmatrix} -2 & 2 & 2 & 0 \\ -1 & 1 & 1 & 0 \\ -1 & 1 & 1 & 0 \\ -1 & 1 & 1 & 0 \end{bmatrix} \begin{bmatrix} x_1 \\ x_2 \\ x_3 \\ x_4 \end{bmatrix} = \begin{bmatrix} 0 \\ 0 \\ 0 \\ 0 \end{bmatrix}.$$

Thus, $\ker(A-I_4)^2$ has a basis

$$\left\{ u_1 = \begin{bmatrix} 1 & 1 & 0 & 0 \end{bmatrix}^t, u_2 = \begin{bmatrix} 1 & 0 & 1 & 0 \end{bmatrix}^t, u_3 = e_4 \right\}.$$

Now $q_3 \in \mathbb{R}^4 \setminus \ker(A-I_4)^2$ if and only if q_3, u_1, u_2, u_3 are linearly independent. Thus, we can take $q_3 = e_1$. Note that there are many choices for q_3, hence for q_3, we should first try with vectors like e_1, e_2, e_3 etc; the computations are easy with such vectors. Now, for $q_3 = e_1$, we have

$$\begin{aligned} q_2 &= (A-I_4)q_3 = (A-I_4)e_1 = \begin{bmatrix} -3 & -2 & -1 & -1 \end{bmatrix}^t, \\ q_1 &= (A-I_4)q_2 = (A-I_4)^2 e_1 = \begin{bmatrix} -2 & -1 & -1 & -1 \end{bmatrix}^t. \end{aligned}$$

Hence,

$$Q = \begin{bmatrix} -2 & -3 & 1 & 0 \\ -1 & -2 & 0 & 0 \\ -1 & -1 & 0 & 0 \\ -1 & -1 & 0 & 1 \end{bmatrix}.$$

We now consider another matrix:

$$A = \begin{bmatrix} 1 & 1 & -2 & -2 \\ 0 & 1 & 1 & 1 \\ 1 & 2 & 1 & 0 \\ -1 & -2 & 0 & 1 \end{bmatrix} \in \mathbb{R}^{4\times 4}.$$

$c_A(x) = (x-1)^4$ and

$$A - I_4 = \begin{bmatrix} 0 & 1 & -2 & -2 \\ 0 & 0 & 1 & 1 \\ 1 & 2 & 0 & 0 \\ -1 & -2 & 0 & 0 \end{bmatrix}$$

is of rank 3. Therefore, the geometric multiplicity of eigenvalue 1 is $4-3=1$, and the Jordan form of A will have only one Jordan block. Thus, A is similar to $J_4(1)$. Let $Q = \begin{bmatrix} q_1 & q_2 & q_3 & q_4 \end{bmatrix}$ such that $Q^{-1}AQ = J_4(1)$, that is, $AQ = QJ_4(1)$. On equating the corresponding columns of AQ and $QJ_4(1)$, we have

$$\begin{aligned} Aq_1 &= q_1, \\ Aq_i &= q_i + q_{i-1},\ i = 2,3,4. \end{aligned}$$

Now q_4 can be any element in $\mathbb{R}^4 \setminus \ker(A - I_4)^3$. On solving the system of equations $(A-I_4)^3 x = 0$, we obtain a basis of $\ker(A-I_4)^3$ given by $\{e_1, e_2, e_3 - e_4\}$. We can thus take $q_4 = e_3$. Then

$$\begin{aligned} q_3 &= (A-I_4)\,q_4 = (A-I_4)\,e_3 = \begin{bmatrix} -2 & 1 & 0 & 0 \end{bmatrix}^t, \\ q_2 &= (A-I_4)\,q_3 = (A-I_4)^2\,e_3 = e_1, \\ q_1 &= (A-I_4)\,q_2 = (A-I_4)^3\,e_3 = \begin{bmatrix} 0 & 0 & 1 & -1 \end{bmatrix}^t. \end{aligned}$$

Hence,

$$Q = \begin{bmatrix} 0 & 1 & -2 & 0 \\ 0 & 0 & 1 & 0 \\ 1 & 0 & 0 & 1 \\ -1 & 0 & 0 & 0 \end{bmatrix}.$$

If we have

$$A = \begin{bmatrix} 1 & 1 & -2 & 0 \\ 2 & 1 & 0 & 2 \\ 1 & 0 & 1 & 1 \\ 0 & -1 & 2 & 1 \end{bmatrix} \in \mathbb{R}^{4\times 4},$$

then $c_A(x) = (x-1)^4$ and the matrix

$$A - I_4 = \begin{bmatrix} 0 & 1 & -2 & 0 \\ 2 & 0 & 0 & 2 \\ 1 & 0 & 0 & 1 \\ 0 & -1 & 2 & 0 \end{bmatrix}$$

is of rank 2. Thus, there are two Jordan blocks. Also as $(A - I_4)^2 = 0$, the Jordan form of A is $J = \operatorname{diag}(J_2(1), J_2(1))$. Let $Q = \begin{bmatrix} q_1 & q_2 & q_3 & q_4 \end{bmatrix}$ such that $AQ = QJ$. Then q_1 and q_3 are eigenvectors of A and $(A - I_4)q_i = q_{i-1}$, for $i = 2, 4$. Now $\left\{u_1 = \begin{bmatrix} 0 & 2 & 1 & 0 \end{bmatrix}^t, u_2 = \begin{bmatrix} 1 & 0 & 0 & -1 \end{bmatrix}^t\right\}$ is a basis of $\ker(A - I_4)$, so q_2 be such that q_2, u_1, u_2 are linearly independent. We let $q_2 = e_1$. Then

$$q_1 = (A - I_4)q_2 = \begin{bmatrix} 0 & 2 & 1 & 0 \end{bmatrix}^t.$$

Next we should have q_4 such that q_2, q_4, u_1, u_2 are linearly independent. We let $q_4 = e_2$. Then

$$q_3 = (A - I_4)q_4 = \begin{bmatrix} 1 & 0 & 0 & -1 \end{bmatrix}^t.$$

Hence,

$$Q = \begin{bmatrix} 0 & 1 & 1 & 0 \\ 2 & 0 & 0 & 1 \\ 1 & 0 & 0 & 0 \\ 0 & 0 & -1 & 0 \end{bmatrix}.$$

2. Two triangulable $n \times n$ matrices are similar if and only if they have the same Jordan canonical form.

   If $A, B \in K^{n\times n}$ and $B = S^{-1}AS$ for some $S \in K^{n\times n}$, then both A and B have the same set of eigenvalues, counting multiplicities. Since for every eigenvalue λ, $\operatorname{rank}(A - \lambda I_n)^k = \operatorname{rank}(B - \lambda I_n)^k$ for all $k \geq 0$, it follows from Theorem 4.4 that A and B have the same Jordan form. Conversely, if A and B have the same Jordan form given by the matrix J, then there are matrices P and Q such that $P^{-1}AP = J = Q^{-1}BQ$. Hence, $(PQ^{-1})^{-1} A (PQ^{-1}) = B$, and the matrices A and B are similar.

   In fact, in view of Theorem 4.4, this result can be stated in the following form.

   Triangulable matrices A and B in $K^{n\times n}$ are similar if and only if $c_A(x) = c_B(x)$, and for every eigenvalue λ of A (hence of B), $\operatorname{rank}(A - \lambda I_n)^k = \operatorname{rank}(B - \lambda I_n)^k$ for all $k \geq 0$.

3. If $A \in K^{n\times n}$ and A is triangulable, then for every eigenvalue λ of A, $\operatorname{rank}(A - \lambda I_n)^k = \operatorname{rank}(A^t - \lambda I_n)^k$ for all $k \geq 0$. Therefore, matrices A and A^t are similar.

   In fact, if J is the Jordan form of A and $P^{-1}AP = J$, then $(P^{-1})^t J^t P^t = A^t$. If

$$Q = \begin{bmatrix} 0 & & 1 \\ & \iddots & \\ 1 & & 0 \end{bmatrix},$$

then $Q^{-1} = Q$ and $Q^{-1}JQ = J^t$. Thus, $A^t = (P^t)^{-1}(Q^{-1}AQ)P^t = (QP^t)^{-1}J(QP^t) = (QP^t)^{-1}(P^{-1}AP)(QP^t) = (PQP^t)^{-1}A(PQP^t)$.

4. Let A be a triangulable matrix in $K^{n\times n}$ and let J be its Jordan canonical form. Let D_0 be the diagonal matrix whose diagonal elements are exactly the diagonal elements of J. Let $N_0 = J - D_0$. Then N_0 is nilpotent and $J = D_0 + N_0$. Since, $A = SJS^{-1}$ for some invertible matrix S, we have $A = SD_0S^{-1} + SN_0S^{-1} = D + N$, where D is diagonalizable and N is nilpotent. Since D and N are block diagonal matrices with respective blocks of the same size and blocks in D are scalar matrices, $DN = ND$. This is another proof of the result mentioned in Example 8, p.95.

4.2. Exercises

1. Determine the Jordan canonical form for the following matrices and find the geometric multiplicity of 1.

$$\begin{bmatrix} 1&1&1&1\\0&1&1&1\\0&0&1&1\\0&0&0&1 \end{bmatrix}, \begin{bmatrix} 1&0&1&1\\0&1&0&1\\0&0&1&0\\0&0&0&1 \end{bmatrix}, \begin{bmatrix} 1&1&0&1\\0&1&1&0\\0&0&1&1\\0&0&0&1 \end{bmatrix}, \begin{bmatrix} 1&1&1&0\\0&1&1&1\\0&0&1&1\\0&0&0&1 \end{bmatrix}.$$

2. For each of the following matrices A, find an invertible matrix P such that $P^{-1}AP$ is in the Jordan canonical form.

$$\begin{bmatrix} 1&-1&-1\\1&-1&-1\\0&0&-1 \end{bmatrix}, \begin{bmatrix} 1&-1&-1\\2&-2&-2\\3&-3&-3 \end{bmatrix}, \begin{bmatrix} 0&3&0\\-3&6&0\\4&1&2 \end{bmatrix}, \begin{bmatrix} 12&-6&4\\36&-18&12\\18&-9&6 \end{bmatrix},$$

$$\begin{bmatrix} -6&1&3\\11&-1&-5\\-24&3&11 \end{bmatrix}, \begin{bmatrix} 8&7&-2\\-10&-9&3\\-9&-9&4 \end{bmatrix}, \begin{bmatrix} 8&12&-3\\-3&-5&1\\9&12&-4 \end{bmatrix},$$

$$\begin{bmatrix} 1&-3&0&3\\-2&-6&0&13\\0&-3&1&3\\-1&-4&0&8 \end{bmatrix}, \begin{bmatrix} 3&0&0&-1\\-2&1&1&0\\2&1&1&-1\\6&0&-1&-1 \end{bmatrix}, \begin{bmatrix} 4&-9&-18&-9\\-2&8&14&7\\3&-10&-19&-10\\-3&10&20&11 \end{bmatrix},$$

$$\begin{bmatrix} 1&1&0&-3\\0&1&0&0\\0&5&1&-14\\0&0&0&1 \end{bmatrix}, \begin{bmatrix} 1&-3&-2&1\\-1&-1&-2&1\\1&3&4&-1\\0&0&0&2 \end{bmatrix}, \begin{bmatrix} 1&3&7&9\\1&2&3&4\\2&5&10&13\\0&1&4&5 \end{bmatrix}.$$

3. Write all possible Jordan canonical forms if the characteristic polynomial is $(x-2)^4(x-3)^2$.

4. The characteristic polynomial of a matrix is $(x-1)^3(x-2)^2$ and the minimal polynomial is $(x-1)^2(x-2)$. Write its Jordan canonical form.

5. Show that there is no real 3×3 matrix whose minimal polynomial is x^2+1, but there is a real 2×2 as well as complex 3×3 matrix with minimal polynomial x^2+1.

6. Let λ be an eigenvalue of T, $T \in L(V)$ and V a finite dimensional vector space over K. Prove that $\operatorname{rank}(T-\lambda I) = \operatorname{rank}(T-\lambda I)^2$ if and only if the algebraic and geometric multiplicities of λ are equal.
7. Let T be a triangulable linear operator on a finite dimensional vector space V over K, and let $\lambda_1, \ldots, \lambda_k$ be distinct eigenvalues of T. Show that if m_i is the least value of k such that $\operatorname{rank}(T-\lambda I)^k = \operatorname{rank}(T-\lambda I)^{k+1}$, then $(x-\lambda_1)^{m_1} \cdots (x-\lambda_k)^{m_k}$ is the minimal polynomial of T.
8. Show that for $J_n(\lambda)$ ($\lambda \in \mathbb{R}$ or $\mathbb{C}$), and for $p(x) \in F[x]$ ($F = \mathbb{R}$ or $\mathbb{C}$),

$$p(J_n(\lambda)) = \begin{bmatrix} p(\lambda) & p^{(1)}(\lambda)/1! & p^{(2)}(\lambda)/2! & \cdots & p^{(n-1)}(\lambda)/(n-1)! \\ 0 & p(\lambda) & p^{(1)}(\lambda)/1! & \cdots & p^{(n-2)}(\lambda)/(n-2)! \\ & & \ddots & & \vdots \\ & & & \ddots & \vdots \\ & & & & p(\lambda) \end{bmatrix}.$$

9. Using the above exercise, prove the Cayley Hamilton theorem for triangulable matrices $A \in F^{n\times n}$ ($F = \mathbb{R}$ or $\mathbb{C}$), that is, $c_A(A) = 0$.

5. The rational form

Let V be a finite dimensional vector space over a field K and let T be a linear operator on V. For a vector v in V, the subspace

$$Z(v,T) = \langle T^k v \mid k = 0,1,2,\ldots \rangle$$

is called the T–**cyclic subspace** of V generated by v. Thus, w is in $Z(v,T)$ if and only if $w = p(T)v$ for some $p(x)$ in $K[x]$. Clearly, $Z(v,T)$ is the smallest T–invariant subspace of V which contains v. If the degree of the annihilator of v with respect to T, that is, $\deg m_T^v(x)$, is m then $Z(v,T)$ has a basis $\mathcal{B} = \{v, Tv, \ldots, T^{m-1}v\}$. Further, if T_1 is the linear operator on $Z(v,T)$ induced by T, then $[T_1]_{\mathcal{B}}$ is the companion matrix of $m_T^v(x)$.

PROPOSITION 5.1. *A Jordan subspace for a linear operator T is T–cyclic.*

PROOF. Let $J = \langle x_1, \ldots, x_k \rangle$ be a Jordan subspace corresponding to T, where $x_1, \ldots, x_k$ is a Jordan chain associated with eigenvalue λ. Then

$$x_{k-1} = (T-\lambda I)x_k, \ldots, x_2 = (T-\lambda I)^{k-2} x_k, x_1 = (T-\lambda I)^{k-1} x_k.$$

Using the binomial theorem, it is clear that $J \subseteq Z(x_k, T)$.

Since $Tx_1 = \lambda x_1$ and $Tx_i = \lambda x_i + x_{i-1}$, $i = 2, \ldots, k$, J is a T–invariant subspace of V which contains x_k. Therefore, $Z(x_k,T) \subseteq J$. Hence, $J = Z(x_k,T)$, a T–cyclic subspace. ∎

An immediate consequence of the above proposition and the Theorem 4.2 is that if T is a linear operator on a finite dimensional vector space V over K and the characteristic polynomial of T splits in $K[x]$, then V is a direct sum of T–cyclic subspaces. The following theorem shows that the condition of splitting of the characteristic polynomial is not essential. In fact, it is true for any linear operator on a finite dimensional vector space.

THEOREM 5.2. *Let V be a finite dimensional vector space over K and let T be a linear operator on V. Then V is a direct sum of T–cyclic subspaces.*

PROOF. By induction on $n = \dim V$. The statement is trivially true for one dimensional vector spaces. Assume that $n > 1$.

Let $y \in V$ such that $m_T^y(x) = m_T(x)$ (Proposition 2.1). Then the subspace $W = Z(y, T)$ is T–cyclic of dimension m, where $m = \deg m_T(x)$. If $W = V$, then V itself is T–cyclic and the statement is true in this case. If $W \neq V$, then we will show that there is a T–invariant subspace W' such that $V = W \oplus W'$. Then by the induction hypothesis, W' will be a direct sum of T–cyclic subspaces and this will complete the proof.

Now we construct such a subspace W'. Enlarge the basis

$$\{y_1 = y, y_2 = Ty, \ldots, y_m = T^{m-1}y\}$$

of W to an ordered basis of V: $\mathcal{B} = \{y_1, \ldots, y_m, y_{m+1}, \ldots, y_n\}$. Let $\{y_1^\bullet, \ldots, y_n^\bullet\}$ be the dual basis of $\mathcal{B}$ and let $\widehat{W} = Z(y_m^\bullet, T^\bullet)$. If $\dim \widehat{W} = m$ and $\widehat{W} \cap W^\circ = \{0\}$, then $V^\bullet = \widehat{W} \oplus W^\circ$, a direct sum of $T^\bullet$–invariant subspace of $V^\bullet$. This implies that $V = \widehat{W}^\circ \oplus W^{\circ\circ}$, a direct sum of T–invariant subspace of V. Since $W^{\circ\circ} = W$, we can take $W' = \widehat{W}^\circ$ (see Section 2.8).

To verify that $\widehat{W} \cap W^\circ = \{0\}$ and $\dim \widehat{W} = m$, all we require to show is that there is no nonzero element of the form $w = a_0 y_m^\bullet + \cdots + a_r {T^\bullet}^r y_m^\bullet$, for $a_0, \ldots, a_r \in K$, $a_r \neq 0$ and $0 \leq r \leq m-1$. Suppose that w is a nonzero element of $\widehat{W} \cap W^\circ$. Then we can write $w = a_0 y_m^\bullet + \cdots + a_r {T^\bullet}^r y_m^\bullet$, $a_0, \ldots, a_r \in K$, $a_r \neq 0$ and $0 \leq r \leq m-1$. Since $w \in W^\circ$ and W° is $T^\bullet$–invariant, ${T^\bullet}^{m-r-1} w \in W^\circ$, and so ${T^\bullet}^{m-r-1} w(y) = 0$. But

$$\begin{aligned} {T^\bullet}^{m-r-1}(w(y)) &= \left(a_0 {T^\bullet}^{m-r-1} y_m^\bullet + \cdots + a_r {T^\bullet}^{m-1} y_m^\bullet\right)(y) \\ &= a_0 y_m^\bullet\left(T^{m-r-1}y\right) + \cdots + a_r y_m^\bullet\left(T^{m-1}y\right) \\ &= a_0 y_m^\bullet(y_{m-r}) + \cdots + a_r y_m^\bullet(y_m) \\ &= a_r. \end{aligned}$$

Since $a_r \neq 0$, we have a contradiction. Now to show that $\dim \widehat{W} = m$, we verify that $y_m^\bullet, T^\bullet y_m^\bullet, \ldots, {T^\bullet}^{m-1} y_m^\bullet$ are linearly independent. If not, then $a_0 y_m^\bullet + \cdots + a_r {T^\bullet}^r y_m^\bullet = 0$, for some $a_0, \ldots, a_r \in K$, $a_r \neq 0$ and $0 \leq r \leq m-1$. Now arguing as above, we again have a contradiction. ∎

In the above proof we have obtained a sequence of vectors $v_1, \ldots, v_r$ and polynomials $p_1(x), \ldots, p_r(x)$ such that $V = Z(v_1, T) \oplus \cdots \oplus Z(v_r, T)$, $p_1(x)$ is the minimal polynomial of T and $p_2(x)$ is the minimal polynomial of the restriction of T to the subspace $Z(v_2, T) \oplus \cdots \oplus Z(v_r, T)$ of V, and thus $p_2(x)$ divides $p_1(x)$. Similarly, each $p_{i+1}(x)$ divides $p_i(x)$. Effectively, the proof is the existence part of the following result.

THEOREM 5.3. ***Cyclic decomposition theorem.*** *Let T be a linear operator on a finite dimensional vector space V over K. Then V is a direct sum of T–cyclic subspaces*

$$V = Z(v_1, T) \oplus \cdots \oplus Z(v_r, T),$$

such that if $p_1(x), \ldots, p_r(x)$ *are the annihilators of* $v_1, \ldots, v_r$ *respectively, that is,* $p_i(x) = m_T^{v_i}(x)$, $p_1(x)$ *is the minimal polynomial of* T *and* $p_{i+1}(x)$ *divides* $p_i(x)$ *for* $i = 1, \ldots, r-1$.

Moreover, the number r *and the sequence of polynomials* $p_1(x), \ldots, p_r(x)$ *are uniquely determined by* T *only.*

For the proof of the uniqueness part of this theorem, we require the following lemma.

LEMMA 5.4. *Let* T *be a linear operator on a finite dimensional vector space* V *over* K *and let* $p(x)$ *be a polynomial in* $K[x]$. *Then*

(i) $p(T) Z(v,T) = Z(p(T) v, T)$, $v \in V$;

(ii) *if* $V = W_1 \oplus \cdots \oplus W_k$, *a direct sum of* T*–invariant subspaces, then* $p(T) V = p(T) W_1 \oplus \cdots \oplus p(T) W_k$;

(iii) *if for* $v, w \in V$, $m_T^v(x) = m_T^w(x)$, *then* $m_T^{p(T)v}(x) = m_T^{p(T)w}(x)$. *In particular,* $\dim Z(p(T) v, T) = \dim Z(p(T) w, T)$.

PROOF. (i) $w \in p(T) Z(v,T)$ if and only if $w = p(T) w'$ for some $w' \in Z(v,T)$. Also $w' \in Z(v,T)$ if and only if $w' = q(T) v$ for some $q(x) \in K[x]$. Thus, $w = p(T) q(T) v$ or $w = q(T)(p(T) v)$ and this happens if and only if $w \in Z(p(T) v, T)$.

(ii) Exercise.

(iii) Let $a(x) = m_T^{p(T)v}(x)$ and $b(x) = m_T^{p(T)w}(x)$. Then $a(T) p(T) v = 0$ implies that $m_T^v(x)$ divides $a(x) p(x)$. Thus there exists $q(x)$ in $K[x]$ such that $a(x) p(x) = m_T^v(x) q(x)$. Since $m_T^v(x) = m_T^w(x)$, $a(T) p(T) w = q(T) m_T^w(T) w = 0$. Therefore, $b(x)$ divides $a(x)$. Similarly, $a(x)$ divides $b(x)$. ∎

We now give the proof of the uniqueness part of the Cyclic decomposition theorem.

PROOF. Suppose that there are sequences of vectors $v_1, \ldots, v_r$ and $w_1, \ldots, w_s$ such that $V = Z(v_1,T) \oplus \cdots \oplus Z(v_r,T)$ with $p_i(x) = m_T^{v_i}(x)$, $i = 1, \ldots, r$ and $p_{i+1}(x)$ divides $p_i(x)$ for $i = 1, \ldots, r-1$, and $V = Z(w_1,T) \oplus \cdots \oplus Z(w_s,T)$ with $q_i(x) = m_T^{w_i}(x)$, $i = 1, \ldots, s$ and $q_{i+1}(x)$ divides $q_i(x)$ for $i = 1, \ldots, s-1$.

Note that to establish the uniqueness, it is sufficient to show that $p_i(x) = q_i(x)$, for $i = 1, \ldots, r$. This we shall prove by induction on r. To show that $p_1(x) = q_1(x)$, first write $v_1 = \sum_{i=1}^s a_i(T) w_i$, $a_1(x), \ldots, a_s(x) \in K[x]$. Then $q_1(T) v_1 = \sum_{i=1}^s a_i(T) q_1(T) w_i$. Since each $q_i(x)$ divides $q_1(x)$, $q_1(T) w_i = 0$ for $i = 1, \ldots, s$, and so $q_1(T) v_1 = 0$. Therefore, $p_1(x)$ divides $q_1(x)$. Similarly, $q_1(x)$ divides $p_1(x)$. Hence, $p_1(x) = q_1(x)$. Now if $r = 1$, then , $\dim V = \deg p_1(x) = \deg q_1(x)$, and so $V = Z(w_1, T)$, that is, $s = 1$.

Now assume that up to the index $k-1$, we have $p_i(x) = q_i(x)$. If $r \geq k$, then $\dim Z(v_1,T) + \cdots + \dim Z(v_{k-1},T) < \dim V$. Since for $i = 1, \ldots, k-1$, $\dim Z(w_i,T) = \deg q_i(x) = \deg p_i(x) = \dim Z(v_i,T)$, so $\dim Z(w_1,T) + \cdots + \dim Z(w_{k-1},T) < \dim V$, and so $s \geq k$. Thus, it makes sense to decide whether $p_k(x) = q_k(x)$.

Using Lemma 5.4,we have $p_k(T) V = Z(p_k(T) v_1, T) \oplus \cdots \oplus Z(p_k(T) v_{k-1}, T)$, and also that $p_k(T) V = Z(p_k(T) w_1, T) \oplus \cdots \oplus Z(p_k(T) w_s, T)$. Since $p_i(x) = q_i(x)$, $i = 1, \ldots, k-1$, by Lemma 5.4, $\dim Z(p_k(T) v_i, T) = \dim Z(p_k(T) w_i, T)$ for $i = 1, \ldots, k-1$. This implies that $p_k(T) w_k = 0$. Therefore, $q_k(x)$ divides $p_k(x)$. Similarly, $p_k(x)$ divides $q_k(x)$. Hence, $p_k(x) = q_k(x)$. This establishes the uniqueness part of the Cyclic decomposition theorem. ∎

Thus we have obtained an ordered basis in which the matrix of T is of the block diagonal form with companion matrices (cyclic blocks) of uniquely determined polynomials $p_1(x), \ldots, p_r(x)$ on the diagonal with $p_1(x)$ the minimal polynomial of T and $p_{i+1}(x)$ divides $p_i(x)$ for $i = 1, \ldots, r-1$. In particular, the number of cyclic blocks and the size of each cyclic block is uniquely determined, and is in the descending order of their sizes. This matrix representation of T is called the **rational canonical form** of T. The polynomials $p_1(x), \ldots, p_r(x)$ are called the **invariant factors** of T. Clearly, $c_T(x) = p_1(x) \cdots p_r(x)$.

5.1. Examples

1. Let $T \in L(\mathbb{R}^4)$ whose action on the standard basis of $\mathbb{R}^4$ is

$$\begin{aligned} Te_1 &= e_1 + 2e_2 + e_3, \\ Te_2 &= e_1 + e_2 - e_4, \\ Te_3 &= -2e_1 + e_3 + 2e_4, \\ Te_4 &= 2e_2 + e_3 + e_4. \end{aligned}$$

We want to write $\mathbb{R}^4$ as a direct sum of T–cyclic subspaces, and so the rational canonical form for T.

Verify that $c_T(x) = (x-1)^4$ and $m_T(x) = (x-1)^2$. Also that $m_T^{e_1}(x) = (x-1)^2 = m_T(x)$. Thus, $W = Z(e_1, T)$ is a 2–dimensional T–cyclic subspace of $\mathbb{R}^4$.

We now proceed as in the proof of Theorem 5.2. First, W has a basis $\{e_1, Te_1\}$, and we extend it to a basis of $\mathbb{R}^4$ given by $\{e_1, y = Te_1, e_3, e_4\}$. Then $\widehat{W} = \langle y^\bullet, T^\bullet y^\bullet \rangle$ and $\widehat{W}^\circ = \{v \in V \mid y^\bullet(v) = 0, y^\bullet(Tv) = 0\}$. Next we compute a basis of $\widehat{W}^\circ$. Since for any $v \in V$, $v = \alpha e_1 + \beta y + \gamma e_3 + \delta e_4$, so $y^\bullet(v) = 0$ implies that $\beta = 0$, $y^\bullet(Tv) = 0$ implies that $\alpha + \delta = 0$. Therefore, $\widehat{W}^\circ = \langle e_1 - e_4, e_3 \rangle$. Then $\mathbb{R}^4 = W \oplus \widehat{W}^\circ$. If $\widehat{T} = T|_{\widehat{W}^\circ}$, then $m_{\widehat{T}}(x) = (x-1)^2 = m_{\widehat{T}}^{e_3}(x)$. Therefore, $\widehat{W}^\circ = Z(e_3, T)$ and $\mathbb{R}^4 = Z(e_1, T) \oplus Z(e_3, T)$. Hence, if $\mathcal{B} = \{e_1, Te_1\} \cup \{e_3, Te_3\}$, then $[T]_{\mathcal{B}} = \operatorname{diag}\left(\begin{bmatrix} 0 & -1 \\ 1 & 2 \end{bmatrix}, \begin{bmatrix} 0 & -1 \\ 1 & 2 \end{bmatrix} \right)$.

2. We give a procedure of determining an invertible matrix which transforms the given matrix to its rational canonical form. More precisely, to a given matrix A, we find an invertible matrix P so that $P^{-1}AP$ is the rational canonical form for A.

Let $A \in K^{n \times n}$ and let $m_A(x) = x^m + a_{m-1}x^{m-1} + \cdots + a_1 x + a_0$. Let $y \in K^n$ be such that $m_A(x) = m_A^y(x)$. Then $y, Ay, \ldots, A^{m-1}y$ are linearly independent in K^n. If P_1 is an invertible matrix whose first m columns are $y, Ay, \ldots, A^{m-1}y$, then $P_1^{-1}AP_1 = A_1$, where A_1 is of the form $\begin{bmatrix} A_{11} & A_{12} \\ 0 & A_{22} \end{bmatrix}$, A_{11} is the companion matrix of $m_A(x)$. Next consider the following system

$$\begin{aligned} e_m^\bullet(u) &= 0, \\ e_m^\bullet(A_1 u) &= 0, \\ &\vdots \\ e_m^\bullet(A_1^{m-1} u) &= 0, \end{aligned}$$

where $m = \deg m_A(x)$ and $m_A(x) = m_{A_1}(x)$. The solution space $S \subseteq K^n$ of this system is of dimension $n - m$ (why is this system consistent?). The subspace S is A_1–invariant. Indeed, if $v \in S$, then

$$A_1^m v = -\left(a_{m-1}A_1^{m-1}v + \cdots + a_1 A_1 v + a_0 v\right)$$

implies that $e_m^\bullet(A_1^m v) = 0$, and so $A_1 v \in S$. Let $\{q_{m+1}, \ldots, q_n\}$ be a basis of S. Then $e_1, \ldots, e_m, q_{m+1}, \ldots, q_n$ are linearly independent in K^n. Since if $\sum_{i=1}^m \alpha_i e_i + \sum_{i=m+1}^n \alpha_i q_i = 0$, then operating by $e_m^\bullet, A_1^\bullet e_m^\bullet, \ldots, A_1^{\bullet\, m-1} e_m^\bullet$ respectively, we have $\alpha_m = 0, \alpha_{m-1} = 0, \ldots, \alpha_1 = 0$. Therefore, the matrix $P_2 = \begin{bmatrix} e_1 & \cdots & e_m & q_{m+1} & \cdots & q_n \end{bmatrix}$ is invertible and $A_2 = P_2^{-1}A_1P_2 = \operatorname{diag}\left(A_{11}, \widehat{A}\right)$. Note that $S = \langle q_{m+1}, \ldots, q_n\rangle$, which is invariant under A.

Now, if by induction on the size of matrices, we have invertible matrix Q such that $Q^{-1}\widehat{A}Q$ is the rational canonical form of $\widehat{A}$, then the matrix $P = P_1P_2P_3$, $P_3 = \operatorname{diag}(I_m, Q)$, transforms A to its rational canonical form.

As an illustration of the above, consider the following matrix in $\mathbb{R}^{4\times 4}$:

$$A = \begin{bmatrix} 1 & -2 & 1 & 2 \\ 1 & 3 & 0 & -1 \\ 0 & 2 & 0 & -2 \\ 1 & 0 & 1 & 2 \end{bmatrix}.$$

Then $c_A(x) = (x-1)^2(x-2)^2$ and $m_A(x) = (x-1)(x-2)$. Thus the invariant factors of A are $(x-1)(x-2)$ and $(x-1)(x-2)$ and the rational canonical form for A is

$$R = \operatorname{diag}\left(\begin{bmatrix} 0 & -2 \\ 1 & 3 \end{bmatrix}, \begin{bmatrix} 0 & -2 \\ 1 & 3 \end{bmatrix}\right).$$

We now compute P such that $P^{-1}AP = R$.

Verify that $m_A^{e_1}(x) = m_A(x)$. Then we can have $P_1 = \begin{bmatrix} e_1 \; Ae_1 \; e_3 \; e_4 \end{bmatrix}$, and so

$$A_1 = P_1^{-1}AP_1 = \begin{bmatrix} 0 & -2 & 1 & 3 \\ 1 & 3 & 0 & -1 \\ 0 & 0 & 0 & -2 \\ 0 & 0 & 1 & 3 \end{bmatrix}.$$

The solution space of the system

$$e_2^\bullet(u) = 0,\ e_2^\bullet(A_1 u) = 0$$

for $u = \begin{bmatrix} u_1 & u_2 & u_3 & u_4 \end{bmatrix}^t \in \mathbb{R}^4$, is

$$S = \left\{ u_1 \begin{bmatrix} 1 \\ 0 \\ 0 \\ 1 \end{bmatrix} + u_3 \begin{bmatrix} 0 \\ 0 \\ 1 \\ 0 \end{bmatrix} \mid u_1, u_3 \in \mathbb{R} \right\},$$

and so we can take

$$q_3 = \begin{bmatrix} 1 & 0 & 0 & 1 \end{bmatrix}^t, q_4 = e_3 \text{ and } P_2 = \begin{bmatrix} e_1 & e_2 & q_3 & q_4 \end{bmatrix}.$$

Thus

$$A_2 = P_2^{-1}A_1P_2 = \operatorname{diag}\left(\begin{bmatrix} 0 & -2 \\ 1 & 3 \end{bmatrix}, \begin{bmatrix} 3 & 1 \\ -2 & 0 \end{bmatrix}\right).$$

Now $\widehat{A}=\begin{bmatrix} 3 & 1 \\ -2 & 0 \end{bmatrix}$ and $m_{\widehat{A}}(x)=x^2-3x+2$. Also for $e_2\in\mathbb{R}^2$, $m_{\widehat{A}}^{e_2}(x)=m_{\widehat{A}}(x)$. Thus, for $Q=\begin{bmatrix} 0 & 1 \\ 1 & 0 \end{bmatrix}=\begin{bmatrix} e_2 & \widehat{A}e_2 \end{bmatrix}$, $Q^{-1}\widehat{A}Q=\begin{bmatrix} 0 & -2 \\ 1 & 3 \end{bmatrix}$.

Therefore, if $P_3=\operatorname{diag}(I_2,Q)$, then the matrix $P=P_1P_2P_3$ is such that $P^{-1}AP=R$. Verify that

$$P=\begin{bmatrix} 1 & 1 & 0 & 1 \\ 0 & 1 & 0 & 0 \\ 0 & 0 & 1 & 0 \\ 0 & 1 & 0 & 1 \end{bmatrix}.$$

Note that since $A_1q_4=q_3$, for $P_2=\begin{bmatrix} e_1 & e_2 & q_4 & q_3 \end{bmatrix}$, we will have $P_2^{-1}A_1P_2=R$. Also, $\mathbb{R}^4=Z(e_1,A)\oplus Z(e_3,A)$ is the cyclic decomposition of $\mathbb{R}^4$.

We consider another example. Let

$$A=\begin{bmatrix} 1 & 0 & 0 & 0 & -3 \\ 0 & 0 & 1 & -1 & 0 \\ 0 & 1 & 0 & -1 & 0 \\ 0 & 0 & 1 & 0 & 0 \\ 1 & 0 & 0 & 0 & -2 \end{bmatrix}\in\mathbb{R}^{5\times 5}$$

Then $c_A(x)=(x^3-1)(x^2+x+1)$ and $m_A(x)=(x^3-1)$. Thus invariant factors are x^3-1 and x^2+x+1. So the rational canonical form for A is

$$R=\operatorname{diag}\left(\begin{bmatrix} 0 & 0 & 1 \\ 1 & 0 & 0 \\ 0 & 1 & 0 \end{bmatrix},\begin{bmatrix} 0 & -1 \\ 1 & -1 \end{bmatrix}\right).$$

Since $m_A^{e_2}(x)=m_A(x)$, we can take $P_1=\begin{bmatrix} e_2 & Ae_2 & A^2e_2 & e_1 & e_5 \end{bmatrix}$. Then

$$A_1=P_1^{-1}AP_1=\operatorname{diag}\left(\begin{bmatrix} 0 & 0 & 1 \\ 1 & 0 & 0 \\ 0 & 1 & 0 \end{bmatrix},\begin{bmatrix} 1 & -3 \\ 1 & -2 \end{bmatrix}\right).$$

Now for $\widehat{A}=\begin{bmatrix} 1 & -3 \\ 1 & -2 \end{bmatrix}$, $m_{\widehat{A}}(x)=m_{\widehat{A}}^{e_1}(x)$, $e_1\in\mathbb{R}^2$. Thus, for $Q=\begin{bmatrix} e_1 & \widehat{A}e_1 \end{bmatrix}=\begin{bmatrix} 1 & 1 \\ 0 & 1 \end{bmatrix}$,

$$Q^{-1}\widehat{A}Q=\begin{bmatrix} 0 & -1 \\ 1 & -1 \end{bmatrix}.$$

Therefore, for $P_2=\operatorname{diag}(I_3,Q)$ and $P=P_1P_2$, $P^{-1}AP=R$, where $P=\begin{bmatrix} e_2 & Ae_2 & A^2e_2 & e_1 & Ae_1 \end{bmatrix}$. This also shows that

$$\mathbb{R}^5=Z(e_2,A)\oplus Z(e_1,A).$$

3. Matrices A and B in $K^{n\times n}$ are similar if and only if they have the same set of invariant factors.

Clearly, A and B have the same set of invariant factors if and only if they have the same rational canonical form.

If A and B are similar then $B = S^{-1}AS$ for some invertible matrix S. Let R be the rational canonical form for A. Then $R = P^{-1}AP$, and so $\left(S^{-1}P\right)^{-1} B \left(S^{-1}P\right) = R$. Therefore, A and B have the same set of invariant factors. Conversely, if A and B have the same rational canonical form R, then $P^{-1}AP = R = Q^{-1}BQ$ for some $n \times n$ invertible matrices P and Q. Hence, A and B are similar.

5.2. Exercises

1. For each of the matrices A of Exercise 2, p.107, find an invertible matrix P such that $P^{-1}AP$ is in the rational canonical form.

2. Determine the rational canonical form of a matrix whose characteristic polynomial is $(x-1)^3(x-2)^2$ and the minimal polynomial is $(x-1)^2(x-2)$.

3. Determine the rational canonical form of a matrix whose characteristic polynomial is $(x-1)^2(x-2)^3(x-3)^4$ and minimal polynomial is given by $(x-1)(x-2)^2(x-3)^2$.

4. Give all possible rational canonical forms if the characteristic polynomial is:
   (a) $(x^2+2)(x-3)^2$;
   (b) $(x-1)^2(x+1)^4$.

5. Determine whether the following matrices are similar or not:

   (a) $\begin{bmatrix} 8 & 2 & -5 \\ 12 & 6 & -10 \\ 12 & 4 & -8 \end{bmatrix}$ and $\begin{bmatrix} 40 & 20 & -34 \\ 57 & 32 & -51 \\ 76 & 40 & -66 \end{bmatrix}$;

   (b) $\begin{bmatrix} 2 & 4 & -3 \\ 1 & 3 & -3 \\ 1 & 3 & -2 \end{bmatrix}$ and $\begin{bmatrix} -5 & -30 & 39 \\ -4 & -19 & 26 \\ -4 & -20 & 27 \end{bmatrix}$;

   (c) $\begin{bmatrix} 6 & -8 & 1 & 3 \\ 4 & -6 & 1 & 2 \\ 10 & -19 & 4 & 4 \\ -1 & 1 & 0 & 0 \end{bmatrix}$ and $\begin{bmatrix} 41 & -14 & -26 & -7 \\ 14 & -13 & -91 & -18 \\ 40 & -4 & -25 & -8 \\ 0 & 0 & 0 & 1 \end{bmatrix}$;

   (d) $\begin{bmatrix} 3 & 0 & 0 & -1 \\ -2 & 1 & 1 & 0 \\ 2 & 1 & 1 & -1 \\ 6 & 0 & -1 & -1 \end{bmatrix}$ and $\begin{bmatrix} 1 & 2 & 3 & 4 \\ 0 & 1 & 2 & 3 \\ 0 & 0 & 1 & 2 \\ 0 & 0 & 0 & 1 \end{bmatrix}$.

6. Let $A, B \in \mathbb{R}^{n\times n}$. Prove that if $P^{-1}AP = B$ for some $P \in \mathbb{C}^{n\times n}$, then there exists $Q \in \mathbb{R}^{n\times n}$ such that $Q^{-1}AQ = B$.

CHAPTER 4

Inner Product Spaces

In this chapter, we study vector spaces over the field of real or complex numbers on which an additional function is defined called an inner product. If λ is a complex number, then $\overline{\lambda}$ is the complex conjugate of λ. Throughout this chapter, F will denote $\mathbb{R}$ or $\mathbb{C}$.

1. Inner products

Let V be a vector space over F, where $F = \mathbb{R}$ or $F = \mathbb{C}$. An **inner product** on V is a function $(\cdot,\cdot) : V \times V \to F$ which satisfies the following conditions:

1. $(\alpha u + \beta v, w) = \alpha\,(u, w) + \beta\,(v, w)$ for all $u, v, w \in V$, $\alpha, \beta \in F$ (linearity in the first coordinate);
2. $(v, u) = \overline{(u, v)}$ for all $u, v \in V$ (conjugate symmetry);
3. $(v, v) > 0$ for all $v \in V \setminus \{0\}$ (positive definiteness).

A vector space V together with an inner product is called an **inner product space**.

Note that for any v in V, $(0, v) = 0 = (v, 0)$. Indeed, $(0, v) = (0 + 0, v) = (0, v) + (0, v)$, so $(0, v) = 0$, and then $(v, 0) = \overline{(0, v)} = 0$. In particular, $(0, 0) = 0$. Thus, $(u, u) = 0$ if and only if $u = 0$. This implies that an inner product is nondegenerate, that is, if $(u, v) = 0$ for all v in V, then $u = 0$.

If $F = \mathbb{R}$, then the condition 2 is: $(u, v) = (v, u)$, and so in this case 1 and 2 together imply that an inner product defined on a real vector space is linear in both the coordinates.

If $F = \mathbb{C}$, then for $\alpha, \beta \in F$, $u, v, w \in V$, conditions 1 and 2 imply that

$$(u, \alpha v + \beta w) = \overline{(\alpha v + \beta w, u)} = \overline{\alpha}\overline{(v, u)} + \overline{\beta}\overline{(w, u)} = \overline{\alpha}\,(u, v) + \overline{\beta}\,(u, w),$$

that is, an inner product on a complex vector space is conjugate linear in the second coordinate.

Let V be an inner product space. The **norm** or **length** of a vector v in V is defined by:

$$\|v\| = \sqrt{(v, v)}.$$

Clearly, for every v in V, $\|v\| \geq 0$ and $\|v\| = 0$ if and only if $v = 0$. A vector v with norm 1 is called a **unit vector**.

PROPOSITION 1.1. *Let V be an inner product space over F and let $u, v \in V$.*

(i) $\|u \pm v\|^2 = \|u\|^2 \pm 2\,\mathrm{Re}\,(u, v) + \|v\|^2$, *where* $\mathrm{Re}\, z$ *denotes the real part of the complex number* z.

(ii) $\|u + v\|^2 + \|u - v\|^2 = 2\,\|u\|^2 + 2\,\|v\|^2$ *(Parallelogram law).*

(iii) $\|\lambda u\| = |\lambda|\,\|u\|$ *for all* $\lambda \in F$.

(iv) $4(u,v) = \begin{cases} \|u+v\|^2 - \|u-v\|^2 & \text{if } F = \mathbb{R}, \\ \|u+v\|^2 - \|u-v\|^2 + i\|u+iv\| - i\|u-iv\| & \text{if } F = \mathbb{C}. \end{cases}$
(Polarization identities).

(v) $|(u,v)| \leq \|u\| \, \|v\|$ *(Cauchy–Schwarz Inequality).*

(vi) $\|u \pm v\| \leq \|u\| + \|v\|$ *(Triangle inequality).*

(vii) $|\|u\| - \|v\|| \leq \|u - v\|$.

PROOF. (i)

$$\begin{aligned} \|u \pm v\|^2 &= (u \pm v, u \pm v) = (u,u) \pm (u,v) \pm (v,u) + (v,v) \\ &= \|u\|^2 \pm \left((u,v) + \overline{(u,v)}\right) + \|v\|^2 \\ &= \|u\|^2 \pm 2\,\mathrm{Re}\,(u,v) + \|v\|^2 . \end{aligned}$$

(ii) is an immediate consequence of (i).

(iii) $\|\lambda u\|^2 = (\lambda u, \lambda u) = \lambda\overline{\lambda}\,(u,u) = |\lambda|^2 \|u\|^2$. Therefore, $\|\lambda u\| = |\lambda| \, \|u\|$.

(iv) is also an immediate consequence of (i).

(v) If $v = 0$, then the statement is trivially true. If $v \neq 0$, then for $\lambda, \mu \in F$, using (i) and (iii), we have

$$0 \leq \|\lambda u + \mu v\|^2 = |\lambda|^2 \|u\|^2 + 2\,\mathrm{Re}\,\lambda\overline{\mu}\,(u,v) + |\mu|^2 \|v\|^2 .$$

In particular, if $\lambda = \|v\|^2$ and $\mu = -(u,v)$, then

$$0 \leq \|u\|^2 \|v\|^4 - |(u,v)|^2 \|v\|^2 .$$

Since $v \neq 0$, we have $|(u,v)| \leq \|u\| \, \|v\|$.

(vi) From (i) and (v),

$$\begin{aligned} \|u \pm v\|^2 &\leq \|u\|^2 \pm 2\,|\mathrm{Re}\,(u,v)| + \|v\|^2 \\ &\leq \|u\|^2 \pm 2\,|(u,v)| + \|v\|^2 \\ &\leq \|u\|^2 \pm 2\,\|u\| \, \|v\| + \|v\|^2 \\ &= (\|u\| \pm \|v\|)^2 . \end{aligned}$$

Thus, $\|u \pm v\| \leq |\|u\| \pm \|v\|| \leq \|u\| + \|v\|$.

(vii) From (vi) we have $\|u\| = \|u - v + v\| \leq \|u - v\| + \|v\|$, and so

$$\|u\| - \|v\| \leq \|u - v\| .$$

Interchanging u and v, we have

$$\|v\| - \|u\| \leq \|v - u\| = \|u - v\| .$$

Hence, $\pm(\|u\| - \|v\|) \leq \|u - v\|$ and $|\|u\| - \|v\|| \leq \|u - v\|$. ∎

1.1. Examples

1. Let V be the real vector space of all equivalence classes of directed line segments (Example 1, p.26). Then the usual scalar product (dot product)

$$(a,b) = a \cdot b = |a| \, |b| \cos\theta,$$

where θ is the angle between a and b, is an inner product. The length (norm) of vector a is $|a| = \sqrt{a \cdot a}$. Recall that two vectors a and b are added according to law: the tail of b is put on the head of a by parallel translation of b, join the tail of a to the head of b, and then the resulting directed line segment denotes $a + b$. Since $-1 \leq \cos\theta \leq 1$, $|a \cdot b| \leq |a| \, |b|$, which is

Cauchy–Schwarz Inequality. The geometric interpretation of the statements of the above proposition is clear from the following:

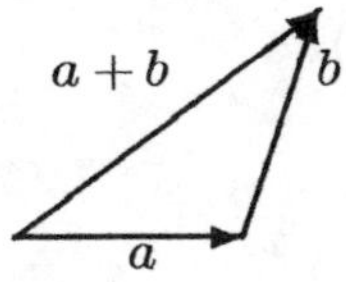

$$\Rightarrow |a+b| \leq |a| + |b|$$

and

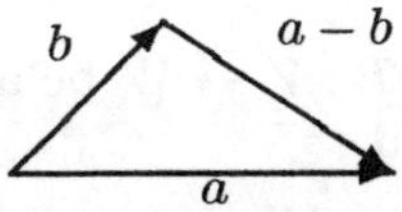

$$\Rightarrow |a-b| \leq |a| + |b|$$

are triangle inequalities. If θ is the angle between a and b, then from the figure

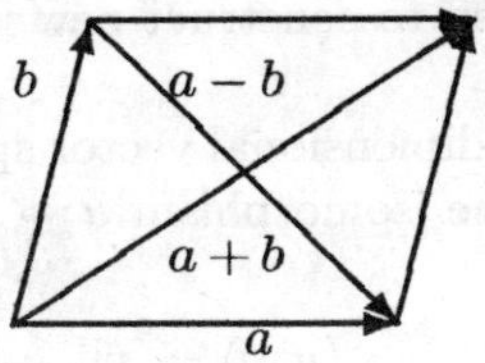

we have

$$\cos\theta = \frac{|a|^2 + |b|^2 - |a-b|^2}{2\,|a|\,|b|}$$

and

$$\cos(\pi - \theta) = \frac{|a|^2 + |b|^2 - |a+b|^2}{2\,|a|\,|b|}.$$

Therefore,

$$|a-b|^2 = |a|^2 + |b|^2 - 2a \cdot b,$$

$$|a+b|^2 = |a|^2 + |b|^2 + 2a \cdot b,$$

and on adding, we have

$$|a+b|^2 + |a-b|^2 = 2\,|a|^2 + 2\,|b|^2\,,$$

the parallelogram law.

2. Let $V = F^n$ and define $(x, y) = \sum_{i=1}^{n} x_i \overline{y_i} = y^* x$. This is an inner product on F^n called the **standard inner product** on F^n. The corresponding norm is defined by

$$\|x\| = \left(\sum_{i=1}^{n} |x_i|^2 \right)^{\frac{1}{2}}.$$

The standard inner product space $\mathbb{R}^n$ (respectively, $\mathbb{C}^n$) is called an n–dimensional **Euclidean** (respectively, **unitary**) space.

3. An inner product on $F^{m\times n}$ is given by $(A, B) = \operatorname{tr}(AB^*)$.

4. Let $V = C[a,b]$, the space of all continuous complex valued functions defined on $[a,b]$, $-\infty < a < b < \infty$. Then an inner product on V is given by

$$(f,g) = \int_a^b f(t)\,\overline{g(t)}\,dt,$$

and the corresponding norm is

$$\|f\| = \left(\int_a^b |f(t)|^2\,dt\right)^{\frac{1}{2}}.$$

5. Let W be an inner product space and let $T : V \to W$ be an isomorphism. Then an inner product on V can be defined:

$$(u,v) = (Tu, Tv)\,,\ u,v \in V.$$

The parenthesis on the right side of above equation are for the inner product on W.

This can be used to construct new inner product spaces from the given one.

Let V be an n–dimensional vector space over F and let $\mathcal{B}$ be an ordered basis of V. Then the isomorphism $u \mapsto [u]_{\mathcal{B}}$ from V to F^n defines an inner product on V:

$$(u,v) = [v]_{\mathcal{B}}^*\,[u]_{\mathcal{B}}\,.$$

If Q is an invertible matrix on $F^{n\times n}$, then the mapping $u \mapsto Qu$ is an isomorphism from F^n to itself. Thus we have an inner product on F^n:

$$(u,v) = (Qv)^*\,Qu = v^*Q^*Qu.$$

Let p be a continuous function on $[a,b], -\infty < a < b < \infty$ and let $p(t) > 0$ for all $t \in [a,b]$. Then the mapping $f \mapsto pf$ is an isomorphism from $C[a,b]$ to itself, and so we have an inner product on $C[a,b]$:

$$(f,g) = \int_a^b p(t)^2\,f(t)\,\overline{g(t)}\,dt.$$

In fact, this means that if $\mu \in C[a,b]$ and $\mu(t) > 0$ for all $t \in [a,b]$, then we can define an inner product on $C[a,b]$:

$$(f,g) = \int_a^b \mu(t)\,f(t)\,\overline{g(t)}\,dt.$$

6. The inner product space l^2 of all complex sequences $\{s_n\}$ with the property that $\sum_{n=1}^{\infty} |s_n|^2 < \infty$ with inner product:

$$(s,t) = \sum_{n=1}^{\infty} s_n\,\overline{t_n},\ s = \{s_n\}\,, t = \{t_n\}\,.$$

Note that since $(|s_n| - |t_n|)^2 \geq 0$, $2\,|s_n|\,|t_n| \leq |s_n|^2 + |t_n|^2$, and so

$$2\left|\sum_{n=1}^{\infty} s_n\,\overline{t_n}\right| \leq 2\sum_{n=1}^{\infty} \left|s_n\,\overline{t_n}\right| \leq \sum_{n=1}^{\infty} |s_n|^2 + \sum_{n=1}^{\infty} |t_n|^2 < \infty,$$

and this shows that this inner product makes sense.

1.2. Exercises

1. Decide which of the following functions define an inner product on $\mathbb{R}^2$. For $x = \begin{bmatrix} x_1 & x_2 \end{bmatrix}^t$, $y = \begin{bmatrix} y_1 & y_2 \end{bmatrix}^t$:
   (a) $(x,y) = x_1 y_2 + x_2 y_1$;
   (b) $(x,y) = x_1 y_2 - x_2 y_1$;
   (c) $(x,y) = x_1 y_1 + x_2 y_2$;
   (d) $(x,y) = x_1 x_2 y_1 y_2$;
   (e) $(x,y) = 2x_1 y_1 + x_1 y_2 + x_2 y_1 + 2x_2 y_2$.

2. Decide which of the following functions define an inner product on $\mathbb{C}^2$. For $x = \begin{bmatrix} x_1 & x_2 \end{bmatrix}^t$, $y = \begin{bmatrix} y_1 & y_2 \end{bmatrix}^t$:
   (a) $(x,y) = x_1 \overline{y_2}$;
   (b) $(x,y) = x_1 \overline{y_1} + x_2 \overline{y_2}$;
   (c) $(x,y) = x_1 y_1 + x_2 y_2$;
   (d) $(x,y) = x_1 \overline{y_2} - x_2 \overline{y_1}$;
   (e) $(x,y) = 2x_1 \overline{y_1} + i\,(x_2 \overline{y_1} - x_1 \overline{y_2}) + 2x_2 \overline{y_2}$.

3. Let u and v be vectors in an inner product space such that $\|u+v\| = 8$, $\|u-v\| = 6$, $\|u\| = 7$. What must be $\|v\|$?

4. Let $A \in \mathbb{C}^{n\times n}$ be such that $(x,y) = y^* Ax$ is an inner product on $\mathbb{C}^n$. Prove that A is a Hermitian matrix with positive diagonal entries.

5. Let V be a finite dimensional inner product space. Prove that for $v \in V\backslash\{0\}$, the set $W = \{w \in W \mid (w,v) = 0\}$ is a subspace of V of dimension $\dim V - 1$.

6. Show that for $A, B \in \mathbb{C}^{n\times n}$:
$$|\operatorname{tr}(AB^*)| \le \sqrt{\operatorname{tr}(AA^*)\operatorname{tr}(BB^*)} \le \frac{\operatorname{tr}(AA^*) + \operatorname{tr}(BB^*)}{2}.$$

7. Let $\theta_1, \ldots, \theta_n$ and $\lambda_1, \ldots \lambda_n$ be positive numbers such that $\theta_1 + \cdots + \theta_n = 1$. Prove that
$$\left(\sum_{i=1}^{n} \theta_i \lambda_i\right)\left(\sum_{i=1}^{n} \theta_i \lambda_i^{-1}\right) \ge 1.$$

8. Show that for x, y in an inner product space V over F: $\|x+y\| = \|x\| + \|y\|$ if and only if $x = 0$ or $y = \lambda x$, for some $\lambda \in F$.

9. Prove that for u, v and w in an inner product space V:
$$\|u-v\|\,\|w\| \le \|v-w\|\,\|u\| + \|w-u\|\,\|v\|.$$
Hint: let $x = u/\|u\|^2$, $y = v/\|v\|^2$ and $z = w/\|w\|^2$; use triangle inequality $\|x-y\| \le \|x-z\| + \|z-y\|$.

2. Orthogonality

Let V be an inner product space. Two vectors u and v in V are said to be **orthogonal** if $(u,v)=0$. If u and v are orthogonal, we write $u \perp v$. If S and T are subsets of V, then S is said to be orthogonal to T, written as $S \perp T$, if for each $s \in S$ and $t \in T$, $(s,t)=0$. For a subset S of V the **orthogonal complement** of S is

$$S^{\perp} = \{v \in V \mid (v,s) = 0 \text{for all } s \in S\}.$$

The reader can verify that $S^{\perp}$ is a subspace of V, and if S is a subspace of V then $S \cap S^{\perp} = \{0\}$. Also, that if $S \perp T$, then $T \subseteq S^{\perp}$. For a vector v in V, by $v \perp S$, we mean that $v \in S^{\perp}$.

A nonempty subset S of an inner product space V is said to be an **orthogonal set** if any two vectors in S are orthogonal, that is, for $x, y \in S$, $(x,y)=0$. Recall that a vector with norm 1 is called a unit vector. An orthogonal set consisting of unit vectors is called an **orthonormal set**. It is easy to construct an orthonormal set from a given set S of orthogonal vectors: multiply each vector of S by the reciprocal of its norm, that is, $\left\{\frac{u}{\|u\|} \mid u \in S\right\}$ is an orthonormal set from S.

PROPOSITION 2.1. *An orthogonal set of nonzero vectors is linearly independent.*

PROOF. Let S be an orthogonal set of nonzero vectors of an inner product space V over F. Let $x_1, \ldots, x_m \in S$ and let $\sum_{i=1}^{m} \alpha_i x_i = 0$, $\alpha_i \in F$. Then for any $k = 1, \ldots, m$:

$$0 = \left(\sum_{i=1}^{m} \alpha_i x_i, x_k\right) = \sum_{i=1}^{m} \alpha_i \left(x_i, x_k\right) = \alpha_i \left(x_k, x_k\right).$$

Since $x_k \neq 0$, $(x_k, x_k) \neq 0$ and so $\alpha_k = 0$. Hence, S is a linearly independent subset of V. ∎

A maximal orthonormal set of vectors in an inner product space V is called a **complete orthonormal set**. The existence of such a set is an application of Zorn's Lemma. Equivalently, an orthonormal subset M of V is a complete orthonormal set if for v in V, $v \perp M$ implies that $v = 0$, that is, there is no nonzero vector in V which is orthogonal to M.

A basis of an inner product space which is also an orthonormal set (respectively, orthogonal set) is called an **orthonormal basis** (respectively, **orthogonal basis**). Our next result shows that in a finite dimensional inner product space a complete orthonormal set is precisely an orthonormal basis. For infinite dimensional inner product spaces a complete orthonormal set need not be a basis. See Example 5 at the end of this section.

PROPOSITION 2.2. *In a finite dimensional inner product space, a complete orthonormal set is a basis.*

PROOF. Let M be a complete orthonormal subset of a finite dimensional inner product space V. Since V is finite dimensional and an orthogonal set is linearly independent, M is a finite set. Let $M = \{x_1, \ldots, x_m\}$. Then for $w = v - \sum_{i=1}^{m} (v, x_i)\, x_i$, $(w, x_k) = 0$ for $k = 1, \ldots, m$. Thus, $w \perp M$. Hence $w = 0$, and $v = \sum_{i=1}^{m} (v, x_i)\, x_i$. Therefore, M is a basis of V. ∎

The proof of the above proposition also shows that if $\{x_1, \ldots, x_n\}$ is an orthonormal basis of V, then for any v in V:

$$v = \sum_{i=1}^{n} (v, x_i)\, x_i.$$

The following result is a procedure of replacing a sequence of linearly independent vectors by a sequence of orthonormal vectors so that the subspace generated by these two sets is the same. In particular, for a finite dimensional inner product space, a method of constructing an orthonormal basis from a given basis.

THEOREM 2.3. ***Gram–Schmidt orthonormalization.*** *Let $\{x_1, \ldots, x_n\}$ be a sequence of linearly independent vectors in an inner product space V. Then there is a sequence of orthonormal vectors $\{y_1, y_2, \ldots\}$ such that for every n:*

$$\langle x_1, \ldots, x_n \rangle = \langle y_1, \ldots, y_n \rangle .$$

PROOF. Let $W_n = \langle x_1, \ldots, x_n \rangle$. We inductively construct a sequence of orthonormal vectors $\{y_1, y_2, \ldots\}$ such that $W_n = \langle y_1, \ldots, y_n \rangle$, $n = 1, 2, \ldots$.

Define $u_1 = x_1$, and $y_1 = \frac{1}{\|u_1\|} u_1$. Clearly, $W_1 = \langle y_1 \rangle$. Assume that orthonormal vectors $y_1, \ldots, y_m$ have been constructed so that for $n = 1, \ldots, m$, $W_n = \langle y_1, \ldots, y_n \rangle$. Define

$$u_{m+1} = x_{m+1} - \sum_{i=1}^{m} (x_{m+1}, y_i)\, y_i.$$

Then $u_{m+1} \neq 0$; otherwise $x_{m+1} \in \langle y_1, \ldots, y_m \rangle = W_m$, a contradiction. Now let

$$y_{m+1} = \frac{1}{\|u_{m+1}\|} u_{m+1}.$$

Then for $k = 1, \ldots, m$,

$$\begin{aligned} (y_{m+1}, y_k) &= \frac{1}{\|u_{m+1}\|} \left[(x_{m+1}, y_k) - \sum_{i=1}^{m} (x_{m+1}, y_i)\,(y_i, y_k) \right] \\ &= \frac{1}{\|u_{m+1}\|} \left[(x_{m+1}, y_k) - \sum_{i=1}^{m} (x_{m+1}, y_i)\,\delta_{ik} \right] = 0. \end{aligned}$$

Thus, $\{y_1, \ldots, y_{m+1}\}$ is an orthonormal set. Since $x_{m+1} \in \langle y_1, \ldots, y_{m+1} \rangle$, it follows that $W_{m+1} = \langle y_1, \ldots, y_{m+1} \rangle$. ∎

PROPOSITION 2.4. *Let W be a subspace of a finite dimensional inner product space V. Then $V = W \oplus W^{\perp}$.*

PROOF. Let $\{x_1, \ldots, x_m\}$ be an orthonormal basis of W and let $v \in V$. Then for $w = v - \sum_{i=1}^{m} (v, x_i)\, x_i$, $(w, x_k) = 0$ for $k = 1, \ldots, m$. Thus, $w \in W^{\perp}$ and so $v = \sum_{i=1}^{m} (v, x_i)\, x_i + w \in W + W^{\perp}$. Clearly $W \cap W^{\perp} = \{0\}$. Hence $V = W \oplus W^{\perp}$. ∎

COROLLARY 2.5. *Let W be a subspace of finite dimensional inner product space V. Then $W^{\perp\perp} = W$.*

PROOF. Clearly $W \subseteq W^{\perp\perp}$. Let $v \in W^{\perp\perp}$. Then $v = w + w'$, $w \in W$, $w' \in W^{\perp}$ and so $0 = (v, w') = (w, w') + (w', w') = (w', w')$. Hence $w' = 0$ and $v \in W$. ∎

Example 5 at the end of this section shows that the statement of Proposition 2.4 is not true for infinite dimensional inner product spaces.

Let V be a vector space and let $V = W_1 \oplus W_2$. Recall that the mapping $P : V \to V$ defined by $P(w_1 + w_2) = w_1$, where $w_1 \in W_1$ and $w_2 \in W_2$ is linear, called the projection of W_1 along W_2. If W is a subspace of an inner product space V such that $V = W \oplus W^\perp$, then the projection of V onto W along $W^\perp$ is called the **orthogonal projection** of V onto W. We will denote it by Pr_W. Thus, for $v \in V$, $\mathrm{Pr}_W(v) \in W$ and $v - \mathrm{Pr}_W(v) \in W^\perp$.

THEOREM 2.6. *Let W be a subspace of finite dimensional inner product space V and let $v \in V$.*

(i) $\|v - \mathrm{Pr}_W(v)\| \le \|v - w\|$ *for all $w \in W$. The equality holds if and only if $w = \mathrm{Pr}_W(v)$.*

(ii) *If $\{x_1, \ldots, x_m\}$ is an orthonormal basis of W, then $\mathrm{Pr}_W(v) = \sum_{i=1}^m (v, x_i)\, x_i$.*

PROOF. (i) Since $v - \mathrm{Pr}_W(v) \in W^\perp$ for any $w \in W$, $v - \mathrm{Pr}_W(v) \perp \mathrm{Pr}_W(v) - w$. Therefore, $\|v - w\|^2 = \|v - \mathrm{Pr}_W(v) + \mathrm{Pr}_W(v) - w\|^2 = \|v - \mathrm{Pr}_W(v)\|^2 + \|\mathrm{Pr}_W(v) - w\|^2 \ge \|v - \mathrm{Pr}_W(v)\|^2$. Clearly, equality holds if and only if $\mathrm{Pr}_W(v) = w$.

(ii) Since $(v - \sum_{i=1}^m (v, x_i)\, x_i, x_k) = 0$ for $k = 1, \ldots, m$, so $v - \sum_{i=1}^m (v, x_i)\, x_i \in W^\perp$. Therefore, $\mathrm{Pr}_W(v - \sum_{i=1}^m (v, x_i)\, x_i) = 0$. Now by the definition of orthogonal projection, it follows that $\mathrm{Pr}_W(v) = \sum_{i=1}^m (v, x_i)\, x_i$. ∎

Theorem 2.6(i) means that orthogonal projection of v onto W is the unique vector $\mathrm{Pr}_W(v)$ in W such that the function $f : W \to \mathbb{R}$ defined by $f(w) = \|v - w\|$ attains its minimum value at $\mathrm{Pr}_W(v)$. If we interpret $\|v - w\|$ as the distance between v and w, then $\mathrm{Pr}_W(v)$ is an element of W which is nearest to v, that is, the distance between v and $\mathrm{Pr}_W(v)$ is the least.

If x is a vector in an inner product space V, then the mapping $f_x : V \to F$ defined by $f_x(v) = (v, x)$ is a **linear functional** on V. The following result shows that linear functionals on a finite dimensional inner product space are of this form. The corresponding version of the following statement, for an infinite dimensional inner product space is not true (Exercise 19).

THEOREM 2.7. ***Riesz representation theorem.*** *Let V be a finite dimensional inner product space and let f be a linear functional on V. Then there exists a unique vector x in V such that $f(v) = (v, x)$, for all v in V.*

PROOF. If $f \equiv 0$, then we may take $x = 0$. Assume that $f \ne 0$. If such a vector x exists then for $v \in \ker f$, we must have $(v, x) = 0$. Thus, we should look for x in $(\ker f)^\perp$. According to Proposition 2.4, $V = \ker f \oplus (\ker f)^\perp$. Since $f \ne 0$, $\dim \ker f = \dim V - 1$, and so $\dim \ker f^\perp = 1$. Thus, if u is a nonzero unit vector in $\ker f^\perp$, then $\ker f^\perp = \langle u \rangle$ and so $V = \langle u \rangle \oplus \ker f$. Since we are searching for x in $\ker f^\perp$, so now our goal is to find $\alpha \in F$ such that $x = \alpha u$ and $f(v) = (v, \alpha u)$ for all $v \in V$. In particular, $f(u) = (u, \alpha u) = \overline{\alpha}$. Therefore, $\alpha = \overline{f(u)}$. Now we verify that $x = \overline{f(u)}\, u$ is the required vector.

For any $v \in V$, write (uniquely) $v = \beta u + w$, $\beta \in F$, $w \in \ker f$. Then

$$(v, x) = \left(v, \overline{f(u)}u\right) = \beta f(u) = f(\beta u) = f(\beta u + w) = f(v).$$

Finally, for uniqueness, suppose that there also exists y in V such that $f(v) = (v, y)$ for all $v \in V$. Then $f(v) = (v, x) = (v, y)$ and so $(v, x - y) = 0$ for all $v \in V$. Therefore, $x = y$. ∎

2.1. Examples.

1. The standard basis $\{e_1, \ldots, e_n\}$ of the standard n–dimensional inner product space F^n is an orthonormal basis.

2. An orthonormal basis for $F^{n \times n}$ with the inner product $(A, B) = \operatorname{tr}(AB^*)$ is the set of matrix units
$$\{E_{ij} \mid i = 1, \ldots, n, j = 1, \ldots, n\}.$$
Since $(E_{rs}, E_{pq}) = \operatorname{tr}\left(E_{rs}E_{pq}^*\right) = \operatorname{tr}(E_{rs}E_{qp}) = \delta_{sq}\delta_{rp}$, that is, for $(r, s) \neq (p, q)$, $(E_{rs}, E_{pq}) = 0$ and $(E_{rs}, E_{rs}) = 1$.

3. Consider the standard inner product space $\mathbb{R}^3$ with basis
$$\left\{ x_1 = \begin{bmatrix} 2 \\ 1 \\ 2 \end{bmatrix}, x_2 = \begin{bmatrix} 4 \\ 1 \\ 0 \end{bmatrix}, x_3 = \begin{bmatrix} 3 \\ 1 \\ -1 \end{bmatrix} \right\}.$$
Applying Gram–Schmidt orthonormalization process, we get an orthonormal basis $\{y_1, y_2, y_3\}$:
$$y_1 = \frac{x_1}{\|x_1\|} = \frac{1}{3}\begin{bmatrix} 2 \\ 1 \\ 2 \end{bmatrix}.$$
$$u_2 = x_2 - (x_2, y_1)\, y_1 = x_2 - \left(y_1^t x_2\right) y_1 = \begin{bmatrix} 4 \\ 1 \\ 0 \end{bmatrix} - \begin{bmatrix} 2 \\ 1 \\ 2 \end{bmatrix} = \begin{bmatrix} 2 \\ 0 \\ -2 \end{bmatrix}.$$
$$y_2 = \frac{u_2}{\|u_2\|} = \frac{1}{\sqrt{2}}\begin{bmatrix} 1 \\ 0 \\ -1 \end{bmatrix}.$$
$$\begin{aligned} u_3 &= x_3 - (x_3, y_1)\, y_1 - (x_3, y_2)\, y_2 = x_3 - \left(y_1^t x_3\right) y_1 - \left(y_2^t x_3\right) y_2 \\ &= \begin{bmatrix} 3 \\ 1 \\ -1 \end{bmatrix} - \frac{5}{9}\begin{bmatrix} 2 \\ 1 \\ 2 \end{bmatrix} - 2\begin{bmatrix} 1 \\ 0 \\ -1 \end{bmatrix} = \begin{bmatrix} -1/9 \\ 4/9 \\ -1/9 \end{bmatrix}. \end{aligned}$$
$$y_3 = \frac{u_3}{\|u_3\|} = \frac{1}{3\sqrt{2}}\begin{bmatrix} -1 \\ 4 \\ -1 \end{bmatrix}.$$
Now if $a = \begin{bmatrix} a_1 & a_2 & a_3 \end{bmatrix}^t \in \mathbb{R}^3$, then $a = (a, y_1)\, y_1 + (a, y_2)\, y_2 + (a, y_3)\, y_3$, and so
$$a = \frac{2a_1 + a_2 + 2a_3}{3}\, y_1 + \frac{a_1 - a_3}{\sqrt{2}}\, y_2 + \frac{-a_1 + 4a_2 - a_3}{3\sqrt{2}}\, y_3.$$

4. Consider the inner product space of polynomials $\mathbb{R}[x]$ with the inner product $(p(x), q(x)) = \int_{-1}^{1} p(x)\, q(x)\, dx$. Applying Gram–Schmidt orthonormalization process to its usual basis $\{1, x, x^2, \dots\}$, we get orthonormal polynomials $\{\alpha_0 P_0(x), \alpha_1 P_1(x), \alpha_2 P_2(x), \dots\}$, where

$$\frac{1}{\alpha_k} = \|P_k(x)\| = \left(\int_{-1}^{1} P_k(x)^2\, dx\right)^{\frac{1}{2}} = \sqrt{\frac{2}{2n+1}}.$$

Polynomials $P_0(x), P_1(x), P_2(x), \dots$ are called the Legendre polynomials. The first four Legendre polynomials are $P_0(x) = 1$, $P_1(x) = x$, $P_2(x) = \frac{1}{2}\left(3x^2 - 1\right)$, $P_3(x) = \frac{1}{2}\left(5x^3 - 3x\right)$.

5. Consider the inner product space l^2 of square summable complex sequences. Let E_k be the sequence whose k–th entry is 1 and all other entries zero, and let $M = \{E_k \mid k = 1, 2, \dots\}$. Then M is a complete orthonormal set. It is clear that $(E_r, E_s) = \delta_{rs}$. To check that M is indeed a complete orthonormal set, let $s = \{s_n\}$ be in l^2 such that $s \perp M$. Then for each $k = 1, 2, \dots$, $s_k = (s, E_k) = 0$, and so s is the zero sequence. The set M is not a basis of l^2 as the sequence $u = \{u_n\}$ with $u_n = 1$ for all $n = 1, 2, \dots$ is such that $u \notin \langle M \rangle$. Indeed, $\langle M \rangle$ is the space of all those sequences whose only finite number of entries are nonzero.

   Further, if $W = \langle M \rangle$, then clearly $W^\perp = \{0\}$. Therefore, for $V = l^2$, $V \neq W \oplus W^\perp$.

6. Let $\mathcal{B} = \{v_1, \dots, v_n\}$ and $\mathcal{B}' = \{w_1, \dots, w_m\}$ be ordered orthonormal bases for inner product spaces V and W, respectively. Let $T \in L(V, W)$ and let the matrix representation of T with respect to these bases be ${}_{\mathcal{B}'}[T]_{\mathcal{B}} = [\alpha_{ij}] \in F^{m \times n}$. Then

$$Tv_j = \sum_{r=1}^{m} \alpha_{rj} w_r,$$

for $j = 1, \dots, n$, and so

$$(Tv_j, w_i) = \sum_{r=1}^{m} \alpha_{rj} (w_r, w_i) = \alpha_{ij}.$$

Therefore, the (i, j)–th entry of $m \times n$ matrix ${}_{\mathcal{B}'}[T]_{\mathcal{B}}$ is (Tv_j, w_i).

   Thus, if $T \in L(V)$ and $\mathcal{B} = \{v_1, \dots, v_n\}$ is an ordered orthonormal basis for inner product space V, then the (i, j)–th entry of $n \times n$ matrix $[T]_{\mathcal{B}}$ is (Tv_j, v_i).

7. Consider the inner product space $\mathbb{R}_3[x]$ with the inner product

$$(p(x), q(x)) = \int_{-1}^{1} p(x)\, q(x)\, dx.$$

The mapping $\varepsilon : \mathbb{R}_3[x] \to \mathbb{R}$ given by $\varepsilon(p(x)) = p(0)$ is linear. Thus by Riesz representation theorem, there exists $r(x) \in \mathbb{R}_3[x]$ such that $\varepsilon(p(x)) = (p(x), r(x))$ for all $p(x) \in \mathbb{R}_3[x]$. According to the proof of this theorem, $r(x)$ is in the one dimensional subspace $\ker \varepsilon^\perp$ such that if $\ker \varepsilon^\perp =$

$\langle u(x)\rangle$, then $r(x) = u(0)\,u(x)$. Clearly, $\ker\varepsilon = \langle x, x^2, x^3\rangle$. Let $\ker\varepsilon^{\perp} = \langle u(x) = a_0 + a_1x + a_2x^2 + a_3x^3\rangle$. Then

$$0 = (u(x), x) = \int_{-1}^{1} \left(a_0 + a_1x + a_2x^2 + a_3x^3\right) x\,dx = \frac{2}{3}\,a_1 + \frac{2}{5}\,a_3;$$

$$0 = (u(x), x^2) = \int_{-1}^{1} \left(a_0 + a_1x + a_2x^2 + a_3x^3\right) x^2\,dx = \frac{2}{3}\,a_0 + \frac{2}{5}\,a_2;$$

$$0 = (u(x), x^3) = \int_{-1}^{1} \left(a_0 + a_1x + a_2x^2 + a_3x^3\right) x^3\,dx = \frac{2}{5}\,a_1 + \frac{2}{7}\,a_3.$$

Solving these equations, we obtain $\ker\varepsilon^{\perp} = \langle u(x) = 1 - \frac{5}{3}\,x^2\rangle$. Therefore, $r(x) = 1 - \frac{5}{3}\,x^2$. Hence, for any $p(x) \in \mathbb{R}_3[x]$:

$$p(0) = \int_{-1}^{1} p(x)\left(1 - \frac{5}{3}\,x^2\right) dx.$$

8. Consider a general system of linear equations: $Ax = b$, $A \in F^{m\times n}$, $x \in F^n$, $b \in F^m$. This system has a solution if and only if $b \in \operatorname{im} A$ and such a solution is unique if and only if $\ker A = \{0\}$.

Assume that $b \in \operatorname{im} A$. If x_0 is a solution of the linear system $Ax = b$, then $x_0 + y$ is also a solution for $y \in \ker A$. However, if x_0 is restricted to $\ker A^{\perp}$, then it is a unique solution. To see this, let x_1 and x_2 be two solutions of the equation $Ax = b$. Since $F^n = \ker A \oplus \ker A^{\perp}$, we have $x_1 = u_1 + v_1$, $x_2 = u_2 + v_2$, $u_1, u_2 \in \ker A$, $v_1, v_2 \in \ker A^{\perp}$. Since $Ax_i = Av_i = b$, $i = 1, 2$, $A(v_1 - v_2) = 0$. Thus, $v_1 - v_2 \in \ker A \cap \ker A^{\perp} = \{0\}$. Therefore, $v_1 = v_2$, and so the component in $\ker A^{\perp}$ of any solution of the system is unique. This result can be stated as follows.

If $A \in F^{m\times n}$ ($F = \mathbb{R}$ or $\mathbb{C}$) and $b \in \operatorname{im} A$, then the system of linear equations $Ax = b$ has a unique solution x_0 in $\ker A^{\perp}$ and any solution can be written as $x_0 + y$, $y \in \ker A$.

2.2. Exercises. In the following exercises, V is a finite dimensional inner product space over F.

1. Verify that the following vectors are orthogonal with respect to the standard inner product, and extend these to form an orthogonal basis.

$$(i)\ \begin{bmatrix}1\\2\\4\\1\end{bmatrix}, \begin{bmatrix}-1\\-2\\1\\1\end{bmatrix};\quad (ii)\ \begin{bmatrix}1\\0\\2\\1\end{bmatrix}, \begin{bmatrix}-2\\0\\1\\0\end{bmatrix};\quad (iii)\ \begin{bmatrix}1+i\\1-i\\2\end{bmatrix}, \begin{bmatrix}2+i\\i\\-1\end{bmatrix}.$$

2. Use Gram–Schmidt orthonormalization process to obtain an orthonormal basis spanned by the following vectors in the standard inner product space.

(i) $\begin{bmatrix} 1 \\ -2 \\ 2 \\ 3 \end{bmatrix}, \begin{bmatrix} 1 \\ 1 \\ 3 \\ -5 \end{bmatrix}, \begin{bmatrix} 3 \\ 2 \\ 1 \\ -1 \end{bmatrix}$; (ii) $\begin{bmatrix} 1 \\ 0 \\ 1 \\ 0 \end{bmatrix}, \begin{bmatrix} 1 \\ 1 \\ 1 \\ 1 \end{bmatrix}, \begin{bmatrix} 1 \\ 0 \\ 0 \\ 1 \end{bmatrix}$;

(iii) $\begin{bmatrix} -1 \\ 0 \\ 1 \\ 2 \end{bmatrix}, \begin{bmatrix} 0 \\ 1 \\ 2 \\ -1 \end{bmatrix}, \begin{bmatrix} 1 \\ 2 \\ -1 \\ 0 \end{bmatrix}, \begin{bmatrix} 2 \\ -1 \\ 0 \\ 1 \end{bmatrix}$; (iv) $\begin{bmatrix} i \\ 1 \\ 1 \end{bmatrix}, \begin{bmatrix} 1 \\ i \\ 1 \end{bmatrix}, \begin{bmatrix} 1 \\ 1 \\ i \end{bmatrix}$.

3. Suppose that W_1 and W_2 are subspaces of V and that $\dim W_1 < \dim W_2$. Prove that there is a nonzero vector in W_2 which is orthogonal to all vectors in W_1.

4. Let $p, q, r \in \mathbb{Z}$. Show that the following vectors are mutually orthogonal in the Euclidean space $\mathbb{R}^3$:

$$u = \begin{bmatrix} p \\ q \\ r \end{bmatrix}, \ v = \begin{bmatrix} q \\ r \\ p \end{bmatrix}, \ w = \begin{bmatrix} r \\ p \\ q \end{bmatrix}.$$

if and only if $pq + qr + rp = 0$. Prove that if this holds, the length of each vector is $|p+q+r|$.

5. Let $x \in \mathbb{C}^n$. Compute eigenvalues and eigenvectors of the matrix

$$A = \begin{bmatrix} 0 & x \\ x^* & 0 \end{bmatrix}.$$

6. Let $x \in F^n$. Find all eigenvalues and eigenvectors of the $n \times n$ matrix xx^*.

7. Let $\mu_1, \ldots, \mu_n$ be nonzero real numbers and let $A \in \mathbb{R}^{n \times n}$ whose (i,j)–th entry is μ_i/μ_j. Obtain eigenvalues and eigenvectors of A.
Hint: verify that $A = \Lambda B \Lambda^{-1}$, $\Lambda = \operatorname{diag}(\mu_1, \ldots, \mu_n)$, B is an $n \times n$ matrix whose all entries are 1.

8. Let $V = \mathbb{R}_5[x]$ with the inner product $(p(x), q(x)) = \int_{-1}^{1} p(x)\, q(x)\, dx$. Find $W^\perp$ if:
   (a) $W = \{p(x) \in \mathbb{R}_5[x] \mid p(x) = p(-x)\}$;
   (b) $W = \{p(x) \in \mathbb{R}_5[x] \mid p(0) = p^{(1)}(0) = p^{(2)}(0) = 0\}$;
   (c) $W = \{p(x) \in \mathbb{R}_5[x] \mid p^{(1)}(x) - p(x) = 0\}$.

9. Prove that for $u, v \in V$, u and v are orthogonal if and only if for all $\mu \in F$, $\|u\| \le \|u - \mu v\|$.

10. Let W_1 and W_2 be subspaces of V. Prove that
    (a) if $W_1 \perp W_2$, then $W_1 \cap W_2 = \{0\}$;
    (b) $W_1 \subseteq W_2$ if and only if $W_2^\perp \subseteq W_1^\perp$;
    (c) $(W_1 + W_2)^\perp = W_1^\perp \cap W_2^\perp$;
    (d) $(W_1 \cap W_2)^\perp = W_1^\perp + W_2^\perp$;
    (e) $\{0\}^\perp = V$ and $V^\perp = \{0\}$.

11. Let $\{x_1, \ldots, x_n\}$ be an orthonormal set of vectors. Prove that for $v \in V$:

$$\sum_{i=1}^{n} |(v, x_i)|^2 \leq \|v\|^2,$$

equality holds if and only if $v \in \langle x_1, \ldots, x_n \rangle$.

In particular, if $\{u_1, \ldots, u_n\}$ is an orthonormal basis, then for any $v \in V$, $\sum_{i=1}^{n} |(v, u_i)|^2 = \|v\|^2$.

12. Show that if $\{u_1, \ldots, u_n\}$ is an orthonormal basis of V, then for $v, w \in V$,

$$(v, w) = \sum_{i=1}^{n} (v, u_i)\, \overline{(w, u_i)}.$$

Deduce that vectors are pairwise orthogonal if and only if for $p \neq q$,

$$\sum_{i=1}^{n} (v_p, u_i)\, \overline{(v_q, u_i)} = 0.$$

13. Two ordered sets of vectors $\{u_1, \ldots, u_r\}$ and $\{v_1, \ldots, v_r\}$ are said to be **biorthogonal** if $(u_i, v_j) = \delta_{ij}$, $i, j = 1, \ldots, r$. Prove that if $\{u_1, \ldots, u_r\}$ and $\{v_1, \ldots, v_r\}$ is a biorthogonal system, then each set consists of linearly independent vectors.

14. Prove that for biorthogonal system $\{u_1, \ldots, u_r\}$ and $\{v_1, \ldots, v_r\}$,

$$\|x\|^2 = \sum_{i=1}^{r} (x, u_i)\, \overline{(x, v_i)},$$

for any $x \in V$ if and only if $\{u_1, \ldots, u_r\}$ is a basis of V.

15. What is $W^\perp$ if W is a subspace of F^4 given by:

(a) $W = \left\{ \begin{bmatrix} x_1 \\ x_2 \\ x_3 \\ x_4 \end{bmatrix} \;\middle|\; \begin{matrix} 2x_1 + 3x_2 + 5x_3 + x_4 = 0 \\ x_1 + x_2 + x_3 = 0 \end{matrix} \right\}$?

(b) $W = \left\langle \begin{bmatrix} 2 \\ 1 \\ 0 \\ 2 \end{bmatrix}, \begin{bmatrix} 3 \\ 1 \\ 1 \\ 1 \end{bmatrix} \right\rangle$?

16. Find an orthogonal complement of the subspace of $\mathbb{C}^{n\times n}$ consisting of all diagonal matrices, with the inner product on $\mathbb{C}^{n\times n}$ given by $(A, B) = \operatorname{tr}(AB^*)$.

17. Find the orthogonal projection of vector x onto subspace W of F^4 if:

(a) $x = \begin{bmatrix} 1 \\ 2 \\ 2 \\ 9 \end{bmatrix}$ and $W = \left\langle \begin{bmatrix} 2 \\ 1 \\ -1 \\ 2 \end{bmatrix}, \begin{bmatrix} 1 \\ 2 \\ 0 \\ 1 \end{bmatrix} \right\rangle$;

(b) $x = \begin{bmatrix} 3 \\ 2 \\ 2 \\ 5 \end{bmatrix}$ and $W = \left\{ \begin{bmatrix} x_1 \\ x_2 \\ x_3 \\ x_4 \end{bmatrix} \;\middle|\; \begin{matrix} 2x_1 + x_2 + x_3 - 2x_4 = 0 \\ x_1 + 2x_2 + x_3 - 3x_4 = 0 \\ x_1 + x_2 + 2x_3 + x_4 = 0 \end{matrix} \right\}$.

18. Let $V = \mathbb{R}_5[x]$ with the inner product $(p(x), q(x)) = \int_{-1}^{1} p(x)\, q(x)\, dx$, and let $\varepsilon \in \mathbb{R}_5[x]^*$ defined by $\varepsilon(f(x)) = f(1)$. Find $g(x) \in \mathbb{R}_5[x]$ such that $\varepsilon(f(x)) = (f(x), g(x))$ for all $f(x) \in \mathbb{R}_5[x]$.

19. Let $V = \mathbb{R}[x]$ with the inner product $(p(x), q(x)) = \int_{-1}^{1} p(x)\, q(x)\, dx$, and let $\varepsilon \in \mathbb{R}[x]^*$ defined by $\varepsilon(f(x)) = f(1)$. Prove that there does not exist any $g(x) \in \mathbb{R}[x]$ such that $\varepsilon(f(x)) = (f(x), g(x))$ for all $f(x) \in \mathbb{R}[x]$.

20. Apply Gram–Schmidt orthonormalization process to the vectors $\{1, x, x^2, x^3\}$ of $\mathbb{R}[x]$ and obtain corresponding orthonormal set if the inner product on $\mathbb{R}[x]$ is:
    (a) $(p(x), q(x)) = \int_{-1}^{1} \frac{p(x)\, q(x)}{\sqrt{1-x^2}}\, dx$;
    (b) $(p(x), q(x)) = \int_{-\infty}^{\infty} e^{-x^2}\, p(x)\, q(x)\, dx$.

3. The adjoint of a linear transformation

Throughout this section, we will restrict ourselves to finite dimensional inner product spaces.

THEOREM 3.1. *Let V and W be finite dimensional inner product spaces and let $T \in L(V, W)$. There exists a unique linear mapping $T^* : W \to V$ such that for all $v \in V$ and $w \in W$:*

$$(Tv, w) = (v, T^*w).$$

*The mapping T^* is called the **adjoint** of T.*

PROOF. For fixed w in W, the function $f_w : V \to F$ defined by $f_w(v) = (Tv, w)$, is a linear functional on V. By Riesz representation theorem, there exists a unique $u \in V$ such that $(Tv, w) = (v, u)$ for all $v \in V$. If we set $T^*(w) = u$, then $(Tv, w) = (v, T^*w)$ for all $v \in V$. This defines a mapping $T^* : W \to V$ which is obviously unique. We now verify that T^* is linear. For $x, y \in W$ and $\alpha, \beta \in F$:

$$\begin{aligned}(v, T^*(\alpha x + \beta y)) &= (Tv, \alpha x + \beta y)\\ &= \overline{\alpha}(Tv, x) + \overline{\beta}(Tv, y)\\ &= \overline{\alpha}(v, T^*x) + \overline{\beta}(v, T^*y)\\ &= (v, \alpha T^*x + \beta T^*y)\end{aligned}$$

for all $v \in V$. Hence

$$T^*(\alpha x + \beta y) = \alpha T^*x + \beta T^*y,$$

for all $x, y \in W$ and $\alpha, \beta \in F$. ∎

The following result lists some basic properties of the adjoint.

PROPOSITION 3.2. *Let V and W be finite dimensional inner product spaces over F.*

(i) *If $S, T \in L(V, W)$, then $(S+T)^* = S^* + T^*$, and for $\lambda \in F$, $(\lambda S)^* = \overline{\lambda} S^*$.*
(ii) *If $S \in L(V, W)$, then $S^{**} = S$, where $S^{**} = (S^*)^*$.*
(iii) *If $S, T \in L(V)$, then $(ST)^* = T^*S^*$.*
(iv) *If $T \in L(V)$ and T is invertible, then $(T^*)^{-1} = (T^{-1})^*$.*

PROOF. (i) For $v \in V$ and $w \in W$:

$$\begin{aligned}(v, (S+T)^* w) &= ((S+T)v, w) = (Sv, w) + (Tv, w) \\ &= (v, S^* w) + (v, T^* w) = (v, (S^* + T^*) w).\end{aligned}$$

Therefore, by uniqueness of adjoint mapping $(S+T)^* = S^* + T^*$.

(ii) By definition of adjoint mapping, for $v \in V$, $w \in W$,

$$(S^* w, v) = (w, S^{**} v).$$

Also

$$(S^* w, v) = \overline{(v, S^* w)} = \overline{(Sv, w)} = (w, Sv).$$

Thus, $(w, Sv) = (w, S^{**}v)$ for all $v \in V$, $w \in W$. Hence, $S^{**} = S$.

(iii) For $u, v \in V$,

$$(u, (ST)^* v) = ((ST) u, v) = (Tu, S^* v) = (u, T^* S^* v).$$

Hence, $(ST)^* = T^* S^*$.

(iv) Since for $u, v \in V$, $(u, v) = (Iu, v) = (u, I^* v)$, it follows that $I^* = I$, where I is the identity linear operator on V. If T is invertible, then $TT^{-1} = I$, and so using (iii), we have

$$I = I^* = (TT^{-1})^* = (T^{-1})^* T^*.$$

Hence, $(T^*)^{-1} = (T^{-1})^*$. ∎

Recall that if $A \in F^{m \times n}$, then we denote by A^*, the conjugate transpose of the matrix A. If we take standard inner product spaces F^n and F^m, and consider A as a linear transformation from inner product space F^n to the inner product space F^m, $x \mapsto Ax$, then for $v \in F^n$ and $w \in F^m$,

$$(Av, w) = w^* (Av) = (A^* v)^* w = (w, A^* v).$$

In the above statement the superscript $*$ denotes the conjugate transpose of a matrix. Thus, $A^* = \left(\overline{A}\right)^t$ is actually the adjoint of linear transformation A between standard inner product spaces.

Throughout the book, the superscript $*$ on the matrix will mean its conjugate transpose.

PROPOSITION 3.3. *Let V and W be finite dimensional inner product spaces over F and let $T \in L(V, W)$. If $\mathcal{B}$ and $\mathcal{B}'$ are ordered orthonormal bases of V and W respectively, then the matrix representation of T^* with respect to these bases is the conjugate transpose of the matrix representation of T with respect to these bases, that is,*

$$_{\mathcal{B}}[T^*]_{\mathcal{B}'} = {}_{\mathcal{B}'}[T]^*_{\mathcal{B}}.$$

PROOF. Let $\mathcal{B} = \{v_1, \ldots, v_n\}$ and $\mathcal{B}' = \{w_1, \ldots, w_m\}$ be ordered orthonormal bases of V and W, respectively. Let $_{\mathcal{B}'}[T]^*_{\mathcal{B}} = A \in F^{m \times n}$ and let $_{\mathcal{B}}[T^*]_{\mathcal{B}'} = B \in F^{n \times m}$. Then by Example 6, p.124, it follows that

$$[B]_{ij} = (T^* w_j, v_i) = \overline{(v_i, T^* w_j)} = \overline{(Tv_i, w_j)} = \left[\overline{A}\right]_{ji} = [A^*]_{ij}.$$

Hence, $B = A^*$ or $_{\mathcal{B}}[T^*]_{\mathcal{B}'} = {}_{\mathcal{B}'}[T]^*_{\mathcal{B}}$. ∎

COROLLARY 3.4. *Let T be a linear operator on a finite dimensional inner product space V over F. Then* $\det T^* = \overline{\det T}$ *and* $\operatorname{tr} T^* = \overline{\operatorname{tr} T}$.

PROOF. Exercise. ∎

COROLLARY 3.5. *Let V and W be finite dimensional inner product spaces over F. Then $L(V,W)$ becomes an inner product space over F if we define*

$$(S,T) = \operatorname{tr}(ST^*).$$

PROOF. If A and B are matrix representations of S and T with respect to ordered orthonormal bases of V and W, then

$$\operatorname{tr}(ST^*) = \operatorname{tr}(AB^*) = \sum_i \sum_j [A]_{ij} [B^*]_{ji} = \sum_i \sum_j [A]_{ij} \overline{[B]_{ij}}.$$

Now verify that this is indeed an inner product. ∎

PROPOSITION 3.6. *Let V and W be finite dimensional inner product spaces over F and let $T \in L(V,W)$.*

(i) $\ker T^* = (\operatorname{im} T)^\perp$ *and* $\operatorname{im} T^* = (\ker T)^\perp$.
(ii) $V = \ker T \oplus \operatorname{im} T^*$ *and* $W = \operatorname{im} T \oplus \ker T^*$.
(iii) $\ker T^*T = \ker T$ *and* $\operatorname{im} T^*T = \operatorname{im} T^*$.

PROOF. (i) $w \in (\operatorname{im} T)^\perp$ if and only if $(Tv, w) = 0$ for all $v \in V$, that is, $(v, T^*w) = 0$ for all $v \in V$, and so $T^*w = 0$ or $w \in \ker T^*$. Hence, $\ker T^* = (\operatorname{im} T)^\perp$.

Now replacing T by T^* we have $\ker T^{**} = (\operatorname{im} T^*)^\perp$, that is, $\ker T = (\operatorname{im} T^*)^\perp$. Hence, $\operatorname{im} T^* = (\ker T)^\perp$.

(ii) Since $V = \ker T \oplus (\ker T)^\perp$ and by (i) $\operatorname{Im} T^* = (\ker T)^\perp$, $V = \ker T \oplus \operatorname{im} T^*$. Similarly, $W = \operatorname{im} T \oplus \ker T^*$.

(iii) Clearly, $\ker T \subseteq \ker T^*T$. If $v \in \ker T^*T$, then $(v, T^*Tv) = 0$, and so $(Tv, Tv) = \|Tv\|^2 = 0$, that is, $Tv = 0$ or $v \in \ker T$. Hence, $\ker T = \ker T^*T$.

For the second part, obviously $\operatorname{im} T^*T \subseteq \operatorname{im} T^*$. From (i),

$$\begin{aligned} \dim \operatorname{im} T^* &= \dim T^\perp = \dim V - \dim \ker T \\ &= \dim V - \dim \ker T^*T = \dim \operatorname{im} T^*T. \end{aligned}$$

Therefore, $\operatorname{im} T^* = \operatorname{im} T^*T$. ∎

3.1. Examples

1. Consider the inner product space $F^{n\times n}$ with the inner product

$$(A, B) = \operatorname{tr}(AB^*).$$

For $P \in F^{n\times n}$, let λ_P be the linear operator on $F^{n\times n}$, the left multiplication by P. Then for $A, B \in F^{n\times n}$, we have

$$\begin{aligned} (\lambda_P(A), B) &= (PA, B) = \operatorname{tr}(PAB^*) = \operatorname{tr}(AB^*P) \\ &= \operatorname{tr}\left(A(P^*B)^*\right) = (A, \lambda_{P^*}(B)). \end{aligned}$$

Thus, the adjoint of λ_P is λ_{P^*}.

2. Consider the inner product space $\mathbb{R}_3[x]$ with the inner product

$$(p(x), q(x)) = \int_{-1}^{1} p(x)\, q(x)\, dx.$$

We find the adjoint of the differential operator D. For this, it is sufficient to find the action of D^* on the basis $\{1, x, x^2, x^3\}$. To determine $D^*(1)$, we consider the equation

$$(Dp(x), 1) = (p(x), D^*(1))$$

for all $p(x) \in \mathbb{R}_3[x]$. By Riesz representation theorem, $(Dp(x), 1) = (p(x), r(x))$ and then $D^*(1) = r(x)$. To compute $r(x)$, let $f_1(p(x)) = (Dp(x), 1)$. Then $r(x) = f_1(u(x))\, u(x)$, where $\ker f_1^{\perp} = \langle u(x) \rangle$. Also $p(x) = p_0 + p_1 x + p_2 x^2 + p_3 x^3 \in \ker f_1$ if and only if

$$\int_{-1}^{1} (p_1 + 2p_2 x + 3p_3 x^2)\, dx = 0.$$

Thus, $\ker f_1 = \langle 1, x - x^3, x^2 \rangle$. Now $u(x) \in \ker f_1^{\perp}$ if and only if

$$(u(x), 1) = 0 = (u(x), x^2) = (u(x), x - x^3).$$

If $u(x) = u_0 + u_1 x + u_2 x^2 + u_3 x^3$, then these equations imply that we can take $u(x) = 7x^3 - 3x$, and so $r(x) = 8u(x)$. Hence, $D^*(1) = 56x^3 - 24x$.

Similarly, taking the linear functional $f_x(p(x)) = (Dp(x), x)$ and using Riesz representation theorem, we have $(Dp(x), x) = (p(x), 12x^2 - 4)$, $p(x) \in \mathbb{R}_3[x]$, and $D^*(x) = 12x^2 - 4$. Similar calculations will give $D^*(x^2) = \frac{124}{3}(35x^3 - x)$, $D^*(x^3) = \frac{12}{5}(3x^2 - 1)$.

3. Consider the infinite dimensional inner product space $\mathbb{R}[x]$ with the inner product

$$(p(x), q(x)) = \int_{-1}^{1} p(x)\, q(x)\, dx.$$

The adjoint of differential operator D does not exist.

For $p(x), q(x) \in \mathbb{R}[x]$:

$$\begin{aligned} (Dp(x), q(x)) &= \int_{-1}^{1} p'(x)\, q(x)\, dx \\ &= p(x)\, q(x)|_{-1}^{1} - \int_{-1}^{1} p(x)\, q'(x)\, dx \\ &= p(1)\, q(1) - p(-1)\, q(-1) - (p(x), D\, q(x)). \end{aligned}$$

If the adjoint of D exist, then $(Dp(x), q(x)) = (p(x), D^* q(x))$, and so

$$(p(x), Dq(x) + D^* q(x)) = p(1)\, q(1) - p(-1)\, q(-1) = \varepsilon_{q(x)}(p(x)),$$

where $\varepsilon_{q(x)}(p(x)) = p(1)\, q(1) - p(-1)\, q(-1)$, a linear functional on $\mathbb{R}[x]$. This implies that for each $q(x) \in \mathbb{R}[x]$, for linear functional $\varepsilon_{q(x)}$, there exists $r(x) = Dq(x) + D^* q(x)$ such that $\varepsilon_{q(x)}(p(x)) = (p(x), r(x))$ for all $p(x) \in \mathbb{R}[x]$, that is,

$$\int_{-1}^{1} p(x)\, r(x)\, dx = p(1)\, q(1) - p(-1)\, q(-1).$$

In particular, for $p(x) = (x^2 - 1)^2 r(x)$, we have $\int_{-1}^{1} ((x^2 - 1)\, r(x))^2\, dx = 0$. Therefore, $(x^2 - 1)\, r(x) = 0$. Since $x^2 - 1 \neq 0$, $r(x) = 0$. But then $\varepsilon_{q(x)} \equiv 0$. Verify that $\varepsilon_{q(x)} \equiv 0$ if and only if $x^2 - 1$ is a factor of $q(x)$. Thus we have a contradiction. Hence, the adjoint of D does not exist.

4. Consider a general system of linear equations $Ax = b$, $A \in F^{m\times n}$, $x \in F^n$, $b \in F^m$. If $b \notin \operatorname{im} A$, then the system is inconsistent. In this case the best we can do is to find $x_0 \in F^m$ such that $\|Ax_0 - b\|$ is the least. Such a solution vector x_0 gives the best approximation of Ax_0 to b, called the **least square solution** of $Ax = b$.

Let $\widetilde{b} = \Pr_{\operatorname{im} A}(b)$. Then $\widetilde{b} \in \operatorname{im} A$ and according to Theorem 2.6, $\widetilde{b}$ is the vector closest to b among all the vectors from the subspace $\operatorname{im} A$. Let $x_0 \in \mathbb{R}^m$ such that $Ax_0 = \widetilde{b}$. Then x_0 is such that

$$\|Ax_0 - b\| \leq \|Ax - b\| \text{ for all } x \in \mathbb{R}^n.$$

In short, since $F^m = \operatorname{im} A \oplus \ker A^*$, there are unique elements $\widetilde{b} \in \operatorname{im} A$ and $b' \in \ker A^*$ such that $b = \widetilde{b} + b'$ and then a solution of the equation $Ax = \widetilde{b}$ is a least square solution.

The computation of $\widetilde{b}$ from b could be quite difficult as $\widetilde{b} = \Pr_{\operatorname{im} A}(b)$, and according to Theorem 2.6, for this we require to find an orthonormal basis of $\operatorname{im} A$. The following result gives a direct method of finding $\widetilde{b}$.

A vector $x_0 \in \mathbb{R}^m$ is a least square solution of equation $Ax = b$ if and only if x_0 is a solution of the normal equation $A^*Ax = A^*b$.

Let $b = \widetilde{b} + b'$, $\widetilde{b} \in \operatorname{im} A$, $b' \in \ker A^*$ and let x_0 be a least square solution of $Ax = b$. Then $Ax_0 = \widetilde{b}$ and $Ax_0 - b = Ax_0 - \left(\widetilde{b} + b'\right) = -b'$, and so $A^*(Ax_0 - b) = 0$, that is, x_0 is a solution of equation $A^*Ax = A^*b$.

Conversely, let x_0 be a solution of equation $A^*Ax = A^*b$. Then $Ax_0 - b \in \ker A^*$. Since, $b \in F^m$ and $F^m = \operatorname{im} A + \ker A^*$, $b = \widetilde{b} + b'$, $\widetilde{b} \in \operatorname{im} A$ and $b' \in \ker A^*$. Thus, $Ax_0 - \widetilde{b} = (Ax_0 - b) + b' \in \ker A^*$. But $Ax_0 - b \in \operatorname{im} A$ also. Thus, $Ax_0 - b \in \operatorname{im} A \cap \ker A^* = \{0\}$. Therefore, x_0 is a least square solution. This proves the above statement.

Next, if $\operatorname{rank} A = n$, then A^*A is nonsingular, and so the normal equation $A^*Ax = A^*b$ has unique solution $(A^*A)^{-1} A^*b$. Thus, in this case we have a unique least square solution.

If $\operatorname{rank} A < n$, then $\operatorname{rank} A^*A < n$, and so the equation $A^*Ax = A^*b$ has infinitely many solutions. From Example 8, p.125, we see that this equation has a unique solution $\widehat{x} \in \ker (A^*A)^{\perp}$. Since, $\ker A^*A = \ker A$, any vector $\widehat{x} + y$, $y \in \ker A$ is also a least square solution. Hence we have infinitely many solutions. But for any such solution $\widehat{x} + y$, $\widehat{x} \in \ker (A^*A)^{\perp}$, $y \in \ker A = \ker A^*A$, we have $\|\widehat{x} + y\| = \|\widehat{x}\| + \|y\| \geq \|\widehat{x}\|$. This shows that $\widehat{x}$ is a unique least square solution of the least norm, that is,

$$\|\widehat{x}\| = \min_{u \in \mathbb{R}^n} \{\|u\| \mid A^*Au = A^*b\}.$$

3.2. Exercises

1. Let $T \in L\left(\mathbb{R}^4\right)$ whose action on the standard basis is

$$\begin{aligned} Te_1 &= e_1 - e_2, \\ Te_2 &= e_1 + e_2 + e_3, \\ Te_3 &= e_1 + e_2 - e_3 + e_4, \\ Te_4 &= -e_1 + e_2 + e_3 + e_4. \end{aligned}$$

For $w = \begin{bmatrix} 1 & 1 & 1 & 1 \end{bmatrix}^t$, find $w' \in \mathbb{R}^4$ such that $(Tv, w) = (v, w')$ for all $v \in \mathbb{R}^4$.

2. Let

$$\left\{ v_1 = \begin{bmatrix} i \\ 1 \\ 1 \\ 1 \end{bmatrix}, v_2 = \begin{bmatrix} 1 \\ i \\ 1 \\ 1 \end{bmatrix}, v_3 = \begin{bmatrix} 1 \\ 1 \\ i \\ 1 \end{bmatrix}, v_4 = \begin{bmatrix} 1 \\ 1 \\ 1 \\ i \end{bmatrix} \right\}$$

be a basis of $\mathbb{C}^4$ and let $T \in L(\mathbb{C}^4)$:

$$\begin{aligned} Tv_1 &= iv_1 + v_2 - v_3 + v_4, \\ Tv_2 &= v_1 + iv_2 + v_3 - iv_4, \\ Tv_3 &= -iv_1 + v_2 + iv_3 + v_4, \\ Tv_4 &= v_1 - iv_2 + v_3 + iv_4. \end{aligned}$$

What is the matrix of T^* with respect to this basis $\{v_1, v_2, v_3, v_4\}$?

3. Let $V = \mathbb{R}_4[x]$ with the inner product $(p(x), q(x)) = \int_{-1}^{1} p(x)\, q(x)\, dx$. Find the adjoint of a linear operator $T \in L(V)$, where

$$T(p(x)) = p^{(2)}(x) + 2p^{(1)}(x) + p(x).$$

4. Prove that λ is an eigenvalue of T if and only if $\overline{\lambda}$ is an eigenvalue of T^*.

5. Let $A, B \in \mathbb{C}^{n \times n}$. Prove that $\operatorname{im} A \perp \operatorname{im} B$ if and only if $A^*B = 0$.

6. Find the least square solution of the equation:

$$\begin{aligned} x_1 - x_2 + 2x_3 &= 1, \\ -x_1 + 2x_2 &= 1, \\ x_1 + 4x_3 &= 1. \end{aligned}$$

In the following exercises, V and W are finite dimensional inner product spaces over F.

7. Let $T \in L(V, W)$. Do T and T^* have the same nonzero eigenvalues? How about T^*T and TT^*?

8. Let $T \in L(V, W)$ and let $\mathcal{B}$ and $\mathcal{B}'$ be ordered orthonormal bases of V and W, respectively. Show that ${}_{\mathcal{B}}[T^*]_{\mathcal{B}'} = {}_{\mathcal{B}'}[T]^*_{\mathcal{B}}$.

9. Let $T \in L(V)$ and U be a subspace of V. Prove that U is T–invariant if and only if $U^\perp$ is T^*–invariant.

10. Let $T \in L(V)$. Show that if $u_1, \ldots, u_k$ are ordered orthonormal eigenvectors of T^*T corresponding to nonzero eigenvalues, then $Tu_1, \ldots, Tu_k$ are ordered orthonormal eigenvectors of TT^* corresponding to the same eigenvalues.

    A converse statement is obtained by replacing T by T^*.

11. Let $T \in L(V, W)$. Prove that either the equation $Tx = b$ is solvable for any $b \in W$ or the equation $T^*y = 0$ has nonzero solutions.

4. Unitary operators

A linear operator T on an inner product space V is called **unitary** if $TT^* = I = T^*T$, that is, T is invertible and $T^{-1} = T^*$.

If S and T are unitary operators, then ST and S^{-1} are also unitary. Since $(ST)^{-1} = T^{-1}S^{-1} = T^*S^* = (ST)^*$ and $(S^{-1})^{-1} = (S^*)^{-1} = (S^{-1})^*$. The product of two unitary operators is unitary, and the inverse of a unitary operator is also unitary.

The following is a characterization of unitary operators on a finite dimensional inner product space.

PROPOSITION 4.1. *Let V be a finite dimensional inner product space and let $T \in L(V)$. The following statements are equivalent:*

(i) *T is unitary;*
(ii) *$(Tu, Tv) = (u, v)$ for all $u, v \in V$;*
(iii) *$\|Tu\| = \|u\|$ for all $u \in V$;*
(iv) *T maps orthonormal bases to orthonormal bases.*

PROOF. For $u, v \in V$, $(Tu, Tv) = (u, T^*Tv)$. Thus, $(Tu, Tv) = (u, v)$ if and only if $T^*T = I$. This proves the equivalence of (i) and (ii). That (ii) implies (iii) is obvious, and (iii) implies (ii) because of polarization identities (Proposition 1.1(iv)). Hence (ii) and (iii) are equivalent.

Clearly, if (ii) holds, then T maps orthonormal bases to orthonormal bases. Conversely, assume that (iv) holds and let $\{u_1, \ldots, u_n\}$ be an orthonormal basis of V. For $u, v \in V$, write $u = \sum_{i=1}^{n} \alpha_i u_i$ and $v = \sum_{i=1}^{n} \beta_i u_i$. Then

$$\begin{aligned}(Tu, Tv) &= \left(\sum_{i=1}^{n} \alpha_i Tu_i, \sum_{j=1}^{n} \beta_j Tu_j\right) = \sum_{i=1}^{n}\sum_{j=1}^{n} \alpha_i \overline{\beta_j} (Tu_i, Tu_j) \\ &= \sum_{i=1}^{n}\sum_{j=1}^{n} \alpha_i \overline{\beta_j} \delta_{ij} = \sum_{i=1}^{n} \alpha_i \overline{\beta_i} = (u, v).\end{aligned}$$

Thus, (iv) implies (ii). ∎

COROLLARY 4.2. *Eigenvalues of a unitary operator have absolute value 1.*

PROOF. If λ is an eigenvalue of a unitary operator T, and u is a corresponding eigenvector, then by (iii) of the above proposition, $\|u\| = \|Tu\| = |\lambda| \|u\|$, and so $|\lambda| = 1$. ∎

In particular, if T is a linear operator on a real inner product space, then eigenvalues of T are 1 or -1.

A linear operator T on an inner product space V is said to preserve the inner product if $(Tu, Tv) = (u, v)$ for all $u, v \in V$, and is said to preserve the norm if $\|Tu\| = \|u\|$ for all $u \in V$. Note that the proof of equivalence of (ii) and (iii) in Proposition 4.1 does not require the finite dimensionality of the inner product space. Thus, a linear operator on an inner product space, not necessarily finite dimensional, preserves the inner product if and only if it preserves the norm. But on an infinite dimensional inner product space a norm preserving linear operator may not be bijective, and so may not be unitary. For example, consider the inner product space l^2. The right shift operator $R : l^2 \to l^2$ defined by $(x_1, x_2, \ldots) \mapsto (0, x_1, x_2, \ldots)$ is a norm preserving linear map which is not surjective. However, if T is a norm preserving linear operator on V, then T is always injective. Since for such T, $||Tv|| = ||v||$ for all $v \in V$, and so $Tv = 0$ if and only if $v = 0$.

A matrix $A \in F^{n\times n}$ is called **unitary** if A is invertible and $A^{-1} = A^*$, and **orthogonal** if $A^{-1} = A^t$. Clearly, a unitary matrix with entries from $\mathbb{R}$is orthogonal.

PROPOSITION 4.3. *Let T be a linear operator on a finite dimensional inner product space V and let $\mathcal{B}$ be an ordered orthonormal basis of V. Then T is unitary if and only if $[T]_{\mathcal{B}}$ is a unitary matrix. If V is an inner product space over $\mathbb{R}$, then $[T]_{\mathcal{B}}$ is actually an orthogonal matrix.*

PROOF. Exercise. ∎

From Proposition 4.3 it is immediate that $n \times n$ matrices when considered as operators on the standard inner product space F^n, unitary matrices are precisely unitary operators. Proposition 4.1 gives the following characterization for unitary matrices.

PROPOSITION 4.4. *Let $A \in F^{n\times n}$. The following statements are equivalent:*

(i) *A is a unitary matrix;*
(ii) *the columns of A form an orthonormal basis of the standard inner product space F^n;*
(iii) *the columns of A^t form an orthonormal basis of the standard inner product space F^n.*

PROOF. *(i)* $\Rightarrow$ *(ii)* If A is a unitary matrix, then $A : F^n \to F^n$ defined by $u \mapsto Au$ is a unitary operator. Since the standard basis $\{e_1, \ldots, e_n\}$ is orthonormal, $\{Ae_1, \ldots, Ae_n\}$ is an orthonormal basis of F^n. Hence, the columns of A form an orthonormal basis of the standard inner product space F^n.

(ii) $\Rightarrow$ *(i)* First, recall that the (i, j)–th entry of a matrix B is $e_i^t B e_j$. Assume that the columns of A are orthonormal. Then $e_i^t (A^*A) e_j = (Ae_i)^* (Ae_j) = \delta_{ij}$. Hence, the (i, j)–th entry of A^*A is δ_{ij}, that is, $A^*A = I_n$.

The equivalence of (i) and (iii) is now obvious as A is unitary if and only if A^t is unitary. ∎

COROLLARY 4.5. *Let T be a unitary operator on V, $\dim V = n$. If $\mathcal{B}_1$ and $\mathcal{B}_2$ are ordered orthonormal bases of V, then ${}_{\mathcal{B}_2}[T]_{\mathcal{B}_1}$ is a unitary matrix.*

PROOF. Let $\mathcal{B}_1 = \{u_1, \ldots, u_n\}$, $\mathcal{B}_2 = \{v_1, \ldots, v_n\}$ and let $A = {}_{\mathcal{B}_2}[T]_{\mathcal{B}_1}$. Since T is unitary, $\{Tu_1, \ldots, Tu_n\}$ is also an orthonormal basis of V. By definition of ${}_{\mathcal{B}_2}[T]_{\mathcal{B}_1}$:

$$Tu_r = \sum_{i=1}^{n} [A]_{ir}\, v_i\,,\; r = 1, \ldots, n.$$

Therefore,

$$\begin{aligned}\delta_{rs} &= (Tu_r, Tu_s) = \left(\sum_{i=1}^{n} [A]_{ir}\, v_i\,, \sum_{j=1}^{n} [A]_{js}\, v_j\right)\\ &= \sum_{i=1}^{n}\sum_{j=1}^{n} [A]_{ir}\, \overline{[A]}_{js}\, (v_i\,, v_j) = \sum_{i=1}^{n} [A]_{ir}\, \overline{[A]}_{is}\,.\end{aligned}$$

Hence, the columns of A are orthonormal, and A is a unitary matrix. ∎

Our aim is to prove Schur's theorem: a triangulable linear operator on a finite dimensional inner product space can actually be triangularized with respect to an orthonormal basis. We need few facts to prove this result.

PROPOSITION 4.6. *Denote by $T(n)$, the set of all $n \times n$ upper triangular matrices with entries from F, that is,*

$$T(n) = \left\{A \in F^{n\times n} \mid [A]_{ij} = 0 \text{ whenever } i > j\right\}.$$

If $A, B \in T(n)$, then $AB \in T(n)$, that is, $T(n)$ is closed under matrix multiplication. If A is an invertible matrix in $T(n)$, then $A^{-1} \in T(n)$.

PROOF.

$$[AB]_{ij} = \sum_{r=1}^{n} [A]_{ir} [B]_{rj} = \sum_{r>j} [A]_{ir} [B]_{rj} + \sum_{r\leq j} [A]_{ir} [B]_{rj}.$$

To show that $AB \in T(n)$, we need to verify that $[AB]_{ij} = 0$ for $i > j$. Assume that $i > j$, then for $r > j$, $[B]_{rj} = 0$, and for $r \leq j$, $i > j \geq r$, and so $[A]_{ir} = 0$. Thus, $[AB]_{ij} = 0$ for $i > j$.

If A is invertible matrix and is in $T(n)$, then as $\det A$ is the product of the diagonal entries, all diagonal entries of A are nonzero. The j-th column of A^{-1} is precisely the solution of equation

$$Ax = e_j.$$

Thus to show that $A^{-1} \in T(n)$, it is enough to show that the last $n-j$ entries of the solution vector of the above equation are zero. Write this equation in the following manner:

$$\begin{bmatrix} A_1 & A_{12} \\ 0 & A_2 \end{bmatrix} \begin{bmatrix} \widehat{x} \\ x' \end{bmatrix} = \begin{bmatrix} b \\ 0 \end{bmatrix},$$

$A_1 \in F^{j\times j}$, $A_2 \in F^{(n-j)\times(n-j)}$, $\widehat{x} \in F^j$, $b = \begin{bmatrix} 0 & \cdots & 0 & 1 \end{bmatrix}^t \in F^j$. Then $A_2 x' = 0$. Since A_2 is invertible, $x' = 0$. Therefore, the solution vector of the equation $Ax = e_j$ has last $n-j$ entries zero. Hence, $A^{-1} \in T(n)$. ∎

COROLLARY 4.7. *Let $T(n)^+$ be the set of all upper triangular matrices with positive diagonal entries:*

$$T(n)^+ = \{A \in T(n) \mid [A]_{ii} > 0,\ i = 1, \dots, n\}.$$

Then $T(n)^+$ is closed under matrix multiplication and inverse.

PROOF. If $A, B \in T(n)^+$, then by Proposition 4.6, $AB \in T(n)$. Since $[AB]_{ii} = [A]_{ii} [B]_{ii} > 0$, $i = 1, \dots, n$, $AB \in T(n)^+$. Hence, $T(n)^+$ is closed under matrix multiplication.

If $A \in T(n)^+$, then A is clearly invertible and so by Proposition 4.6, $A^{-1} \in T(n)$. Since $\left[A^{-1}\right]_{ii} = \frac{1}{[A]_{ii}}$, $A^{-1} \in T(n)^+$. ∎

A matrix $A \in F^{n\times n}$ is called lower triangular if $A^t \in T(n)$. Clearly, a diagonal matrix is lower as well as upper triangular. If A is a unitary upper triangular matrix, then as $A^{-1} = A^*$, and as $A^{-1} \in T(n)$, so A has to be a diagonal matrix. Since the eigenvalues of a unitary matrix are of modulus 1, we have the following result.

PROPOSITION 4.8. *If $A \in F^{n\times n}$ is a unitary upper triangular matrix, then A is a diagonal matrix with diagonal entries complex numbers with modulus 1. In particular, if $A \in T(n)^+$ and is unitary, then $A = I_n$.*

Now we have the main result.

THEOREM 4.9. ***Schur's theorem.*** *Let T be a triangulable linear operator on a finite dimensional inner product space V. There exists an ordered orthonormal basis $\mathcal{B}$ of V such that the matrix of T with respect to $\mathcal{B}$ is upper triangular.*

PROOF. If T is triangulable, there is an ordered basis $\mathcal{B}_1 = \{a_1, \ldots, a_n\}$ of V such that $[T]_{\mathcal{B}_1}$ is upper triangular. If $\mathcal{B}_1$ is not orthonormal, by Gram–Schmidt orthonormalization process, we have an ordered orthonormal basis $\mathcal{B} = \{u_1, \ldots, u_n\}$ such that

$$\begin{aligned} a_1 &= \alpha_{11}u_1, \\ a_2 &= \alpha_{12}u_1 + \alpha_{22}u_2, \\ &\vdots \\ a_n &= \alpha_{1n}u_1 + \cdots + \alpha_{nn}u_n, \end{aligned}$$

where $\alpha_{ij} \in F$, $i, j = 1, \ldots, n$. If I is the identity operator on V, then

$${}_{\mathcal{B}}[I]_{\mathcal{B}_1} = \begin{bmatrix} \alpha_{11} & \cdots & \alpha_{1n} \\ & \ddots & \vdots \\ 0 & & \alpha_{nn} \end{bmatrix} \in T(n).$$

Since, $[T]_{\mathcal{B}} = {}_{\mathcal{B}}[I]_{\mathcal{B}_1}\, [T]_{\mathcal{B}_1}\, {}_{\mathcal{B}_1}[I]_{\mathcal{B}}$, by Proposition 4.6, $[T]_{\mathcal{B}}$ is an upper triangular matrix and $\mathcal{B}$ is required orthonormal basis. ∎

COROLLARY 4.10. *If T is unitary and triangulable linear operator on a finite dimensional inner product space V. Then there exists an orthonormal basis of V consisting of eigenvectors of T. In particular, T is diagonalizable.*

PROOF. Immediate from Theorem 4.9 and Proposition 4.8. ∎

4.1. Examples

1. The matrix analogue of Schur's theorem states that for $A \in \mathbb{C}^{n\times n}$ there exists a unitary matrix U such that U^*AU is upper triangular. If $A \in \mathbb{R}^{n\times n}$ and all eigenvalues of A are real, then the matrix U is actually orthogonal. We give a procedure of finding this unitary matrix U.

   Let λ_1 be an eigenvalue of A and let v_1 be the corresponding eigenvector. Extend $\{v_1\}$ to $\{v_1, \ldots, v_n\}$, an orthogonal basis of $\mathbb{C}^n$. Then $\{u_1, \ldots, u_n\}$ is an orthonormal basis of $\mathbb{C}^n$, where $u_i = \frac{v_i}{\|v_i\|}$, $i = 1, \ldots, n$. Let $U_1 = \begin{bmatrix} u_1 & \cdots & u_n \end{bmatrix}$. Then U_1 is unitary and

   $$U_1^*AU_1 = \begin{bmatrix} \lambda_1 & ** \\ 0 & A_2 \end{bmatrix},$$

   $A_2 \in \mathbb{C}^{(n-1)\times(n-1)}$. As usual, $**$ denotes those entries of the matrix which are of no interest to us. By induction, there exists a unitary matrix V such that V^*A_2V is upper triangular. Let $U_2 = \begin{bmatrix} 1 & 0 \\ 0 & V \end{bmatrix}$. Then U_2 is unitary and

$$U_2^*\left(U_1^*AU_1\right)U_2=\begin{bmatrix}\lambda_1 & ** \\ 0 & V^*A_2V\end{bmatrix},$$

is an upper triangular matrix. Thus, $U=U_1U_2$ is required unitary matrix.

To illustrate this procedure, let

$$A=\begin{bmatrix}1 & 0 & 2\\ 3 & 4 & 2\\ 1 & 1 & 1\end{bmatrix}.$$

A has eigenvalue 0 with eigenvector $v_1=\begin{bmatrix}2 & -1 & -1\end{bmatrix}^t$. Any nonzero vector $x=\begin{bmatrix}x_1 & x_2 & x_3\end{bmatrix}^t$ orthogonal to v_1 is such that $2x_1-x_2-x_3=0$. A nonzero solution of this equation is $x_1=1$, $x_2=1$, $x_3=1$. Let $v_2=\begin{bmatrix}1 & 1 & 1\end{bmatrix}^t$. Similarly, if $\begin{bmatrix}x_1 & x_2 & x_3\end{bmatrix}^t$ is orthogonal to v_1 and v_2, then $2x_1-x_2-x_3=0$ and $x_1+x_2+x_3=0$. A nonzero solution of these equations is $x_1=0$, $x_2=1$, $x_3=-1$, and so we can take $v_3=\begin{bmatrix}0 & 1 & -1\end{bmatrix}^t$. Thus, we have an orthonormal basis of $\mathbb{R}^3$ given by

$$\left\{\begin{bmatrix}2/\sqrt{6}\\ -1/\sqrt{6}\\ -1/\sqrt{6}\end{bmatrix},\begin{bmatrix}1/\sqrt{3}\\ 1/\sqrt{3}\\ 1/\sqrt{3}\end{bmatrix},\begin{bmatrix}0\\ 1/\sqrt{2}\\ -1/\sqrt{2}\end{bmatrix}\right\}.$$

Let

$$U_1=\begin{bmatrix}2/\sqrt{6} & 1/\sqrt{3} & 0\\ -1/\sqrt{6} & 1/\sqrt{3} & 1/\sqrt{2}\\ -1/\sqrt{6} & 1/\sqrt{3} & -1/\sqrt{2}\end{bmatrix}.$$

Then

$$U_1^*AU_1=\begin{bmatrix}0 & -\sqrt{2} & -\sqrt{3}\\ 0 & 5 & 0\\ 0 & \sqrt{6} & 1\end{bmatrix}.$$

For $A_2=\begin{bmatrix}5 & 0\\ \sqrt{6} & 1\end{bmatrix}$, clearly $\begin{bmatrix}0\\ 1\end{bmatrix}$ is an eigenvector corresponding to eigenvalue 1. A vector orthogonal to $\begin{bmatrix}0\\ 1\end{bmatrix}$ is $\begin{bmatrix}1\\ 0\end{bmatrix}$. Then $\left\{\begin{bmatrix}0\\ 1\end{bmatrix},\begin{bmatrix}1\\ 0\end{bmatrix}\right\}$ is an orthonormal basis of $\mathbb{R}^2$. If $V=\begin{bmatrix}0 & 1\\ 1 & 0\end{bmatrix}$, then $V^*A_2V=\begin{bmatrix}1 & \sqrt{6}\\ 0 & 5\end{bmatrix}$.

Now

$$U_2=\begin{bmatrix}1 & 0 & 0\\ 0 & 0 & 1\\ 0 & 1 & 0\end{bmatrix}$$

is a unitary matrix and

$$U_2^*\left(U_1^*AU_1\right)U_2=\begin{bmatrix}0 & -\sqrt{3} & -\sqrt{2}\\ 0 & 1 & \sqrt{6}\\ 0 & 0 & 5\end{bmatrix}.$$

Therefore,

$$U = U_1U_2 = \begin{bmatrix} 2/\sqrt{6} & 0 & 1/\sqrt{3} \\ -1/\sqrt{6} & 1/\sqrt{2} & 1/\sqrt{3} \\ -1/\sqrt{6} & -1/\sqrt{2} & 1/\sqrt{3} \end{bmatrix}$$

is required unitary matrix.

2. *Let V be a finite dimensional inner product space over $\mathbb{R}$ and let $f : V \to V$ be any mapping such that $f(0) = 0$ and $\|f(x) - f(y)\| = \|x - y\|$, for all $x, y \in V$. Then f is a unitary mapping, that is, $f(v) = Tv$ for all $v \in V$ and T is a unitary operator on V.*

   If $y = 0$, then $\|f(x)\| = \|x\|$, for all $x \in V$. Next, $\|f(x) - f(y)\|^2 = \|x - y\|^2$ implies that $(f(x) - f(y), f(x) - f(y)) = (x - y, x - y)$, for all $x, y \in V$. Thus, $(f(x), f(x)) - (f(y), f(x)) - (f(x), f(y)) + (f(y), f(y)) = (x, x) - (y, x) - (x, y) + (y, y)$, and so for $x, y \in V$, $(f(x), f(y)) = (x, y)$.

   This means that the mapping f preserves the inner product. Now let $\{u_1, \ldots, u_n\}$ be an orthonormal basis of V. Since $(f(u_i), f(u_j)) = (u_i, u_j) = \delta_{ij}$, $\{f(u_1), \ldots, f(u_n)\}$ is also an orthonormal basis of V.

   Let T be a linear mapping defined by $T(u_i) = f(u_i)$, $i = 1, \ldots, n$. Since T maps an orthonormal basis $\{u_1, \ldots, u_n\}$ to an orthonormal basis, T is unitary. For $v \in V$,

$$\begin{aligned} f(v) &= \sum_{i=1}^{n} (f(v), f(u_i))\, f(u_i) = \sum_{i=1}^{n} (v, u_i)\, T(u_i) \\ &= T\left(\sum_{i=1}^{n} (v, u_i)\, u_i\right) = Tv. \end{aligned}$$

   Hence, $f \equiv T$.

3. *An invertible matrix can be uniquely written as a product of a unitary matrix and an upper triangular matrix with positive diagonal entries.*

   Let $A = \begin{bmatrix} a_1 & \cdots & a_n \end{bmatrix}$ be an invertible matrix in $F^{n \times n}$. Since A is invertible, $\mathcal{B}_1 = \{a_1, \ldots, a_n\}$ is an ordered basis of the standard inner product F^n. Using Gram–Schmidt orthonormalization process, we have an orthonormal basis $\mathcal{B} = \{y_1, \ldots, y_n\}$, where

$$\begin{aligned} y_1 &= \frac{a_1}{\|a_1\|}, \\ y_j &= \frac{a_j - \sum_{i=1}^{j-1} (a_j, y_i)\, y_i}{\left\| a_j - \sum_{i=1}^{j-1} (a_j, y_i)\, y_i \right\|}, \quad j = 2, \ldots, n. \end{aligned}$$

   Thus, we can write

$$\begin{aligned} a_1 &= \alpha_{11} y_1 \\ a_2 &= \alpha_{12} y_1 + \alpha_{22} y_2 \\ &\vdots \\ a_n &= \alpha_{1n} y_1 + \cdots + \alpha_{nn} y_n \end{aligned}$$

$\alpha_{ij} \in F$ and $\alpha_{ii} > 0$, $i, j = 1, \ldots, n$. Now let

$$M = \begin{bmatrix} \alpha_{11} & \alpha_{12} & \cdots & \alpha_{1n} \\ 0 & \alpha_{22} & & \vdots \\ \vdots & & \ddots & \vdots \\ 0 & \cdots & \cdots & \alpha_{nn} \end{bmatrix}.$$

Then M is an upper triangular matrix with positive diagonal entries. If $U = \begin{bmatrix} y_1 & \cdots & y_n \end{bmatrix}$, then as columns of U form an orthonormal basis of F^n, U is a unitary matrix. Now it is easy to verify that $A = UM$.

To establish uniqueness of U and M, suppose that there also exist a unitary matrix $U_1 \in F^{n\times n}$ and M_1 an upper triangular matrix with positive diagonal entries such that $A = U_1 M_1$. Then $UM = U_1 M_1$, and so $U_1^* U = M_1 M^{-1}$. This means that $U_1^* U$ is an upper triangular unitary matrix with positive diagonal entries, and so $U_1^* U = I_n$. Hence, $U_1 = U$ and $M_1 = M$.

This result can be used to prove Schur's theorem: a square matrix with entries from $\mathbb{C}$ is unitarily similar to an upper triangular matrix.

If $A \in \mathbb{C}^{n\times n}$, then A is triangulable. Let P be an invertible matrix such that $P^{-1}AP$ is upper triangular. Write $P = UM$, where U is unitary and M is an upper triangular matrix with positive diagonal entries. Then $P^{-1}AP = (UM)^{-1} A (UM) = M^{-1} (U^* AU) M$. Thus, $U^* AU = M \left(P^{-1}AP\right) M^{-1}$. Since the right side is an upper triangular matrix, A is unitarily similar to an upper triangular matrix.

4. In $\mathbb{R}^3$, the standard inner product is precisely the dot product. For $u = \begin{bmatrix} u_1 & u_2 & u_3 \end{bmatrix}^t$, $v = \begin{bmatrix} v_1 & v_2 & v_3 \end{bmatrix}^t$:

$$(u, v) = v^t u = u_1 v_1 + u_2 v_2 + u_3 v_3,$$

and the angle α between u and v is given by:

$$\cos\alpha = \frac{(u, v)}{\|u\| \, \|v\|} = \frac{v^t u}{\sqrt{u^t u}\sqrt{v^t v}}.$$

Consider the following orthogonal matrix

$$R(\theta) = \begin{bmatrix} 1 & 0 & 0 \\ 0 & \cos\theta & -\sin\theta \\ 0 & \sin\theta & \cos\theta \end{bmatrix}.$$

Clearly, $R(\theta)\, e_1 = e_1$, $(R(\theta)\, e_2, e_2) = \cos\theta = (R(\theta)\, e_3, e_3)$. Geometrically this means that all vectors which are scalar multiple of e_1 are fixed by $R(\theta)$, or all vectors in the direction of e_1 are fixed by $R(\theta)$. Other two equations mean that all vectors in the direction of e_2 (respectively, e_3) are rotated by an angle θ in anticlockwise direction. This defines a rotation by an angle θ

and the axis of rotation is along e_1 vector (see figure below).

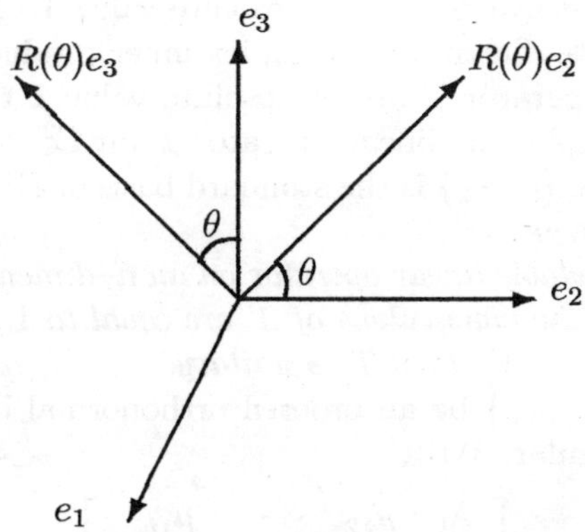

Let A be a 3×3 real orthogonal matrix with determinant 1. Then $\det(I_3 - A) = \det(AA^t - A) = \det\left(A(A - I_3)^t\right)$, and so $\det(A - I_3) = (-1)^3 \det(I_3 - A)$. Thus $\det(I_3 - A) = 0$, and 1 is an eigenvalue of A. If $A \neq I_3$, then the geometric multiplicity of 1 is one. Otherwise A has at least two linearly independent eigenvectors corresponding to 1. This implies that A is a triangulable orthogonal matrix, and so $A = I_3$.

Let x_1 be an eigenvector of A corresponding to 1 with $\|x_1\| = 1$. Then $\mathbb{R}^3 = \langle x_1 \rangle \oplus \langle x_1 \rangle^\perp$ and $\dim \langle x_1 \rangle^\perp = 2$. Since $\langle x_1 \rangle$ is A–invariant, $\langle x_1 \rangle^\perp$ is $A^* = A$–invariant. Let $\{x_2, x_3\}$ be an orthonormal basis of $\langle x_1 \rangle^\perp$. Then $\mathcal{B} = \{x_1, x_2, x_3\}$ is an orthonormal basis of $\mathbb{R}^3$. If $\mathcal{B}_0$ is the standard basis of $\mathbb{R}^3$, then $[A]_{\mathcal{B}} = {}_{\mathcal{B}}[I]_{\mathcal{B}_0} [A]_{\mathcal{B}_0} {}_{\mathcal{B}_0}[I]_{\mathcal{B}}$. By definition, $[A]_{\mathcal{B}_0} = A$ and ${}_{\mathcal{B}}[I]_{\mathcal{B}_0} = \begin{bmatrix} x_1 & x_2 & x_3 \end{bmatrix}$, an orthogonal matrix. Since $\langle x_1 \rangle^\perp$ is A–invariant, $[A]_{\mathcal{B}}$ is of the form

$$[A]_{\mathcal{B}} = \begin{bmatrix} 1 & 0 \; 0 \\ 0 & \\ 0 & B \end{bmatrix},$$

where B is a 2×2 orthogonal matrix. Thus,

$$B = \begin{bmatrix} \cos\theta & -\sin\theta \\ \sin\theta & \cos\theta \end{bmatrix},$$

(Exercise 1). Therefore, $U^t A U = R(\theta)$ for some θ, where $U = {}_{\mathcal{B}_0}[I]_{\mathcal{B}}$. Thus, A defines a rotation about x_1 direction and by an angle θ. Since $\operatorname{tr} A = \operatorname{tr} R(\theta) = 1 + 2\cos\theta$, the angle of rotation is given by $\theta = \cos^{-1}(\operatorname{tr} A - 1)/2$. This proves the following statement.

A 3×3 orthogonal matrix with determinant 1 represents a rotation by angle $\cos^{-1}(\operatorname{tr} A - 1)/2$ and the axis of rotation is along eigenvector corresponding to eigenvalue 1.

The orthogonal matrix $E = \operatorname{diag}(1, 1, -1)$ represents a reflection in the e_1, e_2 plane. If A is a 3×3 orthogonal matrix with determinant 1, then

$\det(EA) = 1$. Thus, $EA = Q^t R(\theta) Q$ for some orthogonal matrix Q, and $A = Q^t (QEQ^t) R(\theta) Q$. Therefore, an orthogonal matrix represents either a rotation or a rotation followed by a reflection.

5. If T is a unitary operator on a finite dimensional complex inner product space then all its eigenvalues are of absolute value 1. The converse of this statement: in a finite dimensional complex inner product space if all eigenvalues of a linear operator T are of absolute value 1 then T is unitary, is not true. For example, the linear operator T on $\mathbb{C}^2$ defined by $Te_1 = e_1$, $Te_2 = e_1 + e_2$, where $\{e_1, e_2\}$ is the standard basis of $\mathbb{C}^2$. However, we prove the following statement.

   Let T be triangulable linear operator on an n–dimensional inner product space V and let all the eigenvalues of T are equal to 1 in absolute value. If $\|Tv\| \leq \|v\|$ for all $v \in V$, then T is unitary.

   Let $\mathcal{B} = \{v_1, \ldots, v_n\}$ be an ordered orthonormal basis of V such that $[T]_{\mathcal{B}}$ is upper triangular. Write

$$[T]_{\mathcal{B}} = \begin{bmatrix} \lambda_1 & \mu_{12} & \cdots & \mu_{1n} \\ & \ddots & \ddots & \vdots \\ & & \ddots & \mu_{n-1\,n} \\ 0 & & & \lambda_n \end{bmatrix}.$$

   Then $|\lambda_i| = 1$, $i = 1, \ldots, n$, $\mu_{ij} \in F$. For $j = 2, \ldots, n$,

$$Tv_j = \mu_{1j}v_1 + \cdots + \mu_{j-1\,j}v_{j-1} + \lambda_j v_j,$$

   and so

$$\|Tv_j\|^2 = |\mu_{1j}|^2 + \cdots + |\mu_{j-1\,j}|^2 + |\lambda_j|^2.$$

   Since $\|Tv_j\|^2 \leq \|v_j\|^2 = 1$ and $|\lambda_j| = 1$, we have

$$|\mu_{1j}|^2 + \cdots + |\mu_{j-1\,j}|^2 = 0.$$

   Therefore, $\mu_{1j} = \cdots = \mu_{j-1\,j} = 0$, and $[T]_{\mathcal{B}} = \operatorname{diag}(\lambda_1, \ldots, \lambda_n)$. Thus, $[TT^*]_{\mathcal{B}} = [T]_{\mathcal{B}} [T^*]_{\mathcal{B}} = \operatorname{diag}(\lambda_1, \ldots, \lambda_n) \operatorname{diag}(\overline{\lambda_1}, \ldots, \overline{\lambda_n}) = I_n$. Hence, $TT^* = I$, and T is unitary.

6. Let $A \in \mathbb{R}^{n \times n}$ and let λ be a root of the characteristic polynomial of A, $c_A(x) = \det(xI_n - A)$. If $\lambda \in \mathbb{R}$, then there exists $x \in \mathbb{R}^n \setminus \{0\}$ such that $Ax = \lambda x$. Let P_1 be an invertible matrix whose first column is x. Then $P_1^{-1} A P_1 = \begin{bmatrix} \lambda_1 & ** \\ 0 & \widetilde{A} \end{bmatrix}$.

   If $\lambda = \alpha + i\beta$, $\alpha, \beta \in \mathbb{R}$, $\beta \neq 0$, is a complex root of $c_A(x)$, then considering A as a linear operator on $\mathbb{C}^n$, A has an eigenvector $z \in \mathbb{C}^n$ corresponding to λ. Write $z = x + iy$, $x, y \in \mathbb{R}^n$, then

(*) $$A(x + iy) = (\alpha + i\beta)(x + iy),$$

   and so

(**) $$Ax = \alpha x - \beta y, \quad Ay = \alpha y + \beta x.$$

   From equations (**), it follows that neither x nor y is zero. If $y = \delta x$, $\delta \in \mathbb{R} \setminus \{0\}$, then from equation (*), $Ax = \lambda x$. Since $x \neq 0$, $x^t Ax = \lambda x^t x$ and so $\lambda \in \mathbb{R}$. Hence, x and y are linearly independent. Let P_1 be an invertible matrix whose first two columns are x and y respectively. Then the first column of $P_1^{-1}AP_1$ is

$$\begin{aligned}\left(P_1^{-1}AP_1\right)e_1 &= P_1^{-1}Ax = P_1^{-1}(\alpha x - \beta y)\\ &= \alpha P^{-1}(Pe_1) - \beta P^{-1}(Pe_2)\\ &= \alpha e_1 - \beta e_2,\end{aligned}$$

and similarly, the second column of $P_1^{-1}AP_1$ is $\left(P_1^{-1}AP_1\right)e_2 = \beta e_1 + \alpha e_2$. Thus,

$$P_1^{-1}AP_1 = \begin{bmatrix} \alpha & \beta & ** \\ -\beta & \alpha & \\ & 0 & \widetilde{A} \end{bmatrix}.$$

Continuing this process inductively on $\widetilde{A}$ as in Example 1, we can obtain an invertible matrix P such that

$$P^{-1}AP = \begin{bmatrix} R_0 & & & ** \\ & R_1 & & \\ & & \ddots & \\ 0 & & & R_k \end{bmatrix},$$

a block upper triangular matrix, R_0 is an $m \times m$ upper triangular matrix, and m is the number of real roots of $c_A(x)$, counted with multiplicity. For $i \geq 1$, each R_i is a 2×2 matrix and $2k$ $(= n - m)$ is the number of nonreal roots of $c_A(x)$, counted with multiplicity. Note that if $\alpha + i\beta$ is a root of $c_A(x)$, then so is $\alpha - i\beta$.

Now $P = QM$, (Example 3), where Q is an orthogonal matrix and M is an upper triangular matrix. Partition M and M^{-1} conformal with the partition of $P^{-1}AP$. Let

$$M = \begin{bmatrix} M_0 & & & ** \\ & M_1 & & \\ & & \ddots & \\ 0 & & & M_k \end{bmatrix} \text{ and } M^{-1} = \begin{bmatrix} M_0' & & & ** \\ & M_1' & & \\ & & \ddots & \\ 0 & & & M_k' \end{bmatrix}.$$

Then

$$Q^tAQ = M\left(P^{-1}AP\right)M^{-1} = \begin{bmatrix} M_0R_0M_0' & & & ** \\ & M_1R_1M_1' & & \\ & & \ddots & \\ 0 & & & M_kR_kM_k' \end{bmatrix}.$$

Since M_0, R_0, M_0' are upper triangular matrices, $M_0R_0M_0'$ is an upper triangular matrix, and for $i \geq 1$, $M_iR_iM_i'$ is a 2×2 matrix. Thus we have the following result.

If $A \in \mathbb{R}^{n \times n}$, then there exists an orthogonal matrix Q such that

$$Q^tAQ = \begin{bmatrix} A_0 & A_{01} & \cdots & A_{0k} \\ & A_1 & \cdots & A_{1k} \\ & & \ddots & \vdots \\ 0 & & & A_k \end{bmatrix},$$

A_0 is an $m \times m$ upper triangular matrix, m is the number of real roots of $c_A(x)$, counted with multiplicity. For $i \geq 1$, each A_i is a 2×2 matrix and $2k$ is the number of nonreal roots of $c_A(x)$, counted with multiplicity.

Note that $c_A(x) = \det(xI_m - A_0)\det(xI_2 - A_1)\cdots\det(xI_2 - A_k)$, and 2×2 matrices corresponds to nonreal roots. Since $\det(xI_2 - A_i) = x^2 - (\operatorname{tr} A_i)x + \det A_i$, and for $i \geq 1$, $\det(xI_2 - A_i)$ has no real roots. This happens if and only if $(\operatorname{tr} A_i)^2 < 4\det A_i$.

Now suppose that T is a linear operator on an n–dimensional inner product space V over F. If $F = \mathbb{C}$, then by Schur's theorem there exists an ordered orthonormal basis $\mathcal{B}$ such that $[T]_{\mathcal{B}}$ is upper triangular. If $F = \mathbb{R}$ and all the roots of the characteristic polynomial of T are real, then again by Schur's theorem there exists an orthonormal basis $\mathcal{B}$ such that $[T]_{\mathcal{B}}$ is upper triangular.

Let $F = \mathbb{R}$ and suppose that T is not triangulable. Let $\mathcal{B}' = \{u_1, \ldots, u_n\}$ be an ordered orthonormal basis of V and let $[T]_{\mathcal{B}'} = A$. Then as discussed above, there exists an orthogonal matrix Q such that

$$Q^t A Q = \begin{bmatrix} A_0 & A_{01} & \cdots & A_{0k} \\ & A_1 & \cdots & A_{1k} \\ & & \ddots & \vdots \\ 0 & & & A_k \end{bmatrix},$$

where A_0 and A_i's are as described above. Let $Q = (q_{ij})$ and let $v_j = \sum_{i=1}^{n} q_{ij} u_i$ so that $[v_j]_{\mathcal{B}'} = Qe_j$. Then $\mathcal{B} = \{v_1, \ldots, v_n\}$ is an ordered basis of V. Since $(v_r, v_s) = [v_s]_{\mathcal{B}'}^* [v_r]_{\mathcal{B}'} = (Qe_s)^* (Qe_r) = \delta_{rs}$, $\mathcal{B}$ is an orthonormal basis of V. Now $Q = {}_{\mathcal{B}'}[I]_{\mathcal{B}}$ and $[T]_{\mathcal{B}} = {}_{\mathcal{B}}[I]_{\mathcal{B}'}\,[T]_{\mathcal{B}'}\,{}_{\mathcal{B}'}[I]_{\mathcal{B}} = Q^t A Q$. This proves the following result.

If T is a linear operator on an n–dimensional inner product space V, then either T is triangulable or there exists an ordered orthonormal basis $\mathcal{B}$ of V such that

$$[T]_{\mathcal{B}} = \begin{bmatrix} A_0 & A_{01} & \cdots & A_{0k} \\ & A_1 & \cdots & A_{1k} \\ & & \ddots & \vdots \\ 0 & & & A_k \end{bmatrix},$$

where $A_0 \in \mathbb{R}^{m \times m}$, an upper triangular matrix and m is the number of real roots of $c_T(x)$, counted with multiplicity and $2k$ $(= n - m)$ is the number of nonreal roots of $c_T(x)$, counted with multiplicity and each $A_i \in \mathbb{R}^{2\times 2}$ such that $(\operatorname{tr} A_i)^2 < 4\det A_i$, $i = 1, \ldots, k$.

4.2. Exercises. In the following exercises, V is a finite dimensional inner product space over F and $T \in L(V)$.

1. Prove that any 2×2 real unitary matrix with determinant equal to 1 is of the form $\begin{bmatrix} \cos\theta & -\sin\theta \\ \sin\theta & \cos\theta \end{bmatrix}$, $\theta \in \mathbb{R}$.
2. Find a unitary matrix whose first two columns are
   (a) $\begin{bmatrix} \frac{1}{\sqrt{3}} & -\frac{1}{\sqrt{3}} & \frac{1}{\sqrt{3}} \end{bmatrix}^t$, $\begin{bmatrix} 0 & \frac{1}{\sqrt{2}} & \frac{1}{\sqrt{2}} \end{bmatrix}^t$;

(b) $\begin{bmatrix} \frac{1}{2} & \frac{i}{2} & \frac{1}{2} & \frac{i}{2} \end{bmatrix}^t$, $\begin{bmatrix} \frac{i}{2} & \frac{1}{2} & \frac{1}{2} & \frac{i}{2} \end{bmatrix}^t$.

3. Prove that if A is a unitary matrix with integer entries, then the only nonzero entries are 1 and -1, and there is exactly one nonzero entry in each row and in each column.

4. For each of the following matrices $A \in \mathbb{C}^{n\times n}$, find a unitary matrix U such that U^*AU is upper triangular.

$$\begin{bmatrix} 3 & -2 & 2 \\ 10 & -9 & 10 \\ 6 & -6 & 7 \end{bmatrix}, \begin{bmatrix} -6 & 1 & 3 \\ 11 & -1 & -5 \\ 24 & 3 & 11 \end{bmatrix}, \begin{bmatrix} 7 & -5 & -4 \\ 2 & -1 & -3 \\ 2 & -2 & -2 \end{bmatrix},$$

$$\begin{bmatrix} 8 & 7 & -2 \\ -10 & -9 & 3 \\ -9 & -9 & 4 \end{bmatrix}, \begin{bmatrix} 6 & -1 & 2 \\ 4 & 1 & 2 \\ -10 & 0 & -3 \end{bmatrix}, \begin{bmatrix} 0 & 2 & -3 \\ 2 & -3 & 6 \\ 2 & -2 & 5 \end{bmatrix}.$$

5. Matrices $A, B \in \mathbb{C}^{n\times n}$ are said to be **unitarily similar** (respectively, **orthogonally similar**) if there exists a unitary (respectively, orthogonal) matrix U such that $B = U^*AU$. Show that if $A = (a_{ij})$ and $B = (b_{ij})$ are unitarily similar $n \times n$ complex matrices, then

$$\sum_{i=1}^{n}\sum_{j=1}^{n} |a_{ij}|^2 = \sum_{i=1}^{n}\sum_{j=1}^{n} |b_{ij}|^2 .$$

6. Give an example to show that two similar matrices may not be unitarily similar.

7. Prove that a Hermitian matrix is unitarily similar to a diagonal matrix. In particular, a symmetric matrix with real entries is orthogonally similar to a diagonal matrix.

8. Prove that two Hermitian matrices are similar if they have the same characteristic polynomial.

9. Verify that $\begin{bmatrix} -1 & i \\ i & 1 \end{bmatrix}$ is a symmetric matrix with nonreal entries, and is not diagonalizable. Thus, a complex symmetric matrix is not always diagonalizable.

10. Show that the matrices $\begin{bmatrix} e^{i\theta} & 0 \\ 0 & e^{-i\theta} \end{bmatrix}$ and $\begin{bmatrix} \cos\theta & -\sin\theta \\ \sin\theta & \cos\theta \end{bmatrix}$, $\theta \in \mathbb{R}$, are similar.

11. Prove that a 2×2 unitary matrix with determinant equal to 1 is similar to a real orthogonal matrix.

12. Show that if A is similar to a unitary matrix then A^* is similar to A^{-1}.

13. Let $a, b \in \mathbb{R}$ such that $a^2 - b^2 = 1$ and $ab \neq 0$. Verify that $\begin{bmatrix} a & bi \\ -bi & a \end{bmatrix}$ is an orthogonal matrix and is not unitary.

14. Let $S \in \mathbb{R}^{n\times n}$ be a skew–symmetric matrix. Prove that $I_n + S$ is invertible. Show that $(I_n - S)(I_n + S)^{-1}$ is orthogonal.

15. Let $A \in \mathbb{C}^{n\times n}$ be Hermitian. Prove that $A - iI_n$ is invertible. Show that $(A - iI_n)^{-1}(A + iI_n)$ is unitary.

16. Let T be a unitary operator. Prove that if W_1 is a T–invariant subspace of V, then there exists a subspace W_2 of V which is also T–invariant and $V = W_1 \oplus W_2$.
Hint: take $W_2 = W_1^{\perp}$.

17. Let T be a unitary operator. Prove that if $F = \mathbb{C}$ or if $F = \mathbb{R}$ and $c_T(x)$ has only real roots, then there exists an ordered orthonormal basis $\mathcal{B}$ such that

$$[T]_{\mathcal{B}} = \operatorname{diag}(\varepsilon_1, \dots, \varepsilon_n), \ \varepsilon_i = \pm 1, \ n = \dim V.$$

Further, prove that if $F = \mathbb{R}$ and $c_T(x)$ has m $(m > 0)$ nonreal roots, counted with multiplicities, then m is even and there is an ordered orthonormal basis $\mathcal{B}$ such that $[T]_{\mathcal{B}} = \operatorname{diag}(A_0, A_1, \dots, A_k)$, where $k = m/2$, $A_0 = \operatorname{diag}(\varepsilon_1, \dots, \varepsilon_r)$, $\varepsilon_i = \pm 1$, $r = n - 2k$, and

$$A_i = \begin{bmatrix} \cos\theta_i & -\sin\theta_i \\ \sin\theta_i & \cos\theta_i \end{bmatrix}, \ \theta_i \in \mathbb{R}, \ i = 1, \dots, k.$$

5. Self adjoint and normal operators

If V is a finite dimensional inner product space over $\mathbb{C}$, then the set $L(V)$ together with the adjoint operation looks very much like the set of complex numbers with the conjugation. For $S, T \in L(V)$, $(a, b \in \mathbb{C})$:

$$(S+T)^* = S^* + T^* \qquad \left(\overline{(a+b)} = \overline{a} + \overline{b}\right);$$
$$(ST)^* = T^* S^* \qquad \left(\overline{ab} = \overline{a}\overline{b}\right);$$
$$(S^*)^* = S \qquad \left(\overline{\overline{a}} = a\right).$$

The algebraic differences between $L(V)$ and $\mathbb{C}$ are that the 'multiplication' in $L(V)$ is the composition of two mappings which is not commutative, and there are nonzero linear operators which do not have multiplicative inverse, that is, which are not invertible.

Given a complex number z, there are unique real numbers a and b such that $z = a + ib$, $a = (z + \overline{z})/2$ and $b = (z - \overline{z})/2i$. Now what should be linear operators corresponding to real numbers. To identify these, first, a complex number is real if and only if $z = \overline{z}$. We say that a linear operator is **self adjoint** if $T^* = T$. Now, if we let $A = (T + T^*)/2$ and $B = (T - T^*)/2i$, then A and B are self adjoint and $T = A + iB$. Moreover, such a representation is unique. Indeed, if $T = C + iD$, C and D are self adjoint, $T^* = C - iD$, and so $C = (T + T^*)/2$ and $D = (T - T^*)/2i$. Also, a complex number is of modulus 1 if $z\overline{z} = 1$. Thus, under this analogy, unitary operators correspond to complex numbers with modulus 1.

Considering $A \in F^{n\times n}$ as a linear operator on the standard inner product space F^n, A is self adjoint if and only if $A^* = A$, that is, A is a Hermitian matrix. If T is a linear operator on a finite dimensional inner product space V, then the matrix of T with respect to an ordered orthonormal basis $\mathcal{B}$, $[T]_{\mathcal{B}}$ is Hermitian. Since $[T]_{\mathcal{B}} = [T^*]_{\mathcal{B}} = [T]_{\mathcal{B}}^*$. Note that a linear operator on a real inner product space is self adjoint if and only if its matrix with respect to an orthonormal basis is symmetric.

PROPOSITION 5.1. *Let S and T be self adjoint operators on an inner product space V. Then*

(i) *$S+T$ is self adjoint;*
(ii) *ST is self adjoint if and only if $ST = TS$;*
(iii) *T^{-1} is self adjoint, if T is invertible.*

PROOF. Exercise. ∎

Statements (i) and (iii) of the above proposition state that, like real numbers, the sum of two self adjoint operators is a self adjoint operator; and the inverse of an invertible self adjoint operator (which corresponds to nonzero real numbers) is also self adjoint. However, (ii) shows that unlike real numbers the product of two self adjoint operators is not always self adjoint.

PROPOSITION 5.2. *Let T be a self adjoint operator on an inner product space V.*

(i) *For all $v \in V$, (Tv, v) is a real number.*
(ii) *If $(Tv, v) = 0$, for all $v \in V$, then $T \equiv 0$.*
(iii) *If $T^k v = 0$ for some $k > 1$, then $Tv = 0$.*

Further, if V is finite dimensional, then

(iv) *all roots of characteristic polynomial of T are real;*
(v) *eigenvectors corresponding to distinct eigenvalues are orthogonal.*

PROOF. (i) $(Tv, v) = \overline{(v, Tv)} = \overline{(v, T^* v)} = \overline{(Tv, v)}$. Hence, $(Tv, v) \in \mathbb{R}$.

(ii) Since $(T(x+y), x+y) = (Tx, x) + (Tx, y) + (Ty, x) + (Ty, y)$, $x, y \in V$, and $(Tv, v) = 0$ for all $v \in V$, we have

$$(*) \qquad 0 = (Tx, y) + (Ty, x).$$

If V is an inner product space over $\mathbb{R}$, then from $(*)$:

$$0 = (Tx, y) + (Ty, x) = (Tx, y) + (y, Tx) = 2(Tx, y).$$

Hence, $(Tx, y) = 0$ for all $x, y \in V$, and $T \equiv 0$.

If V is an inner product space over $\mathbb{C}$, then replacing y by iy in $(*)$, we have $(Tx, iy) + (iTy, x) = 0$. Thus for all $x, y \in V$:

$$(Tx, y) - (Ty, x) = 0.$$

Hence, $(Tx, y) = 0$ for all $x, y \in V$, and $T \equiv 0$.

(iii) If $k = 1$, there is nothing to prove. Assume that the statement is true for all integers up to $k-1$. Choose $m \geq 1$ such that $2^m \geq k$ and $2^{m-1} < k$. Since $T^k v = 0$, $T^{2^m} v = 0$. Thus, $\left(T^{2^{m-1}} v, T^{2^{m-1}} v\right) = \left(T^{2^m} v, v\right) = 0$, and so $T^{2^{m-1}} v = 0$. Therefore, by induction hypothesis, $Tv = 0$.

(iv) First assume that V is an inner product space over $\mathbb{C}$. Let λ be a root of the characteristic polynomial of T, $c_T(x)$, and let v be corresponding eigenvector. Then $\lambda(v, v) = (Tv, v) = (v, Tv) = (v, \lambda v) = \overline{\lambda}(v, v)$. Hence, $\lambda = \overline{\lambda}$ and $\lambda \in \mathbb{R}$.

If V is an inner product space over $\mathbb{R}$, then $c_T(x) \in \mathbb{R}[x]$; it may have complex roots. To prove the statement, we must show that all the roots of $c_T(x)$ are real. Let $\mathcal{B}$ be an ordered orthonormal basis of V and $A = [T]_{\mathcal{B}}$. Then $A \in \mathbb{R}^{n \times n}$, $n = \dim V$. Since T is self adjoint, A is Hermitian (actually, symmetric). Considering A as a linear operator on the standard inner product space $\mathbb{C}^n$, A is self adjoint and $c_A(x) = c_T(x)$. Therefore, all the roots of $c_A(x)$, and hence of $c_T(x)$ are real.

(v) If λ and μ are distinct eigenvalues of T and u and v are corresponding eigenvectors, then $\lambda, \mu \in \mathbb{R}$ and $\lambda(u,v) = (Tu,v) = (u,Tv) = \mu(u,v)$. Since $\lambda \neq \mu$, $(u,v) = 0$. ∎

A linear operator T on an inner product space V is called **orthogonally diagonalizable** if V has orthonormal basis consisting of eigenvectors of T only. An important consequence of the above result is that a self adjoint operator on a finite dimensional inner product space is orthogonally diagonalizable.

THEOREM 5.3. *A self adjoint operator T on a finite dimensional inner product space V is orthogonally diagonalizable.*

PROOF. From Proposition 5.2(iv), the minimal polynomial of T is

$$m_T(x) = (x - \lambda_1)^{m_1} \cdots (x - \lambda_k)^{m_k},$$

where $m_1, \ldots, m_k$ are positive integers and $\lambda_1, \ldots, \lambda_k$ are distinct real numbers. By primary decomposition theorem

$$V = \ker(T - \lambda_1 I)^{m_1} \oplus \cdots \oplus \ker(T - \lambda_k I)^{m_k}.$$

Since each $\lambda_i \in \mathbb{R}$, $(T - \lambda_i I)^* = T^* - \lambda_i I = T - \lambda_i I$, that is, $T - \lambda_i I$ is self adjoint. Thus, for $i = 1, \ldots, k$, $v_i \in \ker(T - \lambda_i I)^{m_i}$ implies that $v_i \in \ker(T - \lambda_i I)$ (Proposition 5.2(iii)). Hence, $\ker(T - \lambda_i I)^{m_i} = \ker(T - \lambda_i I)$ for $i = 1, \ldots, k$. Therefore,

$$V = \ker(T - \lambda_1 I) \oplus \cdots \oplus \ker(T - \lambda_k I),$$

and T is diagonalizable.

Now using Gram–Schmidt orthonormalization process, we can find an orthonormal basis $\mathcal{B}_i$ for $\ker(T - \lambda_i I)$, for $i = 1, \ldots, k$. Then $\mathcal{B} = \mathcal{B}_1 \cup \cdots \cup \mathcal{B}_k$ is an orthonormal basis of V consisting of eigenvectors of T (Proposition 5.2(v)). ∎

A linear operator T on an inner product space V is called **normal** if it commutes with its adjoint, that is, $TT^* = T^*T$. Unitary operators and self adjoint operators are normal operators. $A \in F^{n \times n}$ is called normal if $AA^* = A^*A$. Thus, a linear operator T on a finite dimensional inner product space is normal if and only if for any ordered orthonormal basis $\mathcal{B}$, $[T]_{\mathcal{B}}$ is a normal matrix.

PROPOSITION 5.4. *Let T be a normal operator on an inner product space V over F.*

(i) $\|Tv\| = \|T^*v\|$ *for all* $v \in V$.

(ii) *If for* $v \in V$, $Tv = \lambda v$, $\lambda \in F$, *then* $T^*v = \overline{\lambda} v$.

(iii) *Eigenvectors corresponding to distinct eigenvalues of T are orthogonal.*

(iv) *If* $T^k v = 0$, *for some* $k \geq 1$, *then* $Tv = 0$.

PROOF. (i) $\|Tv\|^2 = (Tv, Tv) = (v, T^*Tv) = (v, TT^*v) = (T^*v, T^*v) = \|T^*v\|^2$. Hence, $\|Tv\| = \|T^*v\|$ for all $v \in V$.

(ii) Since $T - \lambda I$ is a normal operator, for any $\lambda \in F$, so by (i), $\|(T - \lambda I)v\| = \|(T - \lambda I)^* v\| = \|(T^* - \overline{\lambda} I)v\|$. Therefore, if $Tv = \lambda v$, then $T^*v = \overline{\lambda} v$.

(iii) Let v and w be eigenvectors of T corresponding to eigenvalues λ and μ, $\lambda \neq \mu$. Then using (ii), we have $\lambda(v,w) = (Tv,w) = (v,T^*w) = (v,\overline{\mu}w) = \mu(v,w)$. Since $\lambda \neq \mu$, $(v,w) = 0$.

(iv) Since T is normal, $S = T^*T$ is self adjoint. Thus, $T^k v = 0$ implies that $S^k v = 0$. By Proposition 5.2(iii), $Sv = 0$. But then $\|Tv\|^2 = (Tv, Tv) = (v, T^*Tv) = (v, Sv) = 0$. Hence, $Tv = 0$. ∎

In fact, the first statement of the above proposition characterizes normal operators.

PROPOSITION 5.5. *Let T be a linear operator on an inner product space V. Then T is normal if and only if $\|Tv\| = \|T^*v\|$ for all $v \in V$.*

PROOF. We only need to show that the condition is sufficient. If for all $v \in V$, $\|Tv\| = \|T^*v\|$, then $(Tv, Tv) = (T^*v, T^*v)$, and so $((T^*T - TT^*)\, v, v) = 0$. Since $T^*T - TT^*$ is self adjoint, $T^*T = TT^*$ (Proposition 5.2). Hence, T is normal. ∎

Our next result is that normal triangulable linear operators are precisely orthogonally diagonalizable linear operators.

THEOREM 5.6. ***Spectral theorem.*** *Let T be a triangulable linear operator on a finite dimensional inner product space V over F. Then T is normal if and only if V has an orthonormal basis consisting of eigenvectors of T.*

PROOF. If T is triangulable, then its minimal polynomial $m_T(x)$ splits over F. Let $m_T(x) = (x - \lambda_1)^{m_1} \cdots (x - \lambda_k)^{m_k}$, $m_i > 0$ and $\lambda_1, \ldots, \lambda_k$ are distinct roots of $m_T(x)$. Now proceeding exactly as in the proof of Theorem 5.3 and using Proposition 5.4 instead of Proposition 5.2, it can be shown that V has an orthonormal basis consisting of eigenvectors of T.

Conversely, if $\mathcal{B} = \{v_1, \ldots, v_n\}$ is an orthonormal basis of V and each v_i is an eigenvector, then $Tv_i = \mu_i v_i$, $i = 1, \ldots, n$. Also $T^*v_i = \overline{\mu}_i v_i$. Thus, for $i = 1, \ldots, n$,

$$TT^*v_i = T(\overline{\mu}_i v_i) = \overline{\mu}_i \mu_i v_i = \mu_i(\overline{\mu}_i v_i) = \mu_i(T^*v_i) = T^*(\mu_i v_i) = T^*Tv_i.$$

Hence, $TT^* = T^*T$. ∎

COROLLARY 5.7. *Let V be a finite dimensional complex inner product space and let $T \in L(V)$. Then T is normal if and only if T is orthogonally diagonalizable.*

Let V be a finite dimensional inner product space over F and let T be a triangulable normal operator on V. Then by Spectral theorem, T is orthogonally diagonalizable. Let $\lambda_1, \ldots, \lambda_k$ be distinct eigenvalues of T with multiplicities $m_1, \ldots, m_k$ respectively, and let $\mathcal{B} = \{u_{11}, \ldots, u_{1m_1}, \ldots, u_{k1}, \ldots, u_{km_k}\}$ be an ordered basis of V, where $u_{i1}, \ldots, u_{im_1}$ are eigenvectors corresponding to λ_i.

For each $i = 1, \ldots, k$, the mapping $P_i : V \to V$ defined by

$$P_i(v) = \sum_{j=1}^{m_i} (v, u_{ij})\, u_{ij}$$

is a linear operator on V and it is also easy to verify that $P_i(u_{rs}) = \delta_{ir} u_{rs}$. Therefore, $P_i^2 = P_i$, and for $i \neq j$, $P_i P_j = 0$. Now for $v \in V$, we write

$$v = \sum_{i=1}^{k} \sum_{j=1}^{m_i} (v, u_{ij})\, u_{ij} = \sum_{i=1}^{k} P_i(v).$$

Thus, $I = P_1 + \cdots + P_k$, a resolution of identity. Note that each P_i is an orthogonal projection onto $\ker(T - \lambda_i I) = \langle u_{i1}, \ldots, u_{im_i} \rangle$ and is self adjoint as

$$[P_i]_{\mathcal{B}} = \operatorname{diag}\left(0_{m_1}, \ldots, 0_{m_{i-1}}, I_{m_i}, 0_{m_{i+1}}, \ldots, 0_{m_k}\right),$$

a Hermitian matrix, where 0_r denotes the $r \times r$ zero matrix.

Now for $T = TI = TP_1 + \cdots + TP_k$, and for $i = 1, \ldots, k$,

$$TP_i(v) = T\left(\sum_{j=1}^{m_i} (v, u_{ij})\, u_{ij}\right) = \lambda_i \sum_{j=1}^{m_i} (v, u_{ij})\, u_{ij} = \lambda_i P_i(v),$$

for all $v \in V$, Therefore, $TP_i = \lambda_i P_i$ and

$$T = \lambda_1 P_1 + \cdots + \lambda_k P_k.$$

This is called the **spectral resolution** of T. It is easy to verify that for a positive integer m, $T^m = \lambda_1^m P_1 + \cdots + \lambda_k^m P_k$, and more generally for a polynomial $p(x) \in F[x]$,

$$p(T) = p(\lambda_1) P_1 + \cdots + p(\lambda_k) P_k.$$

If T is an invertible matrix, then by Cayley Hamilton theorem, $T^{-1} = q(T)$, for some $q(x) \in F[x]$. Thus,

$$T^{-1} = \lambda_1^{-1} P_1 + \cdots + \lambda_k^{-1} P_k.$$

Therefore, for any integer m,

$$T^m = \lambda_1^m P_1 + \cdots + \lambda_k^m P_k.$$

Let f be any F–valued function, whose domain contains the eigenvalues of T. Define

$$f(T) = f(\lambda_1) P_1 + \cdots + f(\lambda_k) P_k.$$

Since the spectral resolution of T is a finite sum, there exists a polynomial $p(x) \in F[x]$ such that $p(\lambda_i) = f(\lambda_i)$, $i = 1, \ldots, k$, and so

$$f(T) = \sum_{i=1}^{k} f(\lambda_i) P_i = \sum_{i=1}^{k} p(\lambda_i) P_i = p(T).$$

If T is a unitary operator on a finite dimensional complex inner product space, then T has the following spectral resolution:

$$T = \lambda_1 P_1 + \cdots + \lambda_k P_k,$$

with $|\lambda_j| = 1$, $j = 1, \ldots, k$. Thus, $\lambda_j = e^{i\mu_j}$, $\mu_j \in \mathbb{R}$, $j = 1, \ldots, k$. If $S = \mu_1 P_1 + \cdots + \mu_k P_k$, then S is self adjoint operator and $T = e^{iS}$. This is analogous to the fact that a complex number z with unit modulus can be written as $z = e^{i\alpha}$, $\alpha \in \mathbb{R}$.

We now introduce a special class of self adjoint operators. A self adjoint operator T on an inner product space V is called **positive semidefinite** if $(Tv, v) \geq 0$ for all $v \in V$, and **positive definite** if $(Tv, v) > 0$ for all $v \in V \setminus \{0\}$.

The following result characterizes positive definite operators.

THEOREM 5.8. *Let T be a self adjoint operator on a finite dimensional inner product space V. The following statements are equivalent.*

(i) *T is positive definite.*
(ii) *All eigenvalues of T are positive.*

(iii) *There exists a unique positive definite operator A on V such that $T = A^2$. We denote this operator by $\sqrt{T}$.*
(iv) *There exists an invertible linear operator S on V such that $T = S^*S$.*

PROOF. *(i)* $\Rightarrow$ *(ii)* If λ is an eigenvalue of T, then $\lambda \in \mathbb{R}$ (Proposition 5.2). Let v be an eigenvector corresponding to λ, then as $(Tv, v) > 0$, so $\lambda(v, v) > 0$. Hence, $\lambda > 0$.

(ii) $\Rightarrow$ *(i)* Let $\mathcal{B} = \{v_1, \dots, v_n\}$ be an orthonormal basis of V consisting of eigenvectors of T. Then $Tv_i = \mu_i v_i$, $\mu_i > 0$, $i = 1, \dots, n$. For $v \in V$, write $v = \sum_{i=1}^n \alpha_i v_i$, $\alpha_i \in F$. Then

$$(Tv, v) = \left(\sum_{i=1}^n \alpha_i T v_i, \sum_{i=1}^n \alpha_i v_i\right) = \sum_{i=1}^n \sum_{j=1}^n \mu_i \alpha_i \overline{\alpha_j}\,(v_i, v_j) = \sum_{i=1}^n \mu_i |\alpha_i|^2 .$$

Thus, if $v \neq 0$, then $(Tv, v) > 0$.

(ii) $\Rightarrow$ *(iii)* If $T = \lambda_1 P_1 + \cdots + \lambda_k P_k$ is the spectral resolution of T, then $\lambda_i > 0$, and so for $A = \sqrt{\lambda_1} P_1 + \cdots + \sqrt{\lambda_k} P_k$, $T = A^2$. Since the eigenvalues of A are $\sqrt{\lambda_1}, \dots, \sqrt{\lambda_k}$, which are positive, by the equivalence of (i) and (ii), it follows that A is positive definite.

(iii) $\Rightarrow$ *(iv)* Take $S = A$.

(iv) $\Rightarrow$ *(i)* For $v \in V$, $(Tv, v) = (S^*Sv, v) = (Sv, Sv) = \|Sv\|^2 \geq 0$. Since S is invertible, for $v \neq 0$, $Sv \neq 0$, and so $\|Sv\| > 0$. Hence, $(Tv, v) > 0$ for all $v \in V \setminus \{0\}$, and T is positive definite. ∎

Similar is a characterization of positive semidefinite operators.

THEOREM 5.9. *Let T be a self adjoint operator on a finite dimensional inner product space V. The following statements are equivalent.*

(i) *T is positive semidefinite.*
(ii) *All eigenvalues of T are nonnegative.*
(iii) *There exists a unique positive semidefinite operator $\sqrt{T}$ on V such that $T = \left(\sqrt{T}\right)^2$.*
(iv) *There exists $S \in L(V)$ such that $T = S^*S$.*

PROOF. Exercise. ∎

In the beginning of this section we observed that the set of adjoint operators are like the set of real numbers. A real number p is nonnegative if and only if $p = a^2$ for some $a \in \mathbb{R}$, or equivalently, $p = \overline{z}z$ for some $z \in \mathbb{C}$. The above result shows that the set of positive semidefinite operators is analogous to the set of nonnegative real numbers.

5.1. Examples

1. A Hermitian matrix is unitarily similar to a diagonal matrix. If $A \in F^{n\times n}$, A is Hermitian and $\lambda_1 \in \mathbb{R}$ is an eigenvalue of A with corresponding eigenvector u_1, $\|u_1\| = 1$, then $\langle u_1 \rangle$ is an A–invariant subspace of F^n. If $\{u_1, \dots, u_n\}$ is an orthonormal basis of F^n, then $\langle u_1 \rangle^\perp = \langle u_2, \dots, u_n \rangle$ and so $\langle u_2, \dots, u_n \rangle$ is A^*–invariant and hence A–invariant. Therefore, if $U_1 = \begin{bmatrix} u_1 & \cdots & u_n \end{bmatrix}$, then U_1 is a unitary matrix and

$$U_1^* A U_1 = \begin{bmatrix} \lambda_1 & 0 \\ 0 & A_2 \end{bmatrix}.$$

Clearly, $A_2 \in F^{(n-1)\times(n-1)}$ and is Hermitian. Inductively, there exists a unitary matrix V such that V^*AV is a diagonal matrix. Then for $U_2 = \begin{bmatrix} 1 & 0 \\ 0 & V \end{bmatrix}$,

$$U_2^* \left(U_1^* A U_1\right) U_2 = \begin{bmatrix} \lambda_1 & 0 \\ 0 & V^* A_2 V \end{bmatrix},$$

a diagonal matrix, and $U = U_1 U_2$ is a required unitary matrix (orthogonal matrix if $F = \mathbb{R}$) such that U^*AU is a diagonal matrix.

For example, let

$$A = \begin{bmatrix} 2 & 0 & -1 \\ 0 & 3 & 0 \\ -1 & 0 & 2 \end{bmatrix} \in \mathbb{R}^3.$$

Then $Ae_2 = 3e_2$, e_2 is a normalized eigenvector corresponding to eigenvalue 3. Now $\{e_2, e_1, e_3\}$ is an ordered orthonormal basis of $\mathbb{R}^3$ containing e_2. If $U_1 = \begin{bmatrix} e_2 & e_1 & e_3 \end{bmatrix}$, then

$$U_1^* A U_1 = \begin{bmatrix} 3 & 0 & 0 \\ 0 & 2 & -1 \\ 0 & -1 & 2 \end{bmatrix}.$$

Now $A_2 = \begin{bmatrix} 2 & -1 \\ -1 & 2 \end{bmatrix}$ is Hermitian and has eigenvalue 1 (since sum of rows is 1), and corresponding normalized eigenvector is $\frac{1}{\sqrt{2}} \begin{bmatrix} 1 & 1 \end{bmatrix}^t$. An orthonormal basis of $\mathbb{R}^2$ consisting of this eigenvector is

$$\left\{ v_1 = \frac{1}{\sqrt{2}} \begin{bmatrix} 1 & 1 \end{bmatrix}^t, v_2 = \frac{1}{\sqrt{2}} \begin{bmatrix} 1 & -1 \end{bmatrix}^t \right\}.$$

Thus, if $V = \begin{bmatrix} v_1 & v_2 \end{bmatrix}$, then $V^* A_2 V = \operatorname{diag}(1, 3)$. Now if

$$U_2 = \begin{bmatrix} 1 & 0 & 0 \\ 0 & \frac{1}{\sqrt{2}} & \frac{1}{\sqrt{2}} \\ 0 & \frac{1}{\sqrt{2}} & -\frac{1}{\sqrt{2}} \end{bmatrix} \quad \left(= \begin{bmatrix} 1 & 0 \\ 0 & V \end{bmatrix} \right),$$

then $U_2^* \left(U_1^* A U_1\right) U_2 = \operatorname{diag}(3, 1, 3)$, and

$$U = U_1 U_2 = \begin{bmatrix} 0 & \frac{1}{\sqrt{2}} & \frac{1}{\sqrt{2}} \\ 1 & 0 & 0 \\ 0 & \frac{1}{\sqrt{2}} & -\frac{1}{\sqrt{2}} \end{bmatrix}$$

is an orthonormal matrix which transforms A to a diagonal matrix.

If we have

$$A = \begin{bmatrix} 5 & 0 & 4i \\ 0 & 1 & -4i \\ -4i & 4i & 3 \end{bmatrix} \in \mathbb{C}^{3\times 3},$$

then verify that 3 is an eigenvalue of A and corresponding normalized eigenvector is $\begin{bmatrix} \frac{2}{3} & \frac{2}{3} & \frac{i}{3} \end{bmatrix}^t$. An orthonormal basis of $\mathbb{C}^3$ containing this vector is

$$\left\{u_1 = \begin{bmatrix} \frac{2}{3} & \frac{2}{3} & \frac{i}{3} \end{bmatrix}^t, u_2 = \begin{bmatrix} \frac{1}{\sqrt{2}} & -\frac{1}{\sqrt{2}} & 0 \end{bmatrix}^t, u_3 = \begin{bmatrix} \frac{1}{3\sqrt{2}} & \frac{1}{3\sqrt{2}} & \frac{-4i}{3\sqrt{2}} \end{bmatrix}^t\right\}.$$

Thus, we let $U_1 = \begin{bmatrix} u_1 & u_2 & u_3 \end{bmatrix}$, a unitary matrix. Then

$$U_1^* A U_1 = \begin{bmatrix} 3 & 0 & 0 \\ 0 & 3 & 6 \\ 0 & 6 & 3 \end{bmatrix}.$$

Let $A_1 = \begin{bmatrix} 3 & 6 \\ 6 & 3 \end{bmatrix}$. Then 9 is an eigenvalue of A_1 and corresponding normalized eigenvector is $v_1 = \begin{bmatrix} \frac{1}{\sqrt{2}} & \frac{1}{\sqrt{2}} \end{bmatrix}^t$. An orthonormal basis of $\mathbb{C}^3$ containing v_1 is $\left\{v_1, v_2 = \begin{bmatrix} \frac{1}{\sqrt{2}} & -\frac{1}{\sqrt{2}} \end{bmatrix}^t\right\}$. If $V = \begin{bmatrix} v_1 & v_2 \end{bmatrix} \in \mathbb{C}^{2\times 2}$, then $V^* A_1 V = \operatorname{diag}(9, -3)$. Therefore, if $U_2 = \operatorname{diag}(1, V)$, then for $U = U_1 U_2$, we have $U^* A U = \operatorname{diag}(3, 9, -3)$.

2. A matrix $A \in F^{n\times n}$ is called a positive semidefinite matrix if A is Hermitian and A as a linear operator on the standard inner product space F^n is positive semidefinite, that is, $(Ax, x) = x^* A x \geq 0$, for all $x \in F^n$. A is called a positive definite matrix if A is Hermitian and $x^* A x > 0$, for all $x \in F^n \setminus \{0\}$.

   Let V be an inner product space and let $X = \{x_1, \dots, x_n\}$ be a set of vectors in V. Define an $r \times r$ matrix $G(X)$, whose (i, j)–th entry is (x_j, x_i). This matrix $G(X)$ is clearly a Hermitian matrix and is called the **Gram matrix** of X. For $y = \begin{bmatrix} \gamma_1 & \cdots & \gamma_r \end{bmatrix}^t \in F^r$,

$$\begin{aligned} y^* G(X) y &= \sum_{j=1}^{n} \sum_{i=1}^{n} \overline{\gamma_i} \gamma_j (x_j, x_i) = \left(\sum_{j=1}^{n} \gamma_j x_j, \sum_{i=1}^{n} \gamma_i x_i \right) \\ &= \left\| \sum_{i=1}^{n} \gamma_i x_i \right\|^2 \geq 0. \end{aligned}$$

   Therefore, a Gram matrix is always positive semidefinite. From the above equation, clearly, $y^* G(X) y = 0$ if and only if $\sum_{i=1}^{n} \gamma_i x_i = 0$. Therefore, $G(X)$ is positive definite if and only if X is a linearly independent set.

   As a particular case, for $X = \{x_1, x_2\}$, $G(X)$ is a positive semidefinite matrix, and so $\det G(X) \geq 0$. Thus

$$\det \begin{bmatrix} (x_1, x_1) & (x_2, x_1) \\ (x_1, x_2) & (x_2, x_2) \end{bmatrix} \geq 0.$$

   Therefore, $|(x_1, x_2)| \leq \|x_1\| \, \|x_2\|$, and equality holds if and only if x_1 and x_2 are linearly dependent. Thus, we obtain Cauchy–Schwarz inequality.

   If V is a finite dimensional inner product space with a basis $\mathcal{B} = \{u_1, \dots, u_n\}$, then for $x, y \in V$, $x = \sum_{i=1}^{n} \alpha_i u_i$ and $y = \sum_{i=1}^{n} \beta_i u_i$, and so

$$(x, y) = \sum_{i=1}^{n} \sum_{j=1}^{n} \alpha_i \overline{\beta_j} (u_i, u_j) = [y]_{\mathcal{B}}^* \, G(\mathcal{B}) \, [x]_{\mathcal{B}}.$$

Thus the specification of $\mathcal{B}$ and $G(\mathcal{B})$ completely defines the inner product.

If $\mathcal{B}' = \{v_1, \ldots, v_n\}$ is another basis of V, then for $x, y \in V$:

$$\begin{aligned}[y]_{\mathcal{B}'}^* \, G(\mathcal{B}') \, [x]_{\mathcal{B}'} &= (x,y) = [y]_{\mathcal{B}}^* \, G(\mathcal{B}) \, [x]_{\mathcal{B}} \\ &= \left({}_{\mathcal{B}}[I]_{\mathcal{B}'} \, [y]_{\mathcal{B}'} \right)^* G(\mathcal{B}) \left({}_{\mathcal{B}}[I]_{\mathcal{B}'} \, [x]_{\mathcal{B}'} \right) \\ &= [y]_{\mathcal{B}'}^* \; {}_{\mathcal{B}}[I]_{\mathcal{B}'}^* \, G(\mathcal{B}) \; {}_{\mathcal{B}}[I]_{\mathcal{B}'} \, [x]_{\mathcal{B}'} .\end{aligned}$$

Hence, if $\mathcal{B}$ and $\mathcal{B}'$ are bases of V, then there exists an invertible matrix $P = {}_{\mathcal{B}}[I]_{\mathcal{B}'} = \left[\, [v_1]_{\mathcal{B}} \quad \cdots \quad [v_n]_{\mathcal{B}} \, \right] \in F^n$ such that $G(\mathcal{B}') = P^* G(\mathcal{B}) P$.

3. A useful characterization of positive definiteness for matrices is the following.

   Let $A \in F^{n\times n}$ be a Hermitian matrix. For $k = 1, \ldots, n$, let A_k be $k \times k$ submatrix in the left upper corner of A. The following statements are equivalent:

   (i) *A is positive definite;*

   (ii) *$\det A_k > 0$ for $k = 1, \ldots, n$;*

   (iii) *there exists an upper triangular matrix B with positive diagonal entries such that $A = B^* B$* ***(Cholesky decomposition)****.*

   (i) $\Rightarrow$ *(ii)* For $x \in F^k$, let $\widehat{x} \in F^n$ whose first k entries are the entries of x and the next $n-k$ entries zero. Then $x^* A_k x = \widehat{x}^* A \widehat{x}$. Thus, if $x \neq 0$ then $\widehat{x} \neq 0$, and so $x^* A_k x = \widehat{x}^* A \widehat{x} > 0$. Therefore, A_k is positive definite, and so all eigenvalues of A_k are positive. Hence, $\det A_k > 0$.

   (ii) $\Rightarrow$ *(iii)* By induction on n. The case $n = 1$: if $A = (a_{11})$, then take $B = \left(\sqrt{a_{11}}\right)$, and so $B = B^*$ and $A = B^* B$. Now let $A \in F^{n\times n}$, $n > 1$, and write

$$A = \begin{bmatrix} A_{n-1} & b \\ b^* & a_{nn} \end{bmatrix}, \; b \in F^{n-1}.$$

   By induction hypothesis, there exists an upper triangular $(n-1) \times (n-1)$ matrix B_{n-1} with positive diagonal entries such that $A_{n-1} = B_{n-1}^* B_{n-1}$. We now look for B in the desired form:

$$B = \begin{bmatrix} B_{n-1} & u \\ 0 & b_{nn} \end{bmatrix}.$$

   The result will be proved if there exists $u \in F^{n-1}$ such that $b_{nn} > 0$. If such a B exists, then

$$\begin{bmatrix} A_{n-1} & b \\ b^* & a_{nn} \end{bmatrix} = A = B^* B = \begin{bmatrix} B_{n-1}^* & 0 \\ u^* & b_{nn} \end{bmatrix} \begin{bmatrix} B_{n-1} & u \\ 0 & b_{nn} \end{bmatrix}.$$

   Comparing the corresponding entries, we have $B_{n-1}^* u = b$ and $u^* u + b_{nn}^2 = a_{nn}$. Thus, $u = \left(B_{n-1}^*\right)^{-1} b$. Now we only need to see that $b_{nn} > 0$ exists. Clearly, such a b_{nn} exists if and only if $a_{nn} - u^* u > 0$. From the above equation $\det A = b_{nn}^2 \det B_{n-1}^* \det B_{n-1} = b_{nn}^2 \det A_{n-1}$. Thus, $b_{nn}^2 = a_{nn} - u^* u > 0$.

   (iii) $\Rightarrow$ *(i)* If $x \neq 0$, then as B is invertible, $Bx \neq 0$, and so $x^* A x = x^* (B^* B) x = \|Bx\|^2 > 0$. Hence, A is positive definite.

4. If $A = (a_{ij}) \in F^{n\times n}$ is a positive semidefinite matrix, then

$$\det A \leq a_{nn} \det A_{n-1},$$

   where A_{n-1} is as defined in the previous example.

If $\det A = 0$, then as A is positive semidefinite $a_{nn} \geq 0$, and $\det A_{n-1} \geq 0$, and so the above inequality holds.

If $\det A \neq 0$, then all eigenvalues of A are nonzero, and so A is a positive definite matrix. Write

$$A = \begin{bmatrix} A_{n-1} & u \\ u^* & a_{nn} \end{bmatrix}, u \in F^{n-1}.$$

Verify that

$$A = \begin{bmatrix} I_{n-1} & 0 \\ u^* A_{n-1}^{-1} & 1 \end{bmatrix} \begin{bmatrix} A_{n-1} & u \\ 0 & a_{nn} - u^* A_{n-1}^{-1} u \end{bmatrix}.$$

Then $\det A = \left(a_{nn} - u^* A_{n-1}^{-1} u\right) \det A_{n-1}$. Since A_{n-1} is a positive definite matrix, A_{n-1}^{-1} is also a positive definite matrix, and so $u^* A_{n-1}^{-1} u \geq 0$. Therefore, $\det A \leq a_{nn} \det A_{n-1}$.

Now using this result repeatedly, we have the following important result.

If $A = (a_{ij}) \in F^{n\times n}$ *is a positive semidefinite matrix, then*

$$\det A \leq a_{11} \cdots a_{nn}.$$

Finally, if $A = (a_{ij}) \in F^{n\times n}$, then A^*A is always positive semidefinite. Thus, by the above inequality:

$$\begin{aligned} \det(A^*A) &\leq [A^*A]_{11} \cdots [A^*A]_{nn} \\ &= e_1^* (A^*A) e_1 \cdots e_n^* (A^*A) e_n \\ &= (Ae_1)^* (Ae_1) \cdots (Ae_n)^* (Ae_n) \\ &= \prod_{j=1}^{n} \sum_{i=1}^{n} |a_{ij}|^2 . \end{aligned}$$

Hence, we have the following inequality for $A \in F^{n\times n}$,

$$(\det A)^2 \leq \prod_{j=1}^{n} \sum_{i=1}^{n} |a_{ij}|^2 \quad \textbf{(Hadamard inequality)}.$$

5. This example illustrates another method of obtaining the spectral resolution of a diagonalizable linear operator T on a finite dimensional inner product space V. Since T is diagonalizable, the minimal polynomial of T is

$$m_T(x) = (x - \lambda_1) \cdots (x - \lambda_k),$$

where $\lambda_1, \ldots, \lambda_k$ are distinct eigenvalues of T. Writing into partial fractions

$$\frac{1}{m_T(x)} = \sum_{i=1}^{k} \frac{1}{\alpha_i (x - \lambda_i)},$$

where $\alpha_i = \prod_{j=1, j\neq i}^{k} (\lambda_i - \lambda_j) = \frac{d}{dx} m_T(x)\big|_{x=\lambda_i}$. Thus

$$I = \sum_{i=1}^{k} \alpha_i^{-1} \prod_{j=1, j\neq i}^{k} (T - \lambda_j I).$$

If $P_i = \alpha_i^{-1} \prod_{j=1, j\neq i}^{k} (T - \lambda_j I)$, then $P_i \in L(V)$ and we have

$$I = P_1 + \cdots + P_k.$$

This is actually a resolution of identity. Since for $r \neq s$,

$$\begin{aligned} P_r P_s &= \alpha_r^{-1}\alpha_s^{-1} \prod_{j=1, j\neq r}^{k} (T-\lambda_j I) \prod_{j=1, j\neq s}^{k} (T-\lambda_j I) \\ &= (\alpha_r\alpha_s)^{-1} m_T(T) \prod_{j\neq r,s} (T-\lambda_j I) = 0, \end{aligned}$$

and so $P_r = P_r I = P_r(P_1 + \cdots + P_k) = P_r^2$. Now $T = TP_1 + \cdots + TP_k$, and for each i,

$$\begin{aligned} TP_i &= \alpha_i^{-1} T \prod_{j=1, j\neq i}^{k} (T-\lambda_j I) \\ &= \alpha_i^{-1}\left(m_T(T) + \lambda_i \prod_{j=1, j\neq i}^{k} (T-\lambda_j I)\right) \\ &= \lambda_i P_i. \end{aligned}$$

Therefore, $T = \lambda_1 P_1 + \cdots + \lambda_k P_k$, the spectral decomposition of T.

Let

$$A = \begin{bmatrix} 3 & -1 & 1 \\ -1 & 3 & -1 \\ 1 & -1 & 3 \end{bmatrix}.$$

Then A is diagonalizable and $m_A(x) = (x-5)(x-2)$. Now using the above method with $\lambda_1 = 5$ and $\lambda_2 = 2$, we have

$$\frac{1}{m_A(x)} = \frac{1}{3(x-5)} - \frac{1}{3(x-2)}$$

or $1 = \frac{1}{3}(x-2) + \frac{1}{3}(5-x)$. Thus, $P_1 = \frac{1}{3}(A - 2I_3)$ and $P_2 = \frac{1}{3}(5I_3 - A)$, and we have the spectral resolution of A:

$$A = 5P_1 + 2P_2,$$

where

$$P_1 = \frac{1}{3}\begin{bmatrix} 1 & -1 & 1 \\ -1 & 1 & -1 \\ 1 & -1 & 1 \end{bmatrix} \quad \text{and} \quad P_2 = \frac{1}{3}\begin{bmatrix} 2 & 1 & -1 \\ 1 & 2 & 1 \\ -1 & 1 & 2 \end{bmatrix}.$$

6. Let T be a normal operator on a finite dimensional inner product space V over F. If T is not triangulable, then it is clear that $F = \mathbb{R}$. By Example 6, p.142, it follows that there is an ordered orthonormal basis $\mathcal{B}$ of V such that

$$[T]_{\mathcal{B}} = \begin{bmatrix} A_0 & A_{01} & \cdots & A_{0k} \\ & A_1 & \cdots & A_{1k} \\ & & \ddots & \vdots \\ & & & A_k \end{bmatrix},$$

where $A_0 \in \mathbb{R}^{m\times m}$, $m \geq 0$ ($m = 0$ means that T has no real eigenvalues), and each A_i is a 2×2 matrix with $\operatorname{tr} A_i^2 < 4 \det A_i$, $i = 1, \ldots, k$ ($k \geq 1$, otherwise T is triangulable). Since $[T]_{\mathcal{B}}$ is a normal matrix with entries from $\mathbb{R}$,

$$[T]^t_{\mathcal{B}}[T]_{\mathcal{B}} = [T]_{\mathcal{B}}[T]^t_{\mathcal{B}}.$$

Equating the first $m \times m$ main diagonal block entries of the above identity, we get

$$A_0^t A_0 = A_0 A_0^t + A_{01} A_{01}^t + \cdots + A_{0k} A_{0k}^t.$$

Then $\operatorname{tr}(A_0^t A_0) = \operatorname{tr}(A_0 A_0^t) + \operatorname{tr}(A_{01} A_{01}^t) + \cdots + \operatorname{tr}(A_{0k} A_{0k}^t)$. Since $\operatorname{tr}(A_0^t A_0) = \operatorname{tr}(A_0 A_0^t)$, we have

$$\operatorname{tr}(A_{01} A_{01}^t) + \cdots + \operatorname{tr}(A_{0k} A_{0k}^t) = 0.$$

Recall that for $X, Y \in F^{r\times r}$, $(X, Y) = \operatorname{tr}(XY^*)$ is an inner product, so $\operatorname{tr}(XX^*) \geq 0$ and equality holds if and only if $X = 0$. Therefore, we have $A_{01} = \cdots = A_{0k} = 0$. Thus A_0 is normal upper triangular matrix and hence A_0 is diagonal.

Now equating 2×2 main diagonal entries, we have

$$A_1^t A_1 = A_1 A_1^t + A_{12} A_{12}^t + \cdots + A_{1k} A_{1k}^t.$$

Similarly, equating the trace on both sides, we get $A_{12} = \cdots = A_{1k} = 0$. Thus, equating each successive main diagonal 2×2 block of $[T]^t_{\mathcal{B}}[T]_{\mathcal{B}}$ and $[T]_{\mathcal{B}}[T]^t_{\mathcal{B}}$, we conclude that all off diagonal blocks of $[T]_{\mathcal{B}}$ are zero, and each A_i is a 2×2 normal matrix with $\operatorname{tr} A_i^2 < 4 \det A_i$, $i = 1, \ldots, k$. If each A_i is of the form $\begin{bmatrix} a & b \\ c & d \end{bmatrix}$, then the identity $A_i^t A_i = A_i A_i^t$ gives $b^2 = c^2$ and $ac + bd = ab + cd$. Since, $(a+d)^2 < 4(ad - bc)$, we have $c = -b$, $b \neq 0$ and $a = d$. Thus, we have the following situation.

Let V be an n-dimensional inner product space over F and let $T \in L(V)$.

If $F = \mathbb{C}$, then T is normal if and only if there is an ordered orthonormal basis $\mathcal{B}$ of V such that $[T]_{\mathcal{B}} = \operatorname{diag}(\lambda_1, \ldots, \lambda_n)$, $\lambda_i \in \mathbb{C}$.

If $F = \mathbb{R}$, then T is normal if and only if there is an ordered orthonormal basis $\mathcal{B}$ of V such that $[T]_{\mathcal{B}} = \operatorname{diag}(A_1, \ldots, A_r)$, where each A_i is either 1×1 real matrix or a 2×2 real matrix of the form $\begin{bmatrix} a_i & b_i \\ -b_i & a_i \end{bmatrix}$.

For matrices, we have the following.

Let $A \in \mathbb{C}^{n\times n}$. Then A is normal if and only if there exists a unitary matrix U such that $U^*AU = \operatorname{diag}(\lambda_1, \ldots, \lambda_n)$, $\lambda_i \in \mathbb{C}$.

Let $A \in \mathbb{R}^{n\times n}$. Then A is normal if and only if there exists a real orthogonal matrix Q such that $Q^tAQ = \operatorname{diag}(A_1, \ldots, A_r)$, where each A_i is either 1×1 real matrix or a 2×2 real matrix of the form $\begin{bmatrix} a_i & b_i \\ -b_i & a_i \end{bmatrix}$.

5.2. Exercises. In the following exercises, V is a finite dimensional inner product space over F and $T \in L(V)$.

1. Show that if T is a positive definite operator, then so is T^{-1}.

2. Let $A \in F^{n\times n}$ be a positive definite matrix. Prove that for $b \in F^n$, $A - bb^*$ is positive definite if and only if $b^*A^{-1}b < 1$.

3. Show that if T is a positive definite operator, then $\langle x, y\rangle = (Tx, y)$, $x, y \in V$ is an inner product on V.

4. Let $u_1, \ldots, u_n$ be mutually orthogonal nonzero elements of $\mathbb{C}^n$ and let $B = \begin{bmatrix} u_1 & \cdots & u_n \end{bmatrix}$. Prove that for $\sigma \in \mathbb{R}$, the matrix $S = BB^* + \sigma I_n$ is self adjoint and each u_i is an eigenvector of S with eigenvalue $\sigma + \|u_i\|^2$.

5. Using the above exercise, find a symmetric matrix S whose eigenvectors are $\begin{bmatrix} 1 & 1 & 3 \end{bmatrix}^t$, $\begin{bmatrix} 1 & -1 & 0 \end{bmatrix}^t$ and $\begin{bmatrix} 3 & 3 & -2 \end{bmatrix}^t$ and eigenvalue corresponding to $\begin{bmatrix} 1 & 1 & 3 \end{bmatrix}^t$ is 1. Also find other two eigenvalues.

6. Prove that $A \in \mathbb{C}^{n\times n}$ is positive semi definite if and only if there exists an $n \times n$ matrix B whose columns are mutually orthogonal in $\mathbb{C}^n$, such that $A = BB^*$.

7. Prove that $A \in \mathbb{C}^{n\times n}$ is self adjoint if and only if there exists an $n \times n$ matrix B with mutually orthogonal columns and a real number σ such that $A = BB^* + \sigma I_n$.
Hint: Exercises 4 and 6 are useful.

8. Prove that the characteristic polynomial of a self adjoint operator must have real coefficients.

9. Let T be a self adjoint operator and let $\rho_T(u) = \frac{(Tu,u)}{(u,u)}$, $u \neq 0$. Prove that if $\lambda_1 \leq \cdots \leq \lambda_n$ are eigenvalues of T, then $\lambda_1 \leq \rho_T(u) \leq \lambda_n$ for all $u \in V \setminus \{0\}$.

10. Using Example 3, p.139, prove that if A is a positive definite matrix, then there exists an upper triangular matrix B such that $A = B^*B$.

11. Prove that a normal nilpotent operator is a zero operator.

12. Show by example that if S and T are normal operators, then ST may not be normal.

13. Give an example of a real normal matrix that is unitarily similar to a diagonal matrix but not real orthogonally similar to a diagonal matrix.

14. Let $F = \mathbb{C}$ and let $T \in L(V)$. Prove that T is normal if and only if there exists a polynomial $p(x) \in F[x]$ of degree at most $n-1$ ($n = \dim V$) such that $T^* = p(T)$.
Hint: if $\mu_1, \ldots, \mu_k$ are distinct eigenvalues of T, let $p(x) \in F[x]$ such that $p(\mu_i) = \overline{\mu_i}$.

15. Let $F = \mathbb{C}$ and let $T \in L(V)$. Prove that T is normal if and only if there exists a unitary matrix U such that $T^* = TU$.
Hint: for a complex number z, there is $\theta \in \mathbb{R}$ such that $\overline{z} = e^{i\theta}z$. If $\mathcal{B}$ is an ordered orthonormal basis of V consisting of eigenvectors of T, then $[T]_{\mathcal{B}} = \operatorname{diag}(\lambda_1, \ldots, \lambda_n)$, and so $[T^*]_{\mathcal{B}} = \operatorname{diag}(\overline{\lambda_1}, \ldots, \overline{\lambda_n})$.

16. Let T be a triangulable linear operator and let $\{u_1, \cdots, u_n\}$ be an ordered orthonormal basis of V. Let $\lambda_1, \ldots, \lambda_n$ be eigenvalues of T, counted with multiplicities. Prove that the following statements are equivalent:
   (a) T is normal;
   (b) there is an ordered orthonormal basis of V consisting of eigenvectors of T;

(c) $\sum_{i=1}^{n}\sum_{j=1}^{n}|(Tu_i,u_j)|^2=\sum_{i=1}^{n}|\lambda_i|^2$.

17. Prove that two normal matrices are similar if and only if they have the same characteristic polynomial.
18. Let $T_1=(T+T^*)/2$ and $T_2=(T-T^*)/2$. Prove that T is normal if and only if $T_1T_2=T_2T_1$.
19. Prove that a normal operator is self adjoint if and only if all its eigenvalues are real.

    What should be similar equivalent conditions for a normal operator to be unitary?
20. Let $A\in F^{n\times n}$ be a unitarily diagonalizable matrix and let $\mu_1,\dots,\mu_k$ be all distinct eigenvalues of A with multiplicities $n_1,\dots,n_k$, respectively. Suppose that $u_{11},\dots,u_{1n_1},\dots,u_{k1},\dots,u_{kn_k}$ is an ordered orthonormal basis of F^n consisting of eigenvectors of A, u_{ir} is an eigenvector of A corresponding to μ_i. Verify that

$$A=\sum_{r=1}^{k}\mu_rP_r,\quad\text{for }P_r=\sum_{j=1}^{n_r}u_{rj}u_{rj}^*,$$

    is the spectral resolution of A.
21. Write the spectral resolution of following matrices:

$$\begin{bmatrix}1&0&1\\1&0&1\\1&0&1\end{bmatrix},\quad\begin{bmatrix}3&-2&2\\10&-9&10\\6&-6&7\end{bmatrix},\quad\begin{bmatrix}6&-1&2\\4&1&2\\-10&0&-3\end{bmatrix},$$

$$\begin{bmatrix}1&2&2\\2&1&-2\\2&-2&1\end{bmatrix},\quad\begin{bmatrix}4&3&-3\\2&3&-2\\4&4&-3\end{bmatrix},\quad\begin{bmatrix}0&2&1\\2&-3&6\\2&-2&5\end{bmatrix},$$

$$\begin{bmatrix}1&1&1&1\\1&1&-1&-1\\1&-1&1&-1\\1&-1&-1&1\end{bmatrix},\quad\begin{bmatrix}2&1&0&0\\1&2&1&0\\0&1&2&1\\0&0&1&2\end{bmatrix}.$$

6. Polar and singular value decompositions

We have observed an analogy between the linear operators on a finite dimensional inner product space and the complex numbers. Under this analogy, the adjoint of a linear operator corresponds to the conjugate of a complex number, and positive semidefinite operators are like nonnegative real numbers. A complex number is of unit modulus if $z\bar{z}=1$, and so unitary operators are analogous to complex numbers with unit modulus. A complex number z can always be written as $z=\omega\rho$, where $|\omega|=1$ and $\rho\geq 0$. Similar decomposition is possible for a linear operator.

THEOREM 6.1. ***Polar decomposition.*** *A linear operator T on a finite dimensional inner product space V can be represented in the form $T=UP$, where U is unitary and $P=(T^*T)^{1/2}$, a positive semidefinite operator which is uniquely determined by T.*

PROOF. Since TT^* is positive semidefinite operator on V, TT^* has nonnegative eigenvalues. Let $\lambda_1 \geq \cdots \geq \lambda_r > 0 = \lambda_{r+1} = \cdots = \lambda_n$ ($r \leq n$), be eigenvalues of TT^*, repeated according to multiplicities with corresponding eigenvectors $w_1, \ldots, w_n$ that form an orthonormal basis of V (Spectral decomposition theorem). Define

$$u_i = \frac{1}{\sqrt{\lambda_i}} T^* w_i,$$

for $i = 1, \ldots, r$. Verify that $\{u_1, \ldots, u_r\}$ is an orthonormal set and each u_i is an eigenvector of T^*T corresponding to nonzero eigenvalue λ_i, $i = 1, \ldots, r$ (see Exercise 10, p.133). Since T^*T is unitarily diagonalizable, $\dim \ker T^*T = \dim \ker TT^* = n-r$, and so if $\{u_{r+1}, \ldots, u_n\}$ is an orthonormal basis of $\ker T^*T$, then $\{u_1, \ldots, u_n\}$ is an orthonormal basis of V consisting of eigenvectors of T^*T. Let $U \in L(V)$, whose action is given by $Uw_i = u_i$, $i = 1, \ldots, n$. Since U maps an orthonormal basis to an orthonormal basis, U is unitary.

Now we show that $T = UP$, where $P = (T^*T)^{1/2}$. First note that $Pu_i = \sqrt{\lambda_i} u_i$, $i = 1, \ldots, r$ and $Pu_i = 0$, $i = r+1, \ldots, n$. For $i = 1, \ldots, r$,

$$Tu_i = \frac{1}{\sqrt{\lambda_i}} TT^* w_i = \sqrt{\lambda_i} w_i = U\left(\sqrt{\lambda_i} u_i\right) = UPu_i,$$

and for $i = r+1, \ldots, n$,

$$(Tu_i, Tu_i) = (u_i, T^*Tu_i) = 0,$$

$$(Pu_i, Pu_i) = \left(u_i, P^2 u_i\right) = (u_i, T^*Tu_i) = 0,$$

that is, $Tu_i = 0 = Pu_i$. Hence, $T = UP$. ∎

Note that, in the above theorem, if we also assume that T is invertible, then T^*T is positive definite, and so P is invertible. Thus, in this case, $U = P^{-1}T$, which is also uniquely determined by T.

Similar arguments can be used to prove that there exists a unitary operator W and a positive semidefinite operator Q such that $T = QW$.

Let T be a linear operator on an n–dimensional inner product space V over F. The eigenvalues of P $\left(= (T^*T)^{1/2}\right)$ are nonnegative real numbers. The eigenvalues $\sigma_1, \ldots, \sigma_n$ of P, repeated according to their multiplicities, are called **singular values** of T.

Let $\{u_1, \ldots, u_n\}$ be an orthonormal ordered basis of V consisting of eigenvectors of P such that $Pu_i = \sigma_i u_i$, $i = 1, \ldots, n$. Then for $x \in V$, $x = \sum_{i=1}^n (x, u_i)\, u_i$, and

$$Px = \sum_{i=1}^{n} \sigma_i (x, u_i)\, u_i,$$

and so as $T = UP$,

$$Tx = \sum_{i=1}^{n} \sigma_i (x, u_i)\, Uu_i.$$

We write $w_i = Uu_i$, $i = 1, \ldots, n$. Since U is a unitary operator, $\{w_1, \ldots, w_n\}$ is an orthonormal basis of V. This proves the following result.

THEOREM 6.2. ***Singular value decomposition.*** *Let T be a linear operator on an n-dimensional inner product space V. Then there are ordered orthonormal bases $\mathcal{B} = \{u_1, \dots, u_n\}$ and $\mathcal{B}' = \{w_1, \dots, w_n\}$ such that for any $x \in V$,*

$$Tx = \sum_{i=1}^{n} \sigma_i \,(x, u_i)\, w_i,$$

where $\sigma_1, \dots, \sigma_n$ are singular values of T.

We now obtain the matrix analogue of the above result. Let $A \in F^{n\times n}$. Then regarding A as a linear operator on F^n, there are ordered orthonormal bases $\mathcal{B} = \{u_1, \dots, u_n\}$ and $\mathcal{B}' = \{w_1, \dots, w_n\}$ such that for any $x \in V$,

$$Ax = \sum_{i=1}^{n} \sigma_i \,(x, u_i)\, w_i,$$

where $\sigma_1, \dots, \sigma_n$ are singular values of A.

From the above equation, we have for $i = 1, \dots, n$,

$$(*) \qquad Au_i = \sigma_i w_i.$$

This means that ${}_{\mathcal{B}'}[A]_{\mathcal{B}} = \operatorname{diag}(\sigma_1, \dots, \sigma_n)$. Now if $\mathcal{B}_0$ is the standard basis of F^n, then

$${}_{\mathcal{B}'}[A]_{\mathcal{B}} = {}_{\mathcal{B}'}[I]_{\mathcal{B}_0} A\, {}_{\mathcal{B}_0}[I]_{\mathcal{B}}.$$

Recall that $[A]_{\mathcal{B}_0} = A$. Let $U = {}_{\mathcal{B}_0}[I]_{\mathcal{B}} = \begin{bmatrix} u_1 & \cdots & u_n \end{bmatrix}$ and $W = {}_{\mathcal{B}_0}[I]_{\mathcal{B}'} = \begin{bmatrix} w_1 & \cdots & w_n \end{bmatrix}$. Then U and W are unitary matrices and

$$W^* AU = \operatorname{diag}(\sigma_1, \dots, \sigma_n)$$

or $A = W \operatorname{diag}(\sigma_1, \dots, \sigma_n)\, U^*$. Thus, we have the following matrix analogue of singular value decomposition.

THEOREM 6.3. *Let $A \in F^{n\times n}$ and let $\sigma_1, \dots, \sigma_n$ be singular values of A. Then there exist unitary matrices U and W such that $A = W \operatorname{diag}(\sigma_1, \dots, \sigma_n)\, U^*$.*

6.1. Examples

1. Let

$$A = \begin{bmatrix} 1 & 0 & 2 \\ 0 & 1 & 0 \\ 0 & 0 & 0 \end{bmatrix} \in \mathbb{R}^{3\times 3}.$$

We find a polar decomposition of A.

If $A = UP$ is a polar decomposition of A, then $P = (A^t A)^{1/2}$. Now

$$A^t A = \begin{bmatrix} 1 & 0 & 2 \\ 0 & 1 & 0 \\ 2 & 0 & 4 \end{bmatrix}$$

and

$$\left\{ \begin{bmatrix} 0 \\ 1 \\ 0 \end{bmatrix}, \frac{1}{\sqrt{5}} \begin{bmatrix} 1 \\ 0 \\ 2 \end{bmatrix}, \frac{1}{\sqrt{5}} \begin{bmatrix} 2 \\ 0 \\ -1 \end{bmatrix} \right\}$$

is an orthonormal set of eigenvectors of A^tA corresponding to eigenvalues 1, 5 and 0, respectively. Then

$$Q = \begin{bmatrix} 0 & \frac{1}{\sqrt{5}} & \frac{2}{\sqrt{5}} \\ 1 & 0 & 0 \\ 0 & \frac{2}{\sqrt{5}} & \frac{-1}{\sqrt{5}} \end{bmatrix}$$

is an orthogonal matrix and $Q^t\,(A^tA)\,Q = \operatorname{diag}(1,5,0)$.

Thus $P = (A^tA)^{1/2} = Q\operatorname{diag}(1,\sqrt{5},0)\,Q^t$, and so

$$P = \frac{1}{\sqrt{5}}\begin{bmatrix} 1 & 0 & 2 \\ 0 & \sqrt{5} & 0 \\ 2 & 0 & 4 \end{bmatrix}.$$

Next, if $U = [\ u_1 \quad u_2 \quad u_3\ ] \in \mathbb{R}^{3\times 3}$ is a unitary (here orthogonal) matrix such that $A = UP$, then comparing the columns, we get

$$\begin{aligned} e_1 &= Ae_1 = UPe_1 = U\,(e_1 + 2e_3)\,/\sqrt{5} = \frac{1}{\sqrt{5}}\,u_1 + \frac{2}{\sqrt{5}}\,u_3, \\ e_2 &= Ae_2 = UPe_2 = Ue_2 = u_2. \end{aligned}$$

Equating the third columns of A and UP will not yield new equation as $\operatorname{rank} A = 2$. Since $\{u_1, u_2, u_3\}$ is an orthonormal basis of $\mathbb{R}^3$, so $e_3 = \alpha u_1 + \beta u_2 + \gamma u_3$, $\alpha, \beta, \gamma \in F$. Now $(e_1, e_3) = 0$ gives $\alpha + 2\gamma = 0$; $(e_2, e_3) = 0$ implies $\beta = 0$ and $(e_3, e_3) = 1$ gives $\alpha^2 + \gamma^2 = 1$. Therefore, $e_3 = -\frac{2}{\sqrt{5}}\,u_1 + \frac{1}{\sqrt{5}}\,u_3$. Hence

$$U = \frac{1}{\sqrt{5}}\begin{bmatrix} 1 & 0 & 2 \\ 0 & \sqrt{5} & 0 \\ -2 & 0 & 1 \end{bmatrix}.$$

2. We find singular value decomposition of the matrix

$$A = \begin{bmatrix} 1 & 0 & 2 \\ 0 & 1 & 0 \\ 0 & 0 & 0 \end{bmatrix}.$$

The singular values of A are $1, \sqrt{5}$ and 0, and corresponding eigenvector of $(A^tA)^{1/2}$ are

$$u_1 = \begin{bmatrix} 0 \\ 1 \\ 0 \end{bmatrix},\ u_2 = \begin{bmatrix} 1/\sqrt{5} \\ 0 \\ 2/\sqrt{5} \end{bmatrix},\ u_3 = \begin{bmatrix} 2/\sqrt{5} \\ 0 \\ -1/\sqrt{5} \end{bmatrix}.$$

Hence, the unitary matrix U is

$$\begin{bmatrix} 0 & \frac{1}{\sqrt{5}} & \frac{2}{\sqrt{5}} \\ 1 & 0 & 0 \\ 0 & \frac{2}{\sqrt{5}} & \frac{-1}{\sqrt{5}} \end{bmatrix}.$$

Now we only need to find a unitary matrix W so that $A = W\operatorname{diag}(1,\sqrt{5},0)U^*$. From equation (*), we have if $\sigma_i \neq 0$, then $w_i = \frac{1}{\sigma_i}Au_i$. Note that $(AA^t)\,w_i = \frac{1}{\sigma_i}A\,(A^tA)\,u_i = \sigma_i A u_i = \sigma_i^2 w_i$. Thus, for any nonzero singular values we have this equation for w_i and then we can extend this to an orthonormal basis of F^n by choosing an orthonormal basis of $\ker AA^t$.

Now, if we let $\sigma_1 = 1$, $\sigma_2 = \sqrt{5}$ and $\sigma_3 = 0$, then

$$\begin{aligned} w_1 &= Au_1 = \begin{bmatrix} 0 & 1 & 0 \end{bmatrix}^t = e_2, \\ w_2 &= \frac{1}{\sqrt{5}} Au_2 = \begin{bmatrix} 1 & 0 & 0 \end{bmatrix}^t = e_1. \end{aligned}$$

Since $e_3 \in \ker AA^t$, so

$$W = \begin{bmatrix} 0 & 1 & 0 \\ 1 & 0 & 0 \\ 0 & 0 & 1 \end{bmatrix}.$$

3. Let $A \in F^{n\times n}$ and let $\sigma_1, \dots, \sigma_n$ be singular values of A. Then by singular value decomposition theorem, there are unitary matrices U and W such that $A = W \operatorname{diag}(\sigma_1, \dots, \sigma_n) U^*$. Now $A = (WU^*) U \operatorname{diag}(\sigma_1, \dots, \sigma_n) U^*$. Since WU^* is unitary and $U \operatorname{diag}(\sigma_1, \dots, \sigma_n) U^*$ is positive semidefinite, we have polar decomposition of A.

6.2. Exercises. In the following exercises, V is a finite dimensional inner product space over F and $T \in L(V)$.

1. Let $T = UP$, U a unitary and P a positive semi definite operator, be a polar decomposition of T. Prove that T is normal if and only if $PU = UP$.
2. Let $T = UP$ and $T = QW$ be polar decompositions of T, U and W are unitary and P and Q are positive semi definite operators. Prove that if T is invertible then P and Q have the same set of eigenvalues.
3. Show that for a triangulable linear operator:
$$|\lambda_1 \cdots \lambda_n| = \sigma_1 \cdots \sigma_n,$$
where λ_i and σ_i are eigenvalues and singular values of T.
4. Let $A \in \mathbb{C}^{n\times n}$ with eigenvalues $\lambda_1, \dots, \lambda_n$ and singular values $\sigma_1, \dots, \sigma_n$. Prove that
$$\sum_{i=1}^{n} |\lambda_i|^2 \le \operatorname{tr}(AA^*) = \sum_{i=1}^{n} \sigma_i^2.$$
Equality holds if and only if A is normal.
5. What are the singular values of a unitary operator?
6. Let $A \in \mathbb{C}^{n\times n}$ and let $M = \begin{bmatrix} 0 & A \\ A^* & 0 \end{bmatrix}$. Show that the eigenvalues of M are $-\sigma_r, \dots, -\sigma_1, 0, \dots, 0, \sigma_1, \dots, \sigma_r$, where $\sigma_1 \le \cdots \le \sigma_r$ are singular values of A and the multiplicity of eigenvalue 0 is $2n - 2r$.
7. Let $A, B \in \mathbb{C}^{n\times n}$. Do AB and BA have the same set of singular values?

CHAPTER 5

Bilinear Forms

An inner product space V over K ($=\mathbb{R}$ or $\mathbb{C}$) is a mapping from $V\times V$ to K which is linear in the first coordinate and conjugate linear in the second; such a mapping is called a sesquilinear form (sesquilinear means one and a half). In this chapter, we extend the notion of inner product to vector spaces over arbitrary field K.

1. Definition and examples

A **bilinear form** ϕ on a vector space V over K is a mapping $\phi : V\times V\to K$ which is linear in both the coordinates, that is,

$$\begin{aligned}\phi(\alpha u+\beta v,w) &= \alpha\phi(u,w)+\beta\phi(v,w),\\ \phi(u,\alpha v+\beta w) &= \alpha\phi(u,v)+\beta\phi(u,w),\end{aligned}$$

for all $u,v,w\in V$ and $\alpha,\beta\in K$. A vector space V together with a bilinear form ϕ is called a **bilinear space**, denoted by (V,ϕ).

Let ϕ be a bilinear form on a vector space V over K. A vector $u\in V$ is said to be **orthogonal** to $v\in V$, with respect to ϕ, if $\phi(u,v)=0$. The orthogonality relation on V may not be symmetric (Exercise 4). A bilinear form ϕ is called **reflexive** if orthogonality relation (with respect to ϕ) is symmetric, that is, for $u,v\in V$, $\phi(v,u)=0$ whenever $\phi(u,v)=0$. A bilinear form is called

symmetric, if $\phi(u,v)=\phi(v,u)$ for all $u,v\in V$;
skew–symmetric, if $\phi(u,v)=-\phi(v,u)$ for all $u,v\in V$; and
alternating if $\phi(u,u)=0$ for all $u\in V$.

Obviously, a symmetric or a skew–symmetric bilinear form is reflexive. An alternating bilinear form is actually skew–symmetric and hence reflexive. This follows from the following identity:

$$\phi(u+v,u+v)=\phi(u,u)+\phi(u,v)+\phi(v,u)+\phi(v,v),$$

for $u,v\in V$.

The following result is a characterization of reflexive bilinear forms.

THEOREM 1.1. *A bilinear form is reflexive if and only if it is either symmetric or alternating.*

PROOF. Let ϕ be a reflexive bilinear form on a vector space V over K. For $x,y,z\in V$, the identity $\phi(x,\phi(x,y)z-\phi(x,z)y)=0$ implies that

$$\phi(\phi(x,y)z-\phi(x,z)y,x)=0,$$

that is,

$$(*)\qquad \phi(x,y)\phi(z,x)-\phi(x,z)\phi(y,x)=0.$$

In particular, for $z = x$, we have

$$(**) \qquad \phi(x,x)(\phi(x,y) - \phi(y,x)) = 0.$$

Now assume that ϕ is not symmetric. We will prove that ϕ is alternating. If for $u \in V$, there exists $v \in V$ such that $\phi(u,v) \neq \phi(v,u)$, then from $(**)$ it follows that $\phi(u,u) = 0$.

Let $u \in V$ be such that $\phi(x,u) = \phi(u,x)$ for all $x \in V$. Choose $v, w \in V$ such that $\phi(v,w) \neq \phi(w,v)$. Then from $(*)$,

$$\phi(v,w)\phi(u,v) - \phi(v,u)\phi(w,v) = 0,$$

that is, $\phi(u,v)(\phi(v,w) - \phi(w,v)) = 0$, and so $\phi(u,v) = 0 = \phi(v,u)$. Similarly, $\phi(u,w) = 0 = \phi(w,u)$. Now

$$\phi(v,u+w) = \phi(v,w) \neq \phi(w,v) = \phi(u+w,v),$$

and so $\phi(u+w,u+w) = 0$. Hence, $\phi(u,u) = 0$, and ϕ is alternating.

Converse is obvious. ∎

COROLLARY 1.2. *A bilinear form is reflexive if and only if it is either symmetric or skew–symmetric.*

Let ϕ be a bilinear form on V. A vector $v \in V$ is said to be **isotropic** (with respect to ϕ) if $v \neq 0$ and v is orthogonal to itself, that is, $\phi(v,v) = 0$; otherwise v is called **anisotropic**.

1.1. Examples.

1. An inner product on a vector space V over $\mathbb{R}$ is a bilinear form on V, and every nonzero vector of V is anisotropic.

2. Let $\mu : \mathbb{R}^4 \times \mathbb{R}^4 \to \mathbb{R}$ defined by

$$\mu\left(\begin{bmatrix} x_1 & x_2 & x_3 & x_4 \end{bmatrix}^t, \begin{bmatrix} y_1 & y_2 & y_3 & y_4 \end{bmatrix}^t\right) = x_1y_1 + x_2y_2 + x_3y_3 - x_4y_4.$$

   Then μ is a bilinear form on $\mathbb{R}^4$. It is clear that the vector $\begin{bmatrix} 1 & 0 & 0 & 1 \end{bmatrix}^t$ is isotropic. This bilinear space is called **Minkowski's space**.

3. Let $x_1, \ldots, x_{n-2}$, $n \geq 2$, be elements of $\mathbb{R}^n$. The mapping $\delta : \mathbb{R}^n \times \mathbb{R}^n \to \mathbb{R}$ defined by $\delta(x,y) = \det\begin{bmatrix} x_1 & \cdots & x_{n-2} & x & y \end{bmatrix}$ is an alternating bilinear form. Note that if $x_1, \ldots, x_{n-2}$ are linearly dependent, then $\delta \equiv 0$.

4. Let V be a vector space over K and let f, g be linear forms on V. Then the mapping $\phi(x,y) = f(x)g(y)$ is a bilinear form on V.

5. Let V and W be vector spaces over K, $T \in L(V,W)$ and let ξ be a bilinear form on W. This can be used to define a bilinear form on V. Define $\phi : V \times V \to K$ by

$$\phi(x,y) = \xi(Tx,Ty), \quad x, y \in V.$$

   Then ϕ is a bilinear form on V. If ξ is symmetric (respectively, skew–symmetric), so is ϕ.

6. Let $A \in K^{n\times n}$. Define $\theta_A(x,y) = x^t A y$, $x, y \in K^n$. Then θ_A is a bilinear form on K^n. θ_A is symmetric (respectively, skew–symmetric) if and only if A is a symmetric (respectively, skew–symmetric) matrix.
   If $K = \mathbb{R}$, $n = 4$ and $A = \operatorname{diag}(1,1,1,-1)$, the bilinear space $(\mathbb{R}^4, \theta_A)$ is the Minkowski's space.

7. Let ϕ be a bilinear form on K^n. Then there exists a unique matrix A such that $\phi = \theta_A$.
   Let $\{e_1, \ldots, e_n\}$ be the standard basis of K^n. For $x, y \in K^n$, write $x = \sum_{i=1}^n x_i e_i$, $y = \sum_{i=1}^n y_i e_i$. Then
$$\phi(x,y) = \phi\left(\sum_{i=1}^n x_i e_i, \sum_{j=1}^n y_j e_j\right) = \sum_{i=1}^n \sum_{j=1}^n x_i\, y_j\, \phi(e_i, e_j).$$
   Let $A \in K^{n\times n}$ whose (i,j)–th entry is $\phi(e_i, e_j)$, $i, j = 1, \ldots, n$. Then clearly, $\phi(x,y) = x^t A y = \theta_A(x,y)$.
   Now to verify uniqueness, let matrices $A, B \in K^{n\times n}$ be such that $\xi(x,y) = \theta_A(x,y) = \theta_B(x,y)$, for all $x, y \in K^n$. Then $x^t(A-B)y = 0$ for all $x, y \in K^n$. In particular, $e_i^t(A-B)e_j = 0$, $i, j = 1, \ldots, n$. Hence, $A = B$.
   From now onwards, for $A \in K^{n\times n}$, θ_A will always denote the bilinear form on K^n defined by $\theta_A(x,y) = x^t A y$, $x, y \in K^n$.

1.2. Exercises

1. Decide which of the following mappings $\phi : \mathbb{R}^2 \times \mathbb{R}^2 \to \mathbb{R}$ are bilinear, if for $x = \begin{bmatrix} x_1 & x_2 \end{bmatrix}^t$, $y = \begin{bmatrix} y_1 & y_2 \end{bmatrix}^t$ in $\mathbb{R}^2$:
   (a) $\phi(x,y) = x_1 x_2 + y_1 y_2 - x_2 y_1 - x_1 y_2$;
   (b) $\phi(x,y) = (x_1 + x_2)(y_1 + y_2)$;
   (c) $\phi(x,y) = (x_1 + x_2)/(y_1 + y_2)$;
   (d) $\phi(x,y) = x_1 x_2 y_1 y_2$;
   (e) $\phi(x,y) = (3x_1 + 5)y_1 + (3y_2 + 5)x_1$;
   (f) $\phi(x,y) = x_1 y_2 + x_2 y_1^2$;
   (g) $\phi(x,y) = x_1 y_1 + 1$.
2. Prove that if ϕ is a bilinear form on V, then
   (a) $\phi(-x,y) = -\phi(x,y) = \phi(x,-y)$ for all $x, y \in V$;
   (b) $\phi(x,0) = 0 = \phi(0,x)$ for all $x \in V$.
3. If ϕ is a bilinear form on V and $v \in V$ is such that $\phi(x,v) = 0$ for all $x \in V$, then does it imply that $v = 0$?
4. Let $A \in K^{n\times n}$ such that $[A]_{12} = 1$ and $[A]_{21} = 0$. Consider the bilinear form $\theta_A : K^n \times K^n \to K$ given by $\theta_A(x,y) = x^t A y$. Verify that e_2 is orthogonal to e_1 and e_1 is not orthogonal to e_2.
5. Let $A \in K^{m\times m}$. Prove that the mapping $\phi : K^{m\times n} \times K^{m\times n} \to K$ given by $\phi(X,Y) = \operatorname{tr}(X^t A Y)$ is bilinear. Show that if $\operatorname{char} K \neq 2$ and A is skew–symmetric then ϕ is alternating.
6. Let $\phi : K^{n\times n} \times K^{n\times n} \to K$ defined by $\phi(A,B) = \operatorname{tr}(AB) - \operatorname{tr}(A)\operatorname{tr}(B)$. Prove that ϕ is a symmetric bilinear form. Are there isotropic vectors with respect to ϕ?

7. Let $P(X)$ be the family of all subsets of a finite set X. For $A, B \in P(X)$, define addition: $A + B = (A \cup B) \setminus (A \cap B)$, and scalar multiplication: $0A = \emptyset$, $1A = A$. Verify that $P(X)$ is an $|X|$–dimensional vector space over $\mathbb{Z}_2$. Prove that the mapping $P(X) \times P(X) \to \mathbb{Z}_2$ defined by $\langle A, B\rangle = |A \cap B| \bmod 2$ is a bilinear form on $P(X)$. Show that A is an isotropic vector if and only if A has even number of elements.
8. Let K be a field of characteristic 2. Verify that the mapping $\phi : K \times K \to K$ defined by $\phi(x, y) = xy$ is a skew–symmetric bilinear form which is not alternating.
9. Prove that if $\operatorname{char} K \neq 2$, then a bilinear form on V over K is skew–symmetric if and only if it is alternating.
10. Prove that if $\operatorname{char} K \neq 2$ and ϕ is a symmetric and alternating bilinear form on a vector space V over K, then $\phi \equiv 0$.
11. Prove that the set of all bilinear forms on a vector space V over K forms a vector space over K.
12. Prove that in a reflexive bilinear space, if $v_1, \dots, v_n$ are mutually orthogonal anisotropic vectors, then $v_1, \dots, v_n$ are linearly independent.
13. Let V be a vector space over K, $\operatorname{char} K \neq 2$. Prove that a bilinear form on V can be uniquely written as a sum of symmetric and skew–symmetric bilinear forms.

2. The matrix of a bilinear form

If ξ is a bilinear form on K^n, then there exists a matrix $A \in K^{n\times n}$ such that $\xi(x, y) = x^t A y$ for all $x, y \in K^n$ (Example 7, p.167). Since an n–dimensional vector space over K is isomorphic to K^n, it is natural to expect a similar result on finite dimensional vector spaces.

Let V be an n–dimensional vector space over K and let $\mathcal{B} = \{x_1, \dots, x_n\}$ be an ordered basis of V. Then for $u, v \in V$, we have a unique representation $u = \sum_{i=1}^n \alpha_i x_i$, $\alpha_i \in K$ and $v = \sum_{i=1}^n \beta_i x_i$, $\beta_i \in K$, and so $[u]_{\mathcal{B}} = \begin{bmatrix} \alpha_1 & \cdots & \alpha_n \end{bmatrix}^t$ and $[v]_{\mathcal{B}} = \begin{bmatrix} \beta_1 & \cdots & \beta_n \end{bmatrix}^t$. If ϕ is a bilinear form on V, then

$$\phi(u, v) = \phi\left(\sum_{i=1}^n \alpha_i x_i, \sum_{j=1}^n \beta_j x_j\right) = \sum_{i=1}^n \sum_{j=1}^n \alpha_i \beta_j \phi(x_i, x_j).$$

If we denote by $[\phi]_{\mathcal{B}}$ an $n \times n$ matrix whose (i, j)–th entry is $\phi(x_i, x_j)$, then the above equation is

$$\phi(u, v) = [u]_{\mathcal{B}}^t [\phi]_{\mathcal{B}} [v]_{\mathcal{B}}.$$

Obviously, the matrix $[\phi]_{\mathcal{B}}$ is uniquely determined by the bilinear form ϕ and the ordered basis $\mathcal{B}$. This matrix is called the **matrix of the bilinear form** with respect to an ordered basis $\mathcal{B}$.

Denote by $\operatorname{Bil}_K(V)$, the set of all bilinear forms on a vector space V over K. We have seen that $\operatorname{Bil}_K(V)$ is a vector space over K (Exercise 11, p.168). Define a mapping from $\operatorname{Bil}_K(V)$ to $K^{n\times n}$ given by $\phi \mapsto [\phi]_{\mathcal{B}}$. The reader can easily verify that this is a well defined homomorphism of vector spaces which is also an injective mapping. For $A \in K^{n\times n}$, define $f : \mathcal{B}\times\mathcal{B} \to K$ by $f(x_i, x_j) = [A]_{ij}$, $i, j = 1, \ldots, n$. Now for $u = \sum_{i=1}^{n} \alpha_i x_i$, $v = \sum_{i=1}^{n} \beta_i x_i$, define $\phi : V \times V \to K$ by

$$\phi(u, v) = \sum_{j=1}^{n} \sum_{i=1}^{n} \alpha_i \beta_j f(x_i, x_j).$$

Then ϕ is a bilinear form on V and $[\phi]_{\mathcal{B}} = A$. This shows that the mapping $\phi \mapsto [\phi]_{\mathcal{B}}$ is an isomorphism from $\mathrm{Bil}_K(V)$ to $K^{n\times n}$.

The above defined mapping depends on the choice of a basis. If $\mathcal{B}$ and $\mathcal{B}'$ are ordered bases of V, then for $\phi \in \mathrm{Bil}_K(V)$, the matrices $[\phi]_{\mathcal{B}}$ and $[\phi]_{\mathcal{B}'}$ are different. For $u, v \in V$,

$$\phi(u, v) = [u]_{\mathcal{B}}^t [\phi]_{\mathcal{B}} [v]_{\mathcal{B}} \quad \text{and} \quad \phi(u, v) = [u]_{\mathcal{B}'}^t [\phi]_{\mathcal{B}'} [v]_{\mathcal{B}'} .$$

Also

$$\begin{aligned} [u]_{\mathcal{B}}^t [\phi]_{\mathcal{B}} [v]_{\mathcal{B}} &= ({}_{\mathcal{B}}[I]_{\mathcal{B}'} [u]_{\mathcal{B}'})^t [\phi]_{\mathcal{B}} ({}_{\mathcal{B}}[I]_{\mathcal{B}'} [v]_{\mathcal{B}'}) \\ &= [u]_{\mathcal{B}'}^t \left({}_{\mathcal{B}}[I]_{\mathcal{B}'}^t [\phi]_{\mathcal{B}} \, {}_{\mathcal{B}}[I]_{\mathcal{B}'}\right) [v]_{\mathcal{B}'} . \end{aligned}$$

Since this equation holds for all $u, v \in V$,

$$[\phi]_{\mathcal{B}'} = {}_{\mathcal{B}}[I]_{\mathcal{B}'}^t [\phi]_{\mathcal{B}} \, {}_{\mathcal{B}}[I]_{\mathcal{B}'} .$$

Matrices A and B in $K^{n\times n}$ are said to be congruent if there exists an invertible matrix P such that $B = P^t AP$.

This discussion can be stated as the following result.

PROPOSITION 2.1. *Let V be an n–dimensional vector space over K and let $\mathcal{B}$ and $\mathcal{B}'$ be ordered bases of V.*

1. *The mapping defined by $\phi \mapsto [\phi]_{\mathcal{B}}$ is an isomorphism from $\mathrm{Bil}_K(V)$ to $K^{n\times n}$ In particular, $\dim \mathrm{Bil}_K(V) = n^2$.*
2. *For $\phi \in \mathrm{Bil}_K(V)$, matrices $[\phi]_{\mathcal{B}}$ and $[\phi]_{\mathcal{B}'}$ are congruent.*

Thus, for $\phi \in \mathrm{Bil}_K(V)$, $[\phi]_{\mathcal{B}'} = P^t [\phi]_{\mathcal{B}'} P$ for some invertible matrix P, and so $\det [\phi]_{\mathcal{B}'} = (\det P)^2 [\phi]_{\mathcal{B}}$. Therefore, the determinants of a bilinear form with respect to two different ordered bases differ by a nonzero square in K.

Let (V, ϕ) and (W, ξ) be bilinear spaces over K. A linear mapping $T : V \to W$ is called an **isometry** if it preserves the bilinear forms, that is, $\phi(x, y) = \xi(Tx, Ty)$ for all $x, y \in V$. Further, if T is also an isomorphism, T is called an **isometric isomorphism**. If such a mapping exists, we say that the bilinear spaces (V, ϕ) and (W, ξ) are isomorphic.

Let (V, ϕ) be a finite dimensional bilinear space over K, and let $T : (V, \phi) \to (W, \xi)$ be an isometric isomorphism. If $\mathcal{B}$ and $\mathcal{B}'$ are ordered bases of V and W, then for $x, y \in V$,

$$\begin{aligned} \xi(Tx, Ty) &= [Tx]_{\mathcal{B}'}^t [\xi]_{\mathcal{B}'} [Ty]_{\mathcal{B}'} \\ &= ({}_{\mathcal{B}'}[T]_{\mathcal{B}} [x]_{\mathcal{B}})^t [\xi]_{\mathcal{B}'} ({}_{\mathcal{B}'}[T]_{\mathcal{B}} [y]_{\mathcal{B}}) \\ &= [x]_{\mathcal{B}}^t \left({}_{\mathcal{B}'}[T]_{\mathcal{B}}^t [\xi]_{\mathcal{B}'} \, {}_{\mathcal{B}'}[T]_{\mathcal{B}}\right) [y]_{\mathcal{B}} . \end{aligned}$$

Therefore,

$$(*) \qquad [\phi]_{\mathcal{B}} = {}_{\mathcal{B}'}[T]_{\mathcal{B}}^t [\xi]_{\mathcal{B}'} \, {}_{\mathcal{B}'}[T]_{\mathcal{B}}$$

and $[\phi]_{\mathcal{B}}$ and $[\xi]_{\mathcal{B}'}$ are congruent matrices.

PROPOSITION 2.2. *Finite dimensional bilinear spaces (V,ϕ) and (W,ξ) over K are isomorphic if and only if ϕ and ξ admit the same matrix representation with respect to a suitably chosen ordered bases of V and W.*

PROOF. Let $T : (V,\phi) \to (W,\xi)$ be an isometric isomorphism and let $\mathcal{B} = \{x_1,\dots,x_n\}$ be an ordered basis of V. Then $\mathcal{B}' = \{Tx_1,\dots,Tx_n\}$ is an ordered basis of W and ${}_{\mathcal{B}'}[T]_{\mathcal{B}} = I_n$. Thus, from equation (*), it follows that $[\phi]_{\mathcal{B}} = [\xi]_{\mathcal{B}'}$.

Conversely, as ϕ and ξ have the same matrix representations with respect to some ordered bases, $\dim V = \dim W$. Let $\mathcal{B} = \{x_1,\dots,x_n\}$ and $\mathcal{B}' = \{y_1,\dots,y_n\}$ be ordered bases of V and W such that $[\phi]_{\mathcal{B}} = [\xi]_{\mathcal{B}'}$. Let $T : V \to W$ be a mapping whose action on $\mathcal{B}$ is defined by $Tx_i = y_i$, $i = 1,\dots,n$. Then, T is an isomorphism. Now for $u,v \in V$, write $u = \sum_{i=1}^n \alpha_i x_i$, $\alpha_i \in K$ and $v = \sum_{i=1}^n \beta_i x_i$, $\beta_i \in K$. Then

$$\begin{aligned}\phi(u,v) &= \sum_{j=1}^n\sum_{i=1}^n \alpha_i\,\beta_j\,\phi(x_i,x_j) = \sum_{j=1}^n\sum_{i=1}^n \alpha_i\,\beta_j\,\xi(y_i,y_j)\\ &= \xi\left(\sum_{i=1}^n \alpha_i Tx_i, \sum_{j=1}^n \beta_j Tx_j\right) = \xi(Tu,Tv).\end{aligned}$$

Hence, T is an isometry. ∎

2.1. Examples

1. An inner product on an n–dimensional vector space V over $\mathbb{R}$ is a bilinear form on V, and the matrix of inner product with respect to an orthonormal basis is I_n.

2. Let V be an n–dimensional vector space over K and let $\phi \in \mathrm{Bil}_K(V)$. Let $\mathcal{B} = \{x_1,\dots,x_n\}$ be an ordered basis of V and let $[\phi]_{\mathcal{B}} = A$. Let $B \in K^{n\times n}$ be congruent to A. Write $B = P^tAP$, P an invertible matrix. Denote by p_{ij}, the (i,j)–th entry of P, $i,j = 1,\dots,n$. Write

$$y_j = \sum_{i=1}^n p_{ij}x_i, \quad j = 1,\dots,n.$$

Since P is invertible, $\mathcal{B}' = \{y_1,\dots,y_n\}$ is an ordered basis of V. Now $\mathcal{B}'$ is such that $[\phi]_{\mathcal{B}'} = B$. Since for $i,j = 1,\dots,n$,

$$\begin{aligned}[B]_{ij} &= e_i^t\left(P^tAP\right)e_j = (Pe_i)^t\,A\,(Pe_j)\\ &= \begin{bmatrix} p_{1i} & \cdots & p_{ni}\end{bmatrix}\begin{bmatrix}\phi(x_1,x_1) & \cdots & \phi(x_1,x_n)\\ \vdots & & \vdots\\ \phi(x_n,x_1) & \cdots & \phi(x_n,x_n)\end{bmatrix}\begin{bmatrix}p_{1j}\\ \vdots\\ p_{nj}\end{bmatrix}\\ &= \sum_{s=1}^n\sum_{r=1}^n p_{si}\,\phi(x_s,x_r)\,p_{rj}\\ &= \phi\left(\sum_{s=1}^n p_{si}x_s, \sum_{r=1}^n p_{rj}x_r\right)\\ &= \phi(y_i,y_j).\end{aligned}$$

Also note that the j–th column of P is actually $[y_j]_{\mathcal{B}}$. Thus, it also verifies the following statement.

If $\mathcal{B} = \{x_1, \ldots, x_n\}$ and $\mathcal{B}' = \{y_1, \ldots, y_n\}$ are ordered bases of V and $\phi \in \mathrm{Bil}_K(V)$, then $[\phi]_{\mathcal{B}'} = P^t [\phi]_{\mathcal{B}} P$, and the j–th column of P is $[y_j]_{\mathcal{B}}$.

In particular, for $A \in K^{n\times n}$, the matrix of θ_A with respect to the standard basis is A. If $\mathcal{B} = \{x_1, \ldots, x_n\}$ is any other basis of K^n, then $[\theta_A]_{\mathcal{B}} = P^t AP$, where $P \in K^{n\times n}$ and the j–th column of P is x_j.

For example, if

$$A = \begin{bmatrix} 1 & 0 & 3 \\ 3 & -2 & -1 \\ 1 & -1 & 1 \end{bmatrix},$$

then the matrix of θ_A with respect to an ordered basis

$$\left\{ \begin{bmatrix} 1 \\ 1 \\ 0 \end{bmatrix}, \begin{bmatrix} 1 \\ -1 \\ 0 \end{bmatrix}, \begin{bmatrix} 0 \\ 0 \\ 1 \end{bmatrix} \right\}$$

is

$$P^t AP = \begin{bmatrix} 2 & 6 & 2 \\ 0 & -4 & 4 \\ 0 & 2 & 1 \end{bmatrix},$$

where $P = \begin{bmatrix} 1 & 1 & 0 \\ 1 & -1 & 0 \\ 0 & 0 & 1 \end{bmatrix}$.

3. Let (V, ϕ) be a bilinear space over K. A pair of vectors u and v is called a **hyperbolic pair** if

$$\phi(u,v) = 1 = \phi(v,u) \quad \text{and} \quad \phi(u,u) = 0 = \phi(v,v).$$

If $\{u, v\}$ is a hyperbolic pair, then these vectors are linearly independent. Since, if $v = \alpha u$, $\alpha \in K$, then $\phi(v,u) = \alpha\phi(u,u) = 0$, a contradiction. The 2–dimensional subspace $H = \langle u, v\rangle$ is called a **hyperbolic plane.** If $\widehat{\phi} = \phi|_{H\times H}$, then $\widehat{\phi}$ is a symmetric bilinear form on H and the matrix of $\widehat{\phi}$ with respect to the basis $\mathcal{B} = \{u, v\}$ is $\begin{bmatrix} 0 & 1 \\ 1 & 0 \end{bmatrix}$.

If $\operatorname{char} K = 2$, then $\widehat{\phi}$ is alternating. Since for $\alpha, \beta \in K$, we have

$$\widehat{\phi}(\alpha u + \beta v, \alpha u + \beta v) = \alpha^2\phi(u,u) + \beta^2\phi(v,v) + 2\alpha\beta\,\phi(u,v) = 0.$$

If $\operatorname{char} K \neq 2$, then vectors $x = \frac{1}{2}u + v$, $y = \frac{1}{2}u - v$ form an orthogonal basis of H. The matrix of $\widehat{\phi}$ with respect to $\{x, y\}$ is $\begin{bmatrix} 1 & 0 \\ 0 & -1 \end{bmatrix}$. Therefore, if $\operatorname{char} K \neq 2$, $\begin{bmatrix} 0 & 1 \\ 1 & 0 \end{bmatrix}$ and $\begin{bmatrix} 1 & 0 \\ 0 & -1 \end{bmatrix}$ are congruent. In fact,

$$\begin{bmatrix} 1 & 0 \\ 0 & -1 \end{bmatrix} = \begin{bmatrix} 1/2 & 1/2 \\ 1 & -1 \end{bmatrix}^t \begin{bmatrix} 0 & 1 \\ 1 & 0 \end{bmatrix} \begin{bmatrix} 1/2 & 1/2 \\ 1 & -1 \end{bmatrix}.$$

4. For any $n \in \mathbb{Z}^+$, the matrix nI_4 is congruent to I_4. This is based on a nontrivial fact that any positive integer is a sum of four squares, that is, $n = a^2 + b^2 + c^2 + d^2$, $a, b, c, d \in \mathbb{Z}$. Then $P^t I_4 P = P^t P = nI_4$, where

$$P = \begin{bmatrix} a & b & c & d \\ b & -a & -d & c \\ c & d & -a & -b \\ d & -c & b & -a \end{bmatrix}.$$

5. *Let (V, ϕ) be an n–dimensional bilinear space over K and let A be the matrix of ϕ with respect to an ordered basis $\mathcal{B}$ of V. Then (V, ϕ) and (K^n, θ_A) are isomorphic bilinear spaces.*

   Let $T : V \to K^n$ be an isomorphism given by $Tu = [u]_{\mathcal{B}}$. Then for $x, y \in V$,

$$\phi(x, y) = [x]_{\mathcal{B}}^t \, [\phi]_{\mathcal{B}} \, [y]_{\mathcal{B}} = [x]_{\mathcal{B}}^t \, A \, [y]_{\mathcal{B}} = \theta_A(Tx, Ty).$$

6. Let $A, B \in K^{n\times n}$. Bilinear spaces (K^n, θ_A) and (K^n, θ_B) are isomorphic if and only if A and B are congruent matrices.

   If (K^n, θ_A) and (K^n, θ_B) are isomorphic bilinear spaces, then there exists an invertible matrix P such that $P : K^n \to K^n$, defined by $x \mapsto Px$ is an isometric isomorphism, that is, $\theta_B(x, y) = \theta_A(Px, Py)$, for all $x, y \in K^n$. Therefore, $x^t B y = x^t (P^t A P) y$. Hence, for all $i, j = 1, \dots, n$, $[B]_{ij} = e_i^t B e_j = e_i^t (P^t A P) e_j = [P^t A P]_{ij}$, and $B = P^t A P$.

   Conversely, if $B = P^t A P$, for some invertible matrix P, then for all $x, y \in K^n$,

$$\begin{aligned} \theta_B(x, y) &= x^t B y = x^t (P^t A P) y \\ &= (Px)^t A (Py) = \theta_A(Px, Py). \end{aligned}$$

   Hence $P : K^n \to K^n$ given by $x \mapsto Px$ is an isometric isomorphism and bilinear spaces (K^n, θ_A) and (K^n, θ_B) are isomorphic.

7. A real symmetric matrix has only real eigenvalues. By Schur's theorem, if $A \in \mathbb{R}^{n\times n}$ and A is symmetric, then there exists a real orthogonal matrix Q such that $Q^t A Q$ is a diagonal matrix. Thus, a real symmetric matrix is congruent to a diagonal matrix.

   A bilinear form ϕ on a finite dimensional vector space V over $\mathbb{R}$ is called **diagonalizable** if the matrix of ϕ with respect to an ordered basis of V is diagonal. A basis $\mathcal{B} = \{x_1, \dots, x_n\}$ of V is called an **orthogonal basis** with respect to ϕ if $\phi(x_i, x_j) = 0$ for $i \neq j$. Clearly, ϕ is diagonalizable if and only if V has an orthogonal basis with respect to ϕ.

   A diagonalizable bilinear form on a finite dimensional vector space is symmetric (Exercise 10). From the first paragraph of this example and from Example 2, it follows that a symmetric bilinear form on a finite dimensional vector space over $\mathbb{R}$ is diagonalizable.

2.2. Exercises

1. Let $A = \begin{bmatrix} 1 & -1 & 0 \\ 1 & 0 & -1 \\ 0 & 1 & 1 \end{bmatrix}$. Consider the bilinear form ϕ on $K^{n\times n}$ given by $\phi(X,Y) = \operatorname{tr}(X^tAY)$. Find the matrix of ϕ with respect to the ordered basis $\{E_{11}, E_{12}, E_{13}, E_{21}, E_{22}, E_{23}, E_{31}, E_{32}, E_{33}\}$.

2. Find the matrix of a bilinear form ϕ on $K^{2\times 2}$ defined by $\phi(X,Y) = \operatorname{tr}(XY) - \operatorname{tr}(X)\operatorname{tr}(Y)$, with respect to the basis $\{E_{11}, E_{12}, E_{21}, E_{22}\}$.

3. Let ϕ be a bilinear form on a finite dimensional vector space V over K and let $\mathcal{B}$ be an ordered basis of V. Prove that
   (a) ϕ is symmetric if and only if $[\phi]_{\mathcal{B}}$ is symmetric.
   (b) ϕ is alternating if and only if $[\phi]_{\mathcal{B}}$ is an alternating matrix, that is, $[\phi]_{\mathcal{B}}^t = -[\phi]_{\mathcal{B}}$ and diagonal entries are zero.

4. Let A and B be invertible matrices of the same size. Prove that the following are equivalent:
   (a) A and B are congruent.
   (b) A^{-1} and B^{-1} are congruent.
   (c) A^t and B^t are congruent.

5. Let $\operatorname{char} K \neq 2$ and let $a \in K \setminus \{0\}$. Find a matrix Q such that

$$Q^t \begin{bmatrix} 1 & 0 \\ 0 & -1 \end{bmatrix} Q = \begin{bmatrix} a & 0 \\ 0 & -a \end{bmatrix}.$$

6. Let K be a field and let $a, b \in K \setminus \{0\}$. Prove that the following matrices are congruent:

$$\begin{bmatrix} a & 0 \\ 0 & b \end{bmatrix} \quad \text{and} \quad \begin{bmatrix} a+b & 0 \\ 0 & ab(a+b) \end{bmatrix}.$$

7. Prove that $\operatorname{diag}(\lambda_1, \ldots, \lambda_n)$ and $\operatorname{diag}(\lambda_{i_1}, \ldots, \lambda_{i_n})$ are congruent matrices for each permutation $i_1, \ldots, i_n$ of $1, \ldots, n$.

8. Prove that a positive definite matrix is congruent to the identity matrix.

9. Prove that two positive semidefinite matrices A and B are congruent if and only if $\operatorname{rank} A = \operatorname{rank} B$.

10. Prove that a diagonalizable bilinear form is symmetric.

11. Let (V,ϕ) and (W,ξ) be isomorphic bilinear spaces. Prove that ϕ is reflexive if and only if ξ is reflexive.

12. Let V be a finite dimensional vector space over K, let $S, T \in L(V)$ and let $\phi \in \operatorname{Bil}_K(V)$. Show that the matrix of bilinear form ψ on V defined by $\psi(x,y) = \phi(Tx, Sy)$ with respect to an ordered basis $\mathcal{B}$ is $[T]_{\mathcal{B}}^t [\phi]_{\mathcal{B}} [S]_{\mathcal{B}}$.

3. Orthogonality

In this section, we confine ourselves to reflexive bilinear forms. Let ϕ be a reflexive bilinear form on a vector space V over K. Recall that $x, y \in V$ are called orthogonal with respect to ϕ if $\phi(x,y) = 0$. Two subspaces X and Y of V are called orthogonal with respect to ϕ if $\phi(x,y) = 0$ for all $x \in X$ and $y \in Y$. For a subspace S of V, define

$$S^\perp = \{v \in V \mid \phi(x,v) = 0 \text{ for all } x \in S\}.$$

Clearly, $S^\perp$ is a subspace of V. The subspace S is called **isotropic** if $S \cap S^\perp \neq \{0\}$, otherwise **anisotropic**. S is called **totally isotropic** if $S \subseteq S^\perp$. The subspace $V^\perp$ is called the **radical** of ϕ, denoted by $\operatorname{rad}\phi$. Thus,

$$\operatorname{rad}\phi = \{v \in V \mid \phi(x,v) = 0 \text{ for all } x \in V\}.$$

If $\operatorname{rad}\phi = \{0\}$, then ϕ is called **nondegenerate**, otherwise **degenerate**. Thus, ϕ is nondegenerate if ϕ is reflexive and to each $v \in V \setminus \{0\}$, there exists $x \in V$ such that $\phi(x,v) \neq 0$.

PROPOSITION 3.1. *Let ϕ be a reflexive bilinear form on a finite dimensional vector space V over K. Then ϕ is nondegenerate if and only if the matrix of ϕ with respect to an ordered basis of V is invertible.*

PROOF. Let $\mathcal{B}$ be an ordered basis of V. If $\det[\phi]_\mathcal{B} = 0$, then the system of linear equations $[\phi]_\mathcal{B}\, x = 0$ has a nonzero solution. Let $x_0 \in K^n$, $n = \dim V$, be a nonzero solution of this system and let $v \in V$ such that $[v]_\mathcal{B} = x_0$. Then for any $u \in V$,

$$\phi(u,v) = [u]_\mathcal{B}^t\ [\phi]_\mathcal{B}\ [v]_\mathcal{B} = 0.$$

Hence, $v \in \operatorname{rad}\phi \setminus \{0\}$, and ϕ is degenerate.

Conversely, if ϕ is degenerate, there exists $v \in V \setminus \{0\}$ such that $\phi(x,v) = 0$ for all $x \in V$. Let $\mathcal{B} = \{x_1, \dots, x_n\}$ be an ordered basis of V and write $v = \alpha_1 x_1 + \cdots + \alpha_n x_n$. Then $[v]_\mathcal{B} = \begin{bmatrix} \alpha_1 & \cdots & \alpha_n \end{bmatrix}^t$ is a nonzero element of K^n. Now for each $i = 1, \dots, n$, the i–th entry of $[\phi]_\mathcal{B}\,[v]_\mathcal{B}$ is $e_i^t\ [\phi]_\mathcal{B}\ [v]_\mathcal{B} = [x_i]_\mathcal{B}^t\ [\phi]_\mathcal{B}\,[v]_\mathcal{B} = \phi(x_i, v) = 0$. Hence, $[\phi]_\mathcal{B}$ is not invertible. ∎

PROPOSITION 3.2. *Let ϕ be a reflexive bilinear form on a vector space V over K. Let X and Y be subspaces of V.*

(i) *If $X \subseteq Y$, then $Y^\perp \subseteq X^\perp$.*
(ii) $(X+Y)^\perp = X^\perp \cap Y^\perp$.
(iii) $X \subseteq X^{\perp\perp}$.
(iv) $X^\perp = X^{\perp\perp\perp}$.

PROOF. Exercise. ∎

THEOREM 3.3. ***Riesz representation theorem.*** *Let ϕ be a nondegenerate reflexive bilinear form on a finite dimensional vector space V over K. If f is a linear form on V, then there exists a unique vector $v \in V$ such that $f(x) = \phi(x,v)$ for all $x \in V$.*

PROOF. Let $\Phi : V \to V^\bullet$ be a mapping defined by $\Phi(u) = \phi(\cdot, u)$. Clearly Φ is a well defined homomorphism of vector spaces. If $u \in \ker\Phi$, then $\phi(x,u) = 0$ for all $x \in V$, and so $u = 0$ as ϕ is nondegenerate. Therefore, Φ is injective. The mapping Φ is also surjective because V is finite dimensional and $\dim V = \dim V^{\bullet}$. Hence, if $f \in V^{\bullet}$, there exists a unique $v \in V$ such that $f(x) = \phi(x, v)$ for all $x \in V$. ▮

COROLLARY 3.4. *Let ϕ be a nondegenerate reflexive bilinear form on a finite dimensional vector space V over K. For a subspace S of V, $S^{\perp\perp} = S$.*

PROOF. From Proposition 3.2, it follows that $S \subseteq S^{\perp\perp}$. Suppose that the inclusion is proper. Then there exists a linear form f on V such that $f|_S \equiv 0$ and $f|_{S^{\perp\perp}} \neq 0$. Let $v \in V$ such that $f(x) = \phi(x, v)$ for all $x \in V$. Since $\phi(x, v) = 0$ for all $x \in S$ and $\phi(u, v) \neq 0$ for some $u \in S^{\perp\perp}$, so $v \in S^{\perp} \setminus S^{\perp\perp\perp}$, a contradiction (Proposition 3.2(iv)). ▮

PROPOSITION 3.5. *Let ϕ be a reflexive bilinear form on a vector space V over K. If S is a finite dimensional anisotropic subspace of V, then $V = S \oplus S^{\perp}$.*

PROOF. Let $v \in V$. The mapping $f : S \to K$ defined by $f(x) = \phi(x, v)$ is a linear form on S. The bilinear form $\widehat{\phi} = \phi|_{S \times S}$ is nondegenerate on S. Since if $s \in S$ such that $\phi(x, s) = 0$ for all $x \in S$, then $s \in S \cap S^{\perp} = \{0\}$. Therefore, by Riesz representation theorem, there exists $s \in S$ such that $f(x) = \widehat{\phi}(x, s)$ for all $x \in S$. Thus for all $x \in S$, $\phi(x, v) = \widehat{\phi}(x, s) = \phi(x, s)$, and so $\phi(x, v - s) = 0$ for all $x \in S$, that is, $v - s \in S^{\perp}$. Hence, $v = s + v - s \in S + S^{\perp}$. Since v is arbitrary element of V, we have $V = S + S^{\perp}$. Hence, $V = S \oplus S^{\perp}$. ▮

PROPOSITION 3.6. *Let ϕ be a reflexive bilinear form on a vector space V over K. Let W be a finite dimensional subspace of V such that $\phi|_{W \times W}$ is nondegenerate. Then $V = W \oplus W^{\perp}$*

PROOF. Since $\phi_1 = \phi|_{W \times W}$ is nondegenerate, the matrix of ϕ_1 with respect to any ordered basis of W is invertible. If $w \in W \cap W^{\perp}$ and $w \neq 0$, then for any ordered basis $\mathcal{B} = \{x_1 = w, x_2, \ldots, x_m\}$ of W, we have $\phi(x_1, x_i) = 0$, $i = 1, \ldots, m$. But then $[\phi_1]_{\mathcal{B}}$ is not invertible. Therefore, $W \cap W^{\perp} = \{0\}$. Hence, $V = W \oplus W^{\perp}$ (Proposition 3.5). ▮

3.1. Examples

1. Let V be a finite dimensional vector space over K and let ϕ be a nondegenerate bilinear form on V. According to Theorem 3.3, for $f \in V^{\bullet}$ there is a unique $v \in V$ such that $f(x) = \phi(x, v)$ for all $x \in V$.

   Let $\mathcal{B} = \{x_1, \ldots, x_n\}$ be an ordered basis of V. For $x \in V$, write $x = \alpha_1 x_1 + \cdots + \alpha_n x_n$. Then $f(x) = \alpha_1 f(x_1) + \cdots + \alpha_n f(x_n)$. Thus,

$$f(x) = \begin{bmatrix} \alpha_1 & \cdots & \alpha_n \end{bmatrix} \begin{bmatrix} f(x_1) \\ \vdots \\ f(x_n) \end{bmatrix} = [x]_{\mathcal{B}}^t \begin{bmatrix} f(x_1) \\ \vdots \\ f(x_n) \end{bmatrix}.$$

   Since ϕ is nondegenerate, $[\phi]_{\mathcal{B}}$ is an invertible matrix. Thus, $f(x) = [x]_{\mathcal{B}}^t [\phi]_{\mathcal{B}} [\phi]_{\mathcal{B}}^{-1} y$, where $y = \begin{bmatrix} f(x_1) & \cdots & f(x_n) \end{bmatrix}^t$. Let $v \in V$ such that $[v]_{\mathcal{B}} = [\phi]_{\mathcal{B}}^{-1} y$. Then for all $x \in V$,

$$f(x) = [x]_{\mathcal{B}}^t [\phi]_{\mathcal{B}} [v]_{\mathcal{B}} = \phi(x, v).$$

As an illustration of this procedure, let ϕ be a bilinear form on $\mathbb{R}^3$ defined by $\phi(x,y) = x^t Ay$, where

$$A = \begin{bmatrix} 2 & 1 & 1 \\ 1 & 2 & 1 \\ 1 & 1 & 2 \end{bmatrix}.$$

Since A is symmetric and $\det A \neq 0$, ϕ is nondegenerate. Now for the linear form $f : \mathbb{R}^3 \to \mathbb{R}$ defined by

$$f\left(\begin{bmatrix} x_1 & x_2 & x_3 \end{bmatrix}^t\right) = x_1 - x_2 + x_3,$$

we have $f(x) = \phi(x,v)$ for all $x \in V$, where

$$v = \begin{bmatrix} 2 & 1 & 1 \\ 1 & 2 & 1 \\ 1 & 1 & 2 \end{bmatrix}^{-1} \begin{bmatrix} 1 \\ -1 \\ 1 \end{bmatrix} = \frac{1}{4}\begin{bmatrix} 3 \\ -5 \\ 3 \end{bmatrix}.$$

2. Let ϕ be a nondegenerate bilinear form on an n–dimensional vector space V over K and let $\mathcal{B} = \{v_1, \ldots, v_n\}$ be an ordered basis of V. Then $\det [\phi]_{\mathcal{B}} \neq 0$. Thus, for each $i = 1, \ldots, n$, the system of equations $[\phi]_{\mathcal{B}}\, x = e_i$ has a unique solution $y_i \in K^n$, that is, $[\phi]_{\mathcal{B}}\, y_i = e_i$. For $i = 1, \ldots, n$, let $[u_i]_{\mathcal{B}} = y_i$. Then $\{u_1, \ldots, u_n\}$ is an ordered basis of V, and

$$\phi(v_i, u_j) = [v_i]_{\mathcal{B}}^t\, [\phi]_{\mathcal{B}}\, [u_j]_{\mathcal{B}} = e_i^t e_j = \delta_{ij}.$$

Let $v_i^\bullet \in V^\bullet$ defined by $v_i^\bullet(x) = \phi(x, u_i)$, $x \in V$. Then $v_i^\bullet(v_j) = \delta_{ij}$, and $\{v_1^\bullet, \ldots, v_n^\bullet\}$ is a basis of $V^\bullet$ dual to $\{v_1, \ldots, v_n\}$.

3. Let ϕ be a reflexive bilinear form on a finite dimensional vector space V and let V_1 be a subspace complementary to $\operatorname{rad}\phi$, that is, $V = V_1 \oplus \operatorname{rad}\phi$. Let ϕ_1 be the restriction of ϕ to $V_1 \times V_1$. Then ϕ_1 is a nondegenerate bilinear form on V_1. Since if $v \in \operatorname{rad}\phi_1$, then for any $x \in V$, we have $v = x_1 + x_2$, $x_1 \in V_1$, $x_2 \in \operatorname{rad}\phi$ and $\phi(v,x) = \phi_1(x_1,x) + \phi(x_2,x) = 0$.

   Also define $\phi_2 : V/\operatorname{rad}\phi \times V/\operatorname{rad}\phi \to K$ by

$$\phi_2(x + \operatorname{rad}\phi, y + \operatorname{rad}\phi) = \phi(x,y).$$

It is easy to verify that ϕ_2 is a well defined bilinear form on $V/\operatorname{rad}\phi$. Clearly, ϕ_2 is reflexive. If $v + \operatorname{rad}\phi \in V/\operatorname{rad}\phi$ such that $\phi_2(v + \operatorname{rad}\phi, x + \operatorname{rad}\phi) = 0$ for all $x \in V$, then $\phi(v,x) = 0$ for all $x \in V$, and so $v \in \operatorname{rad}\phi$. Thus, ϕ_2 is nondegenerate.

   The mapping $\pi : V_1 \to V/\operatorname{rad}\phi$ defined by $\pi(x) = x + \operatorname{rad}\phi$ is an isomorphism of vector spaces and for $x, y \in V_1$,

$$\phi_1(x,y) = \phi(x,y) = \phi_2(x + \operatorname{rad}\phi, y + \operatorname{rad}\phi) = \phi_2(\pi(x), \pi(y)).$$

Therefore, (V_1, ϕ_1) and $(V/\operatorname{rad}\phi, \phi_2)$ are isomorphic bilinear spaces.

4. Let (V,ϕ) and (W,ξ) be reflexive bilinear spaces and let $T : V \to W$ be an isometry. Then for $x \in \ker T$ and for all $v \in V$,

$$\phi(x,v) = \xi(Tx, Tv) = 0,$$

that is, $x \in \operatorname{rad}\phi$. Therefore, if ϕ is nondegenerate then T is injective.

5. Let ϕ be a reflexive bilinear form on a finite dimensional vector space V over K. The **rank** of ϕ, denoted by $\operatorname{rank}\phi$, is defined by the rank of the matrix $[\phi]_{\mathcal{B}}$, where $\mathcal{B}$ is an ordered basis of V. Since the matrices of ϕ with respect to two different bases are congruent, this definition is independent of the choice of a basis. Thus, if ϕ is nondegenerate, then from Proposition 3.1, $\operatorname{rank}\phi = \dim V$.

   Let V_1 be a subspace of V complementary to $\operatorname{rad}\phi$. Then $V = V_1 \oplus \operatorname{rad}\phi$, and ϕ_1 is a nondegenerate bilinear form on V_1. If $\mathcal{B}_1$ and $\mathcal{B}_2$ are ordered bases of V_1 and $\operatorname{rad}\phi$, then $\mathcal{B} = \mathcal{B}_1 \cup \mathcal{B}_2$ is an ordered basis of V and $[\phi]_{\mathcal{B}} = \operatorname{diag}\left([\phi_1]_{\mathcal{B}_1}, 0\right)$. Since ϕ_1 is nondegenerate, $[\phi_1]_{\mathcal{B}_1}$ is an invertible matrix. Therefore, $\operatorname{rank}\phi = \dim(V/\operatorname{rad}\phi)$.

6. Let ϕ be a reflexive bilinear form on V and let W be a subspace of V such that $V = W \oplus W^{\perp}$. The mapping $S_W : V \to V$ defined by

$$S_W(w + w') = -w + w',$$

   $w \in W$ and $w' \in W^{\perp}$ is an isometric isomorphism from V to itself. If $W = \langle w \rangle$, we write S_w instead of S_W. The mapping S_W is also called the **reflection** along W (or at $W^{\perp}$).

3.2. Exercises. In the following exercises, ϕ is always a reflexive bilinear form on a vector space V over K.

1. Prove that if ϕ is nondegenerate, then for $x, y \in V$:
   (a) if $\phi(x, v) = \phi(y, v)$ for all $v \in V$, then $x = y$;
   (b) if $\phi(v, x) = \phi(v, y)$ for all $v \in V$, then $x = y$.

2. If ϕ is nondegenerate, then does it follow that for every subspace W of V, $\phi|_{W\times W}$ is nondegenerate?

3. Let W be a subspace of V. Prove that the mapping $\widehat{\phi} : V/W \times V/W \to K$ defined by $\widehat{\phi}(x + W, y + W) = \phi(x, y)$ is bilinear if and only if $W \subseteq \operatorname{rad}\phi$.

4. Verify that the mappings from V to $V^\bullet$ defined by $\Phi(v) = \phi(\cdot, v)$ and $\Psi(v) = \phi(v, \cdot)$ for all $v \in V$ are linear. Prove that the following statements are equivalent:
   (a) ϕ is nondegenerate;
   (b) $\operatorname{rank}[\phi]_{\mathcal{B}} = \dim V$, $\mathcal{B}$ is an ordered basis of V;
   (c) Φ is an isomorphism of vector spaces;
   (d) Ψ is an isomorphism of vector spaces.

5. Prove that if $V = V_1 \oplus V_2$ is an orthogonal direct sum and if $\phi_i = \phi|_{V_i\times V_i}$, $i = 1, 2$, then $\operatorname{rad}\phi = \operatorname{rad}\phi_1 \oplus \operatorname{rad}\phi_2$. Also show that if $\mathcal{B}_1$ and $\mathcal{B}_2$ are ordered bases of V_1 and V_2, and $\mathcal{B} = \mathcal{B}_1 \cup \mathcal{B}_2$, then $[\phi]_{\mathcal{B}} = \operatorname{diag}\left([\phi_1]_{\mathcal{B}_1}, [\phi_2]_{\mathcal{B}_2}\right)$.

6. Let (V, ϕ) and (W, ξ) be isomorphic bilinear spaces. Show that ϕ is nondegenerate if and only if ξ is nondegenerate.

7. Show that if $v_1, \ldots, v_m \in V$ such that $(\phi(v_i, v_j)) \in K^{m\times m}$ is invertible, then $v_1, \ldots, v_m$ are linearly independent and $V = W \oplus W^{\perp}$, for $W = \langle v_1, \ldots, v_m \rangle$.

8. Let (V,ϕ) and (W,ξ) be reflexive bilinear spaces over K. Suppose that $T: V \to W$ is a bijective mapping such that $\phi(x,y) = \xi(Tx,Ty)$, for all $x,y \in V$. Show that if ϕ is nondegenerate then T is a linear transformation.
9. Let T be an isometric isomorphism from (V,ϕ) to (W,ξ). Prove that
   (a) $T(\operatorname{rad}\phi) = \operatorname{rad}\xi$; and
   (b) $T(X^\perp) = T(X)^\perp$ for a subspace X of V.
10. Let ϕ be nondegenerate. Prove that if X and Y are subspaces of V, then $(X \cap Y)^\perp = X^\perp + Y^\perp$.

4. Classification of bilinear forms

In this section, we assume that V is a finite dimensional vector space of dimension n over K. Bilinear forms ϕ and ξ on V are said to be **equivalent** if there are ordered bases $\mathcal{B}_1$ and $\mathcal{B}_2$ of V such that $[\phi]_{\mathcal{B}_1} = [\xi]_{\mathcal{B}_2}$. Equivalently, this means that if $\mathcal{B}$ is an ordered basis of V, then $[\phi]_{\mathcal{B}}$ and $[\xi]_{\mathcal{B}}$ are congruent matrices. Thus, classifying bilinear forms up to equivalence is effectively the decomposition of matrices in $K^{n\times n}$ up to congruence.

We start with alternating bilinear forms.

THEOREM 4.1. *Let ϕ be an alternating bilinear form on an n–dimensional vector space V over K. Then there exists an ordered basis $\mathcal{B}$ of V such that the matrix of ϕ is of the form*

$$(*) \qquad \operatorname{diag}\left(\begin{bmatrix} 0 & 1 \\ -1 & 0 \end{bmatrix}, \dots, \begin{bmatrix} 0 & 1 \\ -1 & 0 \end{bmatrix}, 0, \dots, 0\right).$$

PROOF. By induction on $n = \dim V$. If $n = 1$, then $V = \langle v \rangle$, for any $v \neq 0$. Since $\phi(v,v) = 0$, $\phi \equiv 0$ on V and the result is obviously true in this case.

Let ϕ be an alternating bilinear form on V and let the hypothesis be true for all alternating bilinear forms on vector spaces of dimensions up to $n-1$. If $\phi \equiv 0$, then the result is obvious. If $\phi \not\equiv 0$, there are $u_1, u_2 \in V$ such that $\phi(u_1,u_2) \neq 0$. Let $x_1 = u_1$ and $x_2 = \phi(u_1,u_2)^{-1} u_2$. Then $\phi(x_1,x_2) = 1$ and $\phi(x_2,x_1) = -1$. The vectors x_1 and x_2 are linearly independent. If $x_2 = \alpha x_1$, $\alpha \in K$, then $1 = \phi(x_1,x_2) = \phi(x_1,\alpha x_1) = 0$, a contradiction. If $W = \langle x_1, x_2 \rangle$ and $\phi_1 = \phi|_{W\times W}$, then $\mathcal{B}_1 = \{x_1, x_2\}$ is a basis of W and

$$[\phi_1]_{\mathcal{B}_1} = \begin{bmatrix} 0 & 1 \\ -1 & 0 \end{bmatrix}.$$

Since ϕ_1 is nondegenerate, by Proposition 3.6, $V = W \oplus W^\perp$. Let $\phi_2 = \phi|_{W^\perp \times W^\perp}$. Then ϕ_2 is an alternating bilinear form on $W^\perp$ and $\dim W^\perp < n$. Therefore, by induction, there is ordered basis $\mathcal{B}_2$ of $W^\perp$ such that $[\phi_2]_{\mathcal{B}_2}$ is of the form $(*)$. Therefore, if $\mathcal{B} = \mathcal{B}_1 \cup \mathcal{B}_2$, then $\mathcal{B}$ is an ordered basis of V such that $[\phi]_{\mathcal{B}}$ is of the required form. ∎

COROLLARY 4.2. *If ϕ is a nonzero alternating bilinear form on a finite dimensional vector space V over K, then* $\operatorname{rank}\phi$ *is even. In particular, if ϕ is nondegenerate alternating bilinear form on V, then* $\dim V$ *is even.*

PROOF. The rank of ϕ is twice the number of 2×2 blocks appearing in $(*)$. ∎

COROLLARY 4.3. *Two alternating bilinear forms on a finite dimensional vector space are equivalent if and only if they have the same rank.*

The matrix analogue of Theorem 4.1 is the following.

PROPOSITION 4.4. *Let $A \in K^{n\times n}$ be such that $A^t = -A$ and the diagonal entries of A are zero. Then A is congruent to the block diagonal matrix of the form*

$$\operatorname{diag}\left(\begin{bmatrix} 0 & 1 \\ -1 & 0 \end{bmatrix}, \ldots, \begin{bmatrix} 0 & 1 \\ -1 & 0 \end{bmatrix}, 0_{n-2r}\right),$$

where $\operatorname{rank} A = 2r$, *and r is the number of 2×2 blocks.*

PROOF. The bilinear form θ_A on K^n given by $\theta_A(x, y) = x^t A y$ is alternating. According to Theorem 4.1, there is an ordered basis $\mathcal{B} = \{x_1, \ldots, x_n\}$ of K^n such that

$$[\theta_A]_{\mathcal{B}} = \operatorname{diag}\left(\begin{bmatrix} 0 & 1 \\ -1 & 0 \end{bmatrix}, \ldots, \begin{bmatrix} 0 & 1 \\ -1 & 0 \end{bmatrix}, 0_{n-2r}\right),$$

$\operatorname{rank} A = 2r$. Therefore, if $P = \begin{bmatrix} x_1 & \cdots & x_n \end{bmatrix} \in K^{n\times n}$, then

$$P^t A P = \operatorname{diag}\left(\begin{bmatrix} 0 & 1 \\ -1 & 0 \end{bmatrix}, \ldots, \begin{bmatrix} 0 & 1 \\ -1 & 0 \end{bmatrix}, 0_{n-2r}\right).$$

∎

We now classify symmetric bilinear forms. Recall that a bilinear form ϕ on a finite dimensional vector space V is diagonalizable if $[\phi]_{\mathcal{B}}$ is a diagonal matrix with respect to some ordered basis $\mathcal{B}$ of V. Equivalently, V has an orthogonal basis with respect to ϕ, that is, there exists a basis $\mathcal{B} = \{x_1, \ldots, x_n\}$ such that $\phi(x_i, x_j) = 0$ for $i \neq j$. Since a diagonal matrix is symmetric, a diagonalizable bilinear form is symmetric. In fact, the converse is also true.

PROPOSITION 4.5. *A symmetric bilinear form on a finite dimensional vector space V over a field K of characteristic not equal to 2 is diagonalizable.*

PROOF. By induction on the dimension of V. If $\dim V = 1$, then every nonzero vector of v is an orthogonal basis. Assume that the hypothesis holds for all symmetric bilinear forms on vector spaces of dimension less than $\dim V$.

Let ϕ be a symmetric bilinear form on V. If $\phi(u, u) = 0$ for all $u \in V$. Then for $x, y \in V$,

$$0 = \phi(x + y, x + y) = \phi(x, x) + 2\phi(x, y) + \phi(y, y).$$

Since $\operatorname{char} K \neq 2$, this implies that $\phi \equiv 0$. Thus, in this case, every basis of V is orthogonal.

Now assume that for some $x_1 \in V$, $\phi(x_1, x_1) \neq 0$. Let $W = \langle x_1 \rangle$. Then $V = W \oplus W^\perp$(Proposition 3.5). If $\phi_1 = \phi|_{W^\perp \times W^\perp}$, then ϕ_1 is a symmetric bilinear form on $W^\perp$. Since $\dim W^\perp = \dim V - 1$, by induction hypothesis, $W^\perp$ has an orthogonal basis $\mathcal{B}_1$. Hence, $\mathcal{B} = \{x_1\} \cup \mathcal{B}_1$ is an orthogonal basis of V with respect to ϕ. ∎

Now we assume that ϕ is a symmetric bilinear form on a finite dimensional vector space V over a field K of characteristic 2. Since $\operatorname{char} K = 2$, there is no distinction between symmetric and skew-symmetric bilinear forms, and so a symmetric bilinear form can now be alternating. For example, take $\phi(x, y) = x + y$, $x, y \in K^n$. If ϕ is alternating, then we have a classification given by Theorem 4.1. If ϕ is not alternating, then the following result proves that it is diagonalizable.

THEOREM 4.6. *Let ϕ be a symmetric and nonalternating bilinear form on a finite dimensional vector space V over K, $\operatorname{char} K = 2$. Then ϕ is diagonalizable.*

PROOF. Let V_1 be a subspace of V complementary to $\operatorname{rad}\phi$. Then $V = V_1 \oplus \operatorname{rad}\phi$ and $\phi' = \phi|_{V_1\times V_1}$ is a nondegenerate, nonalternating and symmetric bilinear form on V_1. If V_1 has an orthogonal basis $\mathcal{B}_1$ with respect to ϕ', then taking any basis $\mathcal{B}_2$ of $\operatorname{rad}\phi$, we have an orthogonal basis $\mathcal{B} = \mathcal{B}_1 \cup \mathcal{B}_2$ of V. Therefore, without any loss, we can assume that ϕ is also nondegenerate.

Now we prove this theorem by induction on $\dim V$. If $\dim V = 1$, it is trivially true. Assume that the hypothesis is true for all nondegenerate, nonalternating and symmetric bilinear forms on vector spaces of dimension less than $\dim V$. Let ϕ be a nondegenerate, nonalternating and symmetric bilinear form on V. Since ϕ is nonalternating, there exists $x \in V$ such that $\phi(x, x) = a \neq 0$. Let $W = \langle x\rangle$. Then $V = W \oplus W^{\perp}$ (Proposition 3.6). Let $\phi_0 = \phi|_{W^{\perp}\times W^{\perp}}$. If ϕ_0 is nonalternating, then by induction hypothesis, $W^{\perp}$ has an orthogonal basis $\mathcal{B}_0$ with respect to ϕ_0, and so $\mathcal{B} = \{x\} \cup \mathcal{B}_0$ is an orthogonal basis of V with respect to ϕ.

If ϕ_0 is alternating, then as ϕ_0 is nondegenerate, by Theorem 4.1, there is an ordered basis $\mathcal{B}' = \{x_2, x_3, \ldots, x_n\}$ such that

$$[\phi_0]_{\mathcal{B}'} = \operatorname{diag}\left(\begin{bmatrix} 0 & 1 \\ 1 & 0 \end{bmatrix}, \ldots, \begin{bmatrix} 0 & 1 \\ 1 & 0 \end{bmatrix}\right).$$

Note that as $\operatorname{char} K = 2$, $-1 = 1$. Let $W_1 = \langle x, x_2, x_3\rangle$ and let $\phi_1 = \phi|_{W_1\times W_1}$. Then verify that ϕ_1 is a nondegenerate bilinear form on W_1. Also that the vectors $u_1 = x + x_2 + x_3$, $u_2 = x + ax_2$, $u_3 = x + (1 + a)\,x_2 + x_3$ are mutually orthogonal and the matrix of ϕ_1 with respect to the ordered basis $\{u_1, u_2, u_3\}$ is $\operatorname{diag}(a, a, a)$. Now let $W_2 = \langle u_3, x_4, \ldots, x_n\rangle$. Since u_3 is a linear combination of x, x_1 and x_2 and $\phi(u_3, u_3) = a$, the bilinear form $\phi_2 = \phi|_{W_2\times W_2}$ is nondegenerate, nonalternating and symmetric. Since $\dim W_2 < \dim V$, by induction hypothesis, W_2 has an orthogonal basis $\mathcal{B}''$. Hence, $\mathcal{B} = \{u_1, u_2\} \cup \mathcal{B}''$ is an orthogonal basis of V. ∎

The matrix analogue of this theorem is the following.

PROPOSITION 4.7. *Let $A \in K^{n\times n}$ be a symmetric matrix. If $\operatorname{char} K = 2$ and all diagonal entries of A are zero, then A is congruent to a matrix of the form*

$$\operatorname{diag}\left(\underbrace{\begin{bmatrix} 0 & 1 \\ 1 & 0 \end{bmatrix}, \ldots, \begin{bmatrix} 0 & 1 \\ 1 & 0 \end{bmatrix}}_{r\ blocks}, 0_{n-2r}\right),$$

where $2r = \operatorname{rank} A$. Otherwise A is congruent to a diagonal matrix and the number of nonzero entries on the diagonal of A is equal to the rank of A.

If ϕ is a symmetric bilinear form on a finite dimensional vector space V over $\mathbb{R}$, then according to Theorem 4.5, V has an orthogonal basis with respect to ϕ. We can order the elements of this basis $\mathcal{B} = \{x_1, \ldots, x_s, x_{s+1}, \ldots, x_{s+t}, x_{s+t+1}, \ldots, x_n\}$ so that

$$[\phi]_{\mathcal{B}} = \operatorname{diag}(\alpha_1, \ldots, \alpha_s, \alpha_{s+1}, \ldots, \alpha_{s+t}, 0, \ldots, 0),$$

$\alpha_i > 0$ for $i = 1, \ldots, s$, $\alpha_{s+i} < 0$ for $i = 1, \ldots, t$, $s \geq 0$, $t \geq 0$. Clearly, $s = 0$ (respectively, $t = 0$) means that all diagonal entries of $[\phi]_{\mathcal{B}}$ are nonpositive (respectively, nonnegative); and $s + t = n$ means that there are no nonzero entries. Now let

$$y_i = \begin{cases} \frac{1}{\sqrt{|\alpha_i|}} x_i, & i = 1, \ldots, s+t, \\ x_i, & i = s+t+1, \ldots, n. \end{cases}$$

Then

$$\phi(y_i, y_j) = \frac{\phi(x_i, x_j)}{\sqrt{|\alpha_i|\,|\alpha_j|}},$$

and so the matrix of ϕ with respect to the ordered basis $\mathcal{B}' = \{y_1, \cdots, y_n\}$ is

$$\operatorname{diag}(I_s,\ -I_t,\ 0_{n-s-t}).$$

The nonnegative integers s and t are uniquely determined by the symmetric bilinear form ϕ alone. This can be deduced as a consequence of the following result.

THEOREM 4.8. ***Witt's theorem.*** *Let ϕ be a symmetric bilinear form on a finite dimensional vector space V over K, $\operatorname{char} K \neq 2$. Then for each pair of vectors x and y in V such that $\phi(x,x) \neq 0$, $\phi(y,y) \neq 0$ and $\phi(x,x) = \phi(y,y)$, there exists an isometric isomorphism on V which maps x to y.*

PROOF. Since $\phi(x+y, x-y) = \phi(x,x) + \phi(y,x) - \phi(x,y) - \phi(y,y) = 0$, $x+y$ and $x-y$ are orthogonal with respect to ϕ. Also, as

$$\phi(x,x) = \frac{1}{2}\left(\phi(x,x) + \phi(y,y)\right) = \frac{1}{4}\left(\phi(x+y, x+y) + \phi(x-y, x-y)\right),$$

at least one of the values $\phi(x+y, x+y)$ or $\phi(x-y, x-y)$ is not zero.

If $\phi(x-y, x-y) \neq 0$, then $V = \langle x-y\rangle \oplus \langle x-y\rangle^{\perp}$ (Proposition 3.6), and a required isometric isomorphism on V is S_{x-y} (Example 6, p.177). Since $S_{x-y}(x) = S_{x-y}\left(\frac{1}{2}(x-y) + \frac{1}{2}(x+y)\right) = -\frac{1}{2}(x-y) + \frac{1}{2}(x+y) = y$.

If $\phi(x+y, x+y) \neq 0$, then $V = \langle x+y\rangle \oplus \langle x+y\rangle^{\perp}$, and we consider the isometric isomorphism S_{x+y}. Since $S_{x+y}(x) = S_{x+y}\left(\frac{1}{2}(x+y) + \frac{1}{2}(x-y)\right) = -\frac{1}{2}(x+y) + \frac{1}{2}(x-y) = -y$, $S_y \circ S_{x+y}$ is a required isometric isomorphism on V. ∎

PROPOSITION 4.9. *Let $\operatorname{char} K \neq 2$ and let $\lambda_1, \ldots, \lambda_{n-1}, \mu_1, \ldots, \mu_{n-1} \in K$ and $\lambda \in K \setminus \{0\}$. If $\operatorname{diag}(\lambda, \lambda_1, \ldots, \lambda_{n-1})$ and $\operatorname{diag}(\lambda, \mu_1, \ldots, \mu_{n-1})$ are congruent matrices, then so are the matrices $\operatorname{diag}(\lambda_1, \ldots, \lambda_{n-1})$ and $\operatorname{diag}(\mu_1, \ldots, \mu_{n-1})$.*

PROOF. Let ϕ be a symmetric bilinear form on V whose matrix with respect to an ordered basis $\{x, x_1, \ldots, x_{n-1}\}$ is $\operatorname{diag}(\lambda, \lambda_1, \ldots, \lambda_{n-1})$. Then there exists an ordered basis $\{y, y_1, \ldots, y_{n-1}\}$ such that the matrix of ϕ with respect to this basis is $\operatorname{diag}(\lambda, \mu_1, \ldots, \mu_{n-1})$. Clearly, these two are orthogonal bases and $\phi(x,x) = \lambda = \phi(y,y)$. Thus, by Theorem 4.8, there exists an isometric isomorphism S on V such that $Sx = y$. Since S preserves the orthogonality of vectors, $S\left(\langle x\rangle^{\perp}\right) = \langle Sx\rangle^{\perp} = \langle y\rangle^{\perp}$. Therefore, if $S_1 = S|_{\langle x\rangle^{\perp}}$, then S_1 is an isometric isomorphism from $\langle x\rangle^{\perp}$ to $\langle y\rangle^{\perp}$. Since $\{x_1, \ldots, x_{n-1}\}$ is a basis of $\langle x\rangle^{\perp}$, so $\{Sx_1, \ldots, Sx_{n-1}\}$ is an ordered basis of $\langle y\rangle^{\perp}$. Also if $\phi_1 = \phi|_{\langle y\rangle^{\perp}\times\langle y\rangle^{\perp}}$, the matrix of ϕ_1 with respect to an ordered basis $\{Sx_1, \ldots, Sx_n\}$ is $\operatorname{diag}(\lambda_1, \ldots, \lambda_{n-1})$. Since the matrix of ϕ_1 with respect to the ordered basis $\{y_1, \cdots, y_{n-1}\}$ is $\operatorname{diag}(\mu_1, \cdots, \mu_{n-1})$, the matrices $\operatorname{diag}(\lambda_1, \ldots, \lambda_{n-1})$ and $\operatorname{diag}(\mu_1, \cdots, \mu_{n-1})$ are congruent. ∎

An obvious induction argument will yield the following result.

COROLLARY 4.10. *Let* char $K \neq 2$ *and let* $\lambda_1, \ldots, \lambda_m, \mu_1, \ldots, \mu_m \in K$. *If* $\operatorname{diag}(\gamma_1, \ldots, \gamma_k, \lambda_1, \ldots, \lambda_m)$ *and* $\operatorname{diag}(\gamma_1, \ldots, \gamma_k, \mu_1, \ldots, \mu_m)$ *are congruent matrices for* $\gamma_1, \ldots, \gamma_k \in K \setminus \{0\}$, *then so are the matrices* $\operatorname{diag}(\lambda_1, \ldots, \lambda_m)$ *and* $\operatorname{diag}(\mu_1, \ldots, \mu_m)$.

Now we have the following result about symmetric bilinear forms on a real vector space V.

THEOREM 4.11. *Let* ϕ *be a symmetric bilinear form on a finite dimensional vector space* V *over* $\mathbb{R}$. *Then there exists an ordered basis* $\mathcal{B}$ *of* V *such that*

$$[\phi]_{\mathcal{B}} = \operatorname{diag}(I_s, -I_t, 0_{n-s-t}),$$

the numbers s *and* t *are uniquely determined by* ϕ *alone, that is, do not depend upon the choice of the orthogonal basis.*

PROOF. We have already shown the existence of such a basis. Suppose that $\mathcal{B}$ and $\mathcal{B}'$ be ordered bases of V such that $[\phi]_{\mathcal{B}} = \operatorname{diag}(I_s, -I_t, 0_{n-s-t})$ and $[\phi]_{\mathcal{B}'} = \operatorname{diag}(I_{s'}, -I_{t'}, 0_{n-s'-t'})$. Since $\operatorname{rank}[\phi]_{\mathcal{B}} = \operatorname{rank}[\phi]_{\mathcal{B}'}$, $s + t = s' + t'$. If $s > s'$, then by Theorem 4.9, matrices $\operatorname{diag}(I_{s-s'}, -I_t, 0_{n-s-t})$ and $\operatorname{diag}(-I_{t'}, 0_{n-s'-t'})$ are congruent. Thus,

$$P^t \operatorname{diag}(-I_{t'}, 0_{n-s'-t'})\, P = \operatorname{diag}(I_{s-s'}, -I_t, 0_{n-s-t}),$$

for some invertible matrix P. But then the $(1,1)$–th entry of the matrix on the left side of the above equation is negative and the $(1,1)$–th entry of the matrix on the right side of the above equation is positive, a contradiction. Therefore, $s \leq s'$. Similarly, we have a contradiction if $s' > s$. Hence, $s = s'$ and $t = t'$. ∎

Thus, if ϕ is a symmetric bilinear form on an n–dimensional vector space V over $\mathbb{R}$, then there is a unique pair of nonnegative integers s and t. The triplet $(s, t, n-s-t)$ is called the **index** of ϕ; the rank of ϕ is $s+t$ and the number $s-t$ is called the **signature** of ϕ.

COROLLARY 4.12. *Let* V *be an* n*–dimensional vector space over* $\mathbb{R}$. *Two symmetric bilinear forms on* V *are equivalent if and only if they have the same rank and the same signature.*

The following theorem is another application of Witt's theorem, which is also called the **Witt's prolongation theorem**.

THEOREM 4.13. *Let* ϕ *be a symmetric bilinear form defined on a finite dimensional vector space* V *over a field* K, char $K \neq 2$, *and let* U *and* W *be subspaces of* V *such that* $\phi|_{U\times U}$ *and* $\phi|_{W\times W}$ *are nondegenerate bilinear forms. If* $\widehat{T} : U \to W$ *is an isometric isomorphism, then there exists an isometric isomorphism* $T : V \to V$ *such that* $T|_U = \widehat{T}$.

PROOF. By induction on $m = \dim U$. If $m = 1$, then $U = \langle u \rangle$ and $W = \langle w \rangle$ such that $\widehat{T}(u) = w$ and $\phi(u, u) = \phi(w, w) \neq 0$. Thus, there exists an isometric isomorphism $T : V \to V$ such that $Tu = w$ (Theorem 4.9).

Now assume that $m > 1$. From the hypothesis of the theorem it follows that $V = U \oplus U^{\perp}$ and $V = W \oplus W^{\perp}$ (Proposition 3.6). If $\{u_1, \ldots, u_m\}$ is an orthogonal basis of U, then the isometry $\widehat{T}$ carries orthogonal basis $\{u_1, \ldots, u_m\}$ to an orthogonal basis of W: $\left\{w_1 = \widehat{T}u_1, \ldots, w_m = \widehat{T}u_m\right\}$. Extend the ordered basis $\{u_1, \ldots, u_m\}$ of U to an ordered orthogonal basis $\mathcal{B}_1 = \{u_1, \ldots, u_m, u_{m+1}, \ldots, u_n\}$

of V and extend the ordered basis $\{w_1, \ldots, w_m\}$ of W to an ordered orthogonal basis $\mathcal{B}_2 = \{w_1, \ldots, w_m, w_{m+1}, \ldots, w_n\}$ of V. Let $[\phi]_{\mathcal{B}_1} = \operatorname{diag}(\lambda_1, \ldots, \lambda_n)$ and $[\phi]_{\mathcal{B}_2} = \operatorname{diag}(\mu_1, \ldots, \mu_n)$. Since for $i = 1, \ldots, m$: $\mu_i = \phi(w_i, w_i) = \phi\left(\widehat{T}u_i, \widehat{T}u_i\right) = \phi(u_i, u_i) = \lambda_i$, and $[\phi]_{\mathcal{B}_1}$ and $[\phi]_{\mathcal{B}_2}$ are congruent matrices, by Witt's theorem $\operatorname{diag}(\lambda_{m+1}, \ldots, \lambda_n)$ and $\operatorname{diag}(\mu_{m+1}, \ldots, \mu_n)$ are congruent matrices. Therefore, there exists an isometric isomorphism $\overline{T} : U^{\perp} \to W^{\perp}$ (Proposition 2.2).

If $v \in V$, then there are unique elements $u \in U$ and $u' \in U^{\perp}$ such that $v = u + u'$, and so we have a well defined mapping $T : V \to V$ given by $Tv = \widehat{T}u + \overline{T}u'$. It can now be easily verified that T is an isometric isomorphism such that $T|_U = \widehat{T}$. ∎

Corollary 4.14. *Let* $\operatorname{char} K \neq 2$. *If the invertible matrices* $\operatorname{diag}(\beta_1, \ldots, \beta_n)$ *and* $\operatorname{diag}(\gamma_1, \ldots, \gamma_n)$ *are congruent and* $\alpha_1, \ldots, \alpha_m \in K \setminus \{0\}$, *then the matrices* $\operatorname{diag}(\alpha_1, \ldots, \alpha_m, \beta_1, \ldots, \beta_n)$ *and* $\operatorname{diag}(\alpha_1, \ldots, \alpha_m, \gamma_1, \ldots, \gamma_n)$ *are also congruent.*

Proof. Exercise. ∎

4.1. Examples

1. Witt's theorem does not hold when the ground field is allowed to be of characteristic 2.

   Let $\operatorname{char} K = 2$ and let (V, ϕ) be a symmetric 3–dimensional bilinear space over K and let the matrix of ϕ with respect to an ordered basis $\{x, y, z\}$ be

$$\begin{bmatrix} 1 & 0 & 0 \\ 0 & 0 & 1 \\ 0 & 1 & 0 \end{bmatrix}.$$

   Verify that the matrix of ϕ with respect to an ordered basis $\{u, v, w\}$, where $u = x + y + z$, $v = x + y$, $w = x + z$, is

$$\begin{bmatrix} 1 & 0 & 0 \\ 0 & 1 & 0 \\ 0 & 0 & 1 \end{bmatrix}.$$

   For vectors x and u, $\phi(x, x) = 1 = \phi(u, u)$, and yet there is no isometric isomorphism on V such that $Tu = x$. For otherwise, $T\langle u\rangle^{\perp} = \langle x\rangle^{\perp}$ and then $Tv = \alpha y + \beta z$ for some α, β in K, and so $1 = \phi(v, v) = \alpha^2\phi(y, y) + 2\alpha\beta\phi(y, z) + \beta^2\phi(z, z) = 0$.

2. Let ϕ be a nondegenerate symmetric bilinear form on a finite dimensional vector space V over a field K of characteristic different from 2. According to the proof of Witt's theorem if x and y are anisotropic vectors and $\phi(x, x) = \phi(y, y)$, then there exists an isometric isomorphism which is either a reflection or a product of two reflections and which maps x to y. In fact, every isometric isomorphism is a product of reflections.

   Let $T : V \to V$ be an isometric isomorphism and let $\{u_1, \ldots, u_n\}$ be an orthogonal basis of V. Then $\phi(u_1, u_1) = \phi(Tu_1, Tu_1) \neq 0$, and so by Witt's theorem, $R_1(Tu_1) = u_1$, where R_1 is either a reflection or a product of two reflections. Now $R_1 \circ T|_{\langle u_1\rangle^{\perp}}$ is an isometric isomorphism on an $(n-1)$–dimensional bilinear space, and so $R_1 \circ T = R_2' \circ \cdots \circ R_k'$, where each R_i' is a reflection on $\langle u_1\rangle^{\perp}$. For $i = 2, \ldots, k$, let $R_i(u_1) = u_1$ and

$R_i(u_j) = R'_i(u_j)$, then R_i is a reflection on V and $T = R_1 \circ R_2 \circ \cdots \circ R_k$, a product of reflections.

3. Let V be an n–dimensional vector space over a field K. By a **quadratic form** on V determined by a bilinear form ϕ on V we mean a mapping $q : V \to K$ defined by $q(v) = \phi(v, v)$. Thus, if $\mathcal{B}$ is an ordered basis of V, then $q(v) = [v]_{\mathcal{B}}^t [\phi]_{\mathcal{B}} [v]_{\mathcal{B}}$. Equivalently, a quadratic form q on V is a mapping $q : V \to K$ so that to each ordered basis $\mathcal{B}$ of V, there is a matrix A in $K^{n\times n}$ such that $q(v) = [v]_{\mathcal{B}}^t A [v]_{\mathcal{B}}$.

   Assume that characteristic of K is different from 2. If q is a quadratic form determined by a bilinear form ϕ on V, then without any loss, we can assume that ϕ is symmetric. Since the mapping ψ on $V \times V$ defined by $\psi(x, y) = (\phi(x, y) + \phi(y, x))/2$ is symmetric bilinear form and $\psi(x, x) = \phi(x, x)$. Moreover, to a given quadratic form q there corresponds a unique symmetric bilinear form which determines it. Since if q is determined by symmetric bilinear forms ϕ and ψ on V, then $\phi(v, v) = q(v) = \psi(v, v)$ for all $v \in V$, then $\phi - \psi$ is a symmetric and alternating bilinear form on V and so $\phi - \psi \equiv 0$, that is, $\phi \equiv \psi$.

   Let q be a quadratic form on V determined by a symmetric bilinear form ϕ. Let $\mathcal{B} = \{u_1, \ldots, u_n\}$ be an ordered orthogonal basis of V with respect to ϕ. Then $[\phi]_{\mathcal{B}} = \operatorname{diag}(\lambda_1, \ldots, \lambda_n)$, $\lambda_i = \phi(u_i, u_i)$. For $v \in V$, write $v = \alpha_1 u_1 + \cdots + \alpha_n u_n$, so that

$$\begin{aligned} q(v) &= \phi\left(\sum_{i=1}^{n} \alpha_i u_i, \sum_{j=1}^{n} \alpha_j u_j\right) = \sum_{i=1}^{n} \lambda_i \alpha_i^2 \\ &= [v]_{\mathcal{B}}^t \operatorname{diag}(\lambda_1, \ldots, \lambda_n) [v]_{\mathcal{B}}. \end{aligned}$$

   This is called the diagonalization of the quadratic form and the ordered basis $\mathcal{B}$ diagonalizes q.

   Let V be an n–dimensional vector space over $\mathbb{R}$. A quadratic form q on V is called:

   positive semidefinite if $q(v) \geq 0$ for all $v \in V$;
   positive definite if $q(v) > 0$ for all $v \in V \setminus \{0\}$;
   negative semidefinite if $q(v) \leq 0$ for all $v \in V$;
   negative definite if $q(v) < 0$ for all $v \in V \setminus \{0\}$;

   and otherwise indefinite.

   From the above definition it is clear that q is positive semidefinite (respectively, positive definite) if and only if all the diagonal entries in the matrix representation of ϕ with respect to an orthogonal basis are non-negative (respectively, positive); equivalently the index of ϕ is $(s, 0, n - s)$ (respectively, $(n, 0, 0)$).

4. Let q be a quadratic form on K^n determined by a bilinear form ϕ. If A is the matrix of ϕ with respect to the standard basis $\{e_1, \ldots, e_n\}$ of K^n, then $\phi = \theta_A$ and $q(x) = x^t Ax$. From now onwards, whenever we write a quadratic form q by $q(x) = x^t Ax$, we mean A is the matrix of the corresponding bilinear form with respect to the standard basis.

   Also, from the above example if char $K \neq 2$, then A can be taken as a symmetric matrix. This can also be seen directly; if $q(x) = x^t Ax$, then

$q(x) = x^t Ax = (x^t Ax)^t = x^t A^t x$, and so $q(x) = x^t \left(\frac{A+A^t}{2}\right) x$. Thus from now onwards, if the characteristic of the base field is different from 2, whenever we write $q(x) = x^t Ax$, it is understood that A is symmetric.

Now assume that $q(x) = x^t Ax$ is a quadratic form on an n–dimensional Euclidean space $\mathbb{R}^n$. Since A is symmetric $n \times n$ real matrix, A has only real eigenvalues and is diagonalizable. Let $\lambda_1 \geq \cdots \geq \lambda_n$ be eigenvalues of A and let $\{u_1, \ldots, u_n\}$ be an ordered orthonormal basis of $\mathbb{R}^n$ such that $Au_i = \lambda_i u_i$. For $x \neq 0$, write $x = \sum_{i=1}^n \alpha_i u_i$, and so we have diagonalization of q:

$$x^t Ax = \lambda_1 \alpha_1^2 + \cdots + \lambda_n \alpha_n^2.$$

Thus,

$$\lambda_n \left(\alpha_1^2 + \cdots + \alpha_n^2\right) \leq x^t Ax \leq \lambda_1 \left(\alpha_1^2 + \cdots + \alpha_n^2\right).$$

Since $x^t x = \alpha_1^2 + \cdots + \alpha_n^2$, we obtain

$$\lambda_n \leq \frac{x^t Ax}{x^t x} \leq \lambda_1.$$

This proves the following statement.

For any quadratic form $x^t Ax$:

$$\max_{x \neq 0} \frac{x^t Ax}{x^t x} = \max_{\|y\|=1} y^t Ay = \lambda_1,$$

and

$$\min_{x \neq 0} \frac{x^t Ax}{x^t x} = \min_{\|y\|=1} y^t Ay = \lambda_n,$$

where λ_1 and λ_n are the largest and smallest eigenvalues of A. Moreover, these values are obtained at the corresponding eigenvectors of A.

4.2. Exercises

1. Find an orthogonal basis with respect to bilinear form ϕ on $\mathbb{R}^3$ whose matrix with respect to the standard basis is:

$$\begin{bmatrix} 12 & 10 & 3 \\ 10 & 14 & -1 \\ 3 & -1 & 3 \end{bmatrix}, \quad \begin{bmatrix} 1 & 2 & 3 \\ 2 & 1 & 4 \\ 3 & 4 & 1 \end{bmatrix},$$

$$\begin{bmatrix} 2 & 10 & -6 \\ 10 & -14 & 10 \\ -6 & 10 & -7 \end{bmatrix}, \quad \begin{bmatrix} 1 & 1 & 1 \\ 1 & 1 & 1 \\ 1 & 1 & 1 \end{bmatrix}.$$

2. A 2–dimensional symmetric bilinear space (V, ϕ) over a field K is called an **isotropic plane** if ϕ is nondegenerate and there exists a nonzero vector u with $\phi(u, u) = 0$. Prove that (V, ϕ) is an isotropic plane if and only if there is a basis of V with respect to which the matrix of ϕ is $\begin{bmatrix} 0 & 1 \\ 1 & a \end{bmatrix}$ for some $a \in K$.

3. A 2–dimensional symmetric bilinear space (V, ϕ) over a field K is called an **hyperbolic plane** if V has a basis $\{u, v\}$ such that $\phi(u, u) = 0 = \phi(v, v)$ and $\phi(u, v) = 1$. Prove that for $\operatorname{char} K \neq 2$, (V, ϕ) is an hyperbolic plane if and only if (V, ϕ) is an isotropic plane.
4. Prove that if $\operatorname{char} K \neq 2$ and $a \in K \setminus \{0\}$, then $\operatorname{diag}(1, -1)$ and $\operatorname{diag}(a, -a)$ are congruent matrices.
5. Let $\operatorname{char} K = 2$. Prove that (V, ϕ) is an isotropic plane but not a hyperbolic plane if and only if V has an orthogonal basis $\{x, y\}$ such that $\phi(x, x) = \phi(y, y)$.
6. Let ϕ be a symmetric bilinear form on an n–dimensional vector space V over K, and let x be anisotropic vector. Prove that the mapping
$$Ty = -y + 2\frac{\phi(y, x)}{\phi(x, x)} x$$
is an isometry from V to itself and T^2 is the identity. Verify that T is a reflection along $\langle x \rangle$.
7. Let $x^t Ax$ and $x^t Bx$ be two quadratic forms on an n–dimensional Euclidean space $\mathbb{R}^n$. Prove that if A is positive definite, then there exists an orthonormal basis in $\mathbb{R}^n$ which diagonalizes both quadratic forms.
Hint: There exists P such that $P^t AP = I_n$. Let Q be an orthogonal matrix which diagonalizes $P^t BP$.
8. Find an orthonormal basis of $\mathbb{R}^3$ which simultaneously diagonalizes the quadratic forms $x^t Ax$ and $x^t Bx$, where
$$A = \begin{bmatrix} 10 & 2 & 6 \\ 2 & 3 & -8 \\ 6 & -8 & 14 \end{bmatrix} \text{ and } B = \begin{bmatrix} 5 & 1 & 3 \\ 1 & 3 & -1 \\ 3 & -1 & 13 \end{bmatrix}.$$
9. Let B be a positive definite matrix. Prove that
$$\max_{x \neq 0} \frac{x^t Ax}{x^t Bx} = \lambda_1 \left(B^{-1}A\right),$$
where $\lambda_1 \left(B^{-1}A\right)$ is the largest eigenvalue of $B^{-1}A$ and the maximum is attained at an eigenvector corresponding to $\lambda_1 \left(B^{-1}A\right)$.
Hint: Write $B = C^t C$, C invertible and put $Cx = y$ to obtain
$$\frac{x^t Ax}{x^t Bx} = \frac{y^t \left(C^{-1}\right)^t AC^{-1} y}{y^t y}.$$
10. Prove that if B is positive definite and $u \neq 0$, then
$$\max_{x \neq 0} \frac{\left(u^t x\right)^2}{x^t Bx} = u^t B^{-1} u,$$
and the maximum value is attained at x_0 if and only if x_0 is a scalar multiple of $B^{-1}u$.

Bibliography

[1] S. Axler, *Linear Algebra Done Right*, Springer Verlag, New York 1996.

[2] S. Roman, *Advanced Linear Algebra*, Graduate Texts in Mathematics 135, Springer Verlag, New York 1992.

[3] V. Sahai and V. Bist, *Algebra*, Narosa Publishing House, 1999.

[4] K. Spindler, *Abstract Algebra and Applications*, Vol. I and II, Marcel Dekker, New York, 1994.

[5] K. Szymiczek, *Bilinear Algebra: An Introduction to Algebraic Theory of Quadratic Forms*, Gordan and Beach Science Publications, Amsterdam, 1997.

[6] F. Zhang, *Matrix Theory: Basic Results and Techniques*, Springer Verlag, New York 1999

Index